OPTICAL WAVEGUIDES

Here is an authoritative and comprehensive treatment of optical waveguides that will prove an indispensable reference work in connection with the development of optical waveguides for use in future optical communications, data processing, and computer and recognition systems.

The authors begin with the basic theory of optical waveguide propagation along slab waveguides, modifying or translating classic waveguide concepts into the language of the optical physicist. After a chapter on distributed coupling in slab waveguides, they discuss—from an analytical standpoint—waveguide propagation and the coherent interaction in dielectric cylindrical waveguides. They then present experimental studies of mode launching, mode discrimination and identification, radiation characteristics, distributive coupling in passive and active waveguides, and spatial filtering techniques for launching and identifying modes. Finally, they consider the special cases of anisotropic waveguides, noncircular dielectric waveguides, and hollow dielectric waveguides.

This book is intended for optical, computer, microwave, and communications engineers and researchers, and graduate and undergraduate physics and engineering students.

OPTICAL WAVEGUIDES

QUANTUM ELECTRONICS — PRINCIPLES AND APPLICATIONS

A Series of Monographs

EDITED BY
YOH-HAN PAO
Case Western Reserve University
Cleveland, Ohio

Optical Waveguides

N. S. KAPANY
Optics Technology, Inc.
Palo Alto, California

J. J. BURKE
Optical Sciences Center
University of Arizona
Tucson, Arizona

1972

ACADEMIC PRESS New York and London

ACADEMIC PRESS, INC.
111 Fifth Avenue, New York, New York 10003

United Kingdom Edition published by
ACADEMIC PRESS, INC. (LONDON) LTD.
24/28 Oval Road, London NW1

LIBRARY OF CONGRESS CATALOG CARD NUMBER: 72-187230

PRINTED IN THE UNITED STATES OF AMERICA

We dedicate this book to our beloved wives,

SATINDER KAPANY

and

BARBARA BURKE

Contents

Preface

With the emergence of the new technology of optical waveguides, microwave and optical scientists meet at a common frontier, where some evidence of a language barrier has already appeared. Because the literature on waveguides is primarily that of microwave theory and technique, the optical scientist may need a translation. This book, it is hoped, will help to fill that need. Whenever possible, we have attempted to describe waveguide phenomena in classical optical terms. Thus, we use equivalent plane waves, rays, polarization, and intensity distributions, rather than field plots and equivalent circuits, to describe mode propagation. This approach should help the student of optics to make contact with the microwave literature. The microwave investigator, on the other hand, will, we hope, find in it an expanded point of view.

There are essentially two classes of optical waveguides: those in which more or less classical optical elements, placed periodically along the direction of propagation of the wave, serve to confine the wave by successive refocusing in the vicinity of the optical axis; and those in which the guiding mechanism is that of multiple total internal reflections from interfaces parallel to the optical axis. Laser resonators and multiple lens waveguides are examples of the first type. Fiber optical waveguides, slab waveguides, and resonators are examples of the second. In this book, we confine our attention to the latter guides, discussing the former only when their observable properties are very similar to those of some member of the latter class. Furthermore, a third type of waveguide, in which a continuous decrease in the refractive index away from the optical axis serves to confine the wave, is also not treated. Though such guides may be expected to become important for the transmission of optical signals, their basic physical properties are similar to those of guides of the second type.

The book begins with a brief historical sketch (Chapter 1), the purpose of which is to direct the reader to some of the more important literature of dielectric waveguides at microwave frequencies and to the early work at optical frequencies. In Chapter 2, we introduce basic waveguide concepts and terminology in such a way as to relate them to familiar optical models of propagation. We consider in detail the proper and improper TE modes on slab waveguides so as to introduce the variety of waves that are observable on more complicated slab and circular guides. In Chapter 3, we develop the subject of the coupling of optical beams into slab waveguides and the distributed coupling of radiation between parallel guides. The approach is akin to that of optical interferometry, rather than field theoretic, and illustrates the usefulness of optical models in describing waveguide phenomena. Chapter 4 is devoted to a detailed theoretical study of the modes of circular waveguides, whose experimental study is described in the next three chapters. Thus, Chapter 5 describes methods of launching modes on dielectric guides; Chapter 6 discusses their radiation characteristics, together with spatially filtering techniques that provide mode discrimination in the far field; and Chapter 7 is devoted to the theoretical description and measurement of distributed coupling between passive dielectric waveguides and active, fiber-laser resonators. Chapter 8 discusses anisotropic circular waveguides.

Two Appendices, one by C. Yeh on noncircular waveguides and one by T. Sawatari on hollow dielectric waveguides, expand the studies of the text to elliptical and rectangular dielectric waveguides and to fluid-filled guides that may become important as technology develops.

It is hoped that students of physics and optics, as well as optical, electronic, and communications engineers, will find this book of interest.

Acknowledgments

In the past fifteen years, significant advances have been made in the field of optical waveguides, and several of our colleagues have been of invaluable assistance. We are particularly indebted to K. L. Frame, B. G. Phillips, Dr. T. Sawatari, and R. E. Wilcox for their cooperation and efforts.

We are also thankful to Miss E. Miller for her assistance in typing, proofreading, and correcting the manuscript, and to D. Cowen for the illustrations.

Substantial portions of Chapter 2, 3, and 4 are reproduced from a doctoral dissertation submitted by one of the authors (J. J. Burke) to the University of Arizona.

A large portion of the work included in this book was sponsored by the Air Force Office of Scientific Research, and we are particularly appreciative for the continued encouragement and assistance provided by Dr. Marshall C. Harrington of the Physics Branch.

CHAPTER 1

Historical Sketch

The dielectric-sheet (thin-film) and rod waveguides that we will study in this book have been the subjects of a truly massive literature. They are members of a broad class of structures called *surface waveguides* or *surface wave antennas*. These, in turn, form a subset of a broader class of structures, *traveling wave antennas*. Serious engineering studies of many of these began shortly before World War II, just after high-frequency microwave sources were developed. New work is still being produced today, at both microwave and optical frequencies. While it would not be practical for us to attempt to chart a course through all of the microwave literature, we would nevertheless like to bring to the reader's attention several of the excellent texts and comprehensive review papers that are available, particularly those dealing with the theoretical analyses of surface waveguides. Because these analyses are so germane to the many current investigations of optical waveguides, they may well be regarded as part of the optical, as well as the microwave, literature. Therefore, in our attempt to sketch the history of optical waveguides, we will begin with the earliest studies of dielectric guides, defer to texts and review papers for the history and content of the most pertinent work from the microwave literature, and then sketch the optical waveguide work of this and the preceding decade.

Although Rayleigh and Sommerfeld studied guided waves on and within conducting structures near the turn of the century, the first study of an all-dielectric guide was apparently that of Hondros and Debye (*1*) in 1910. They presented the first theoretical description of mode propagation along a dielectric rod. Several low-order modes were later investigated experimentally by Zahn (*2*) in 1916 and Schriever (*3*) in 1920. The subject then seems to have lain dormant for sixteen years until the development of high-frequency sources. In 1936, Barrow (*4*) and Southworth (*5*) reported experimental studies of the circular rod modes, including measurements of attenuation, while Carson *et al.* (*6*) advanced the theory by treating dielectric loss mathematically. In the late 1930's and early 1940's, the dielectric rod came into consideration as an antenna. Because of the evanescent field that propagates outside the rod and radiates away from it at bends, the rod had been completely overshadowed by hollow conducting tubes in transmission systems. The surface wave of the rod, however, provided the possibility of highly directional radiation, so that it became a candidate for microwave antenna systems.

The early theory of dielectric aerials and its experimental verification was described in a monograph by Kiely (*7*) in 1953. The subject of surface wave antennas has developed considerably since that time. These developments have been admirably digested and reviewed by Zucker (*8*) in an outstanding two-volume work *Antenna Theory* edited by him and Collin (*9*). Zucker's paper lists (with titles) more than one hundred references to the microwave literature of the 1950's and 1960's, a great many of which might well be required reading for optical frequency workers. In the same category is Chapter 11 on "Surface Waveguides" in Collin's book *Field Theory of Guided Waves* (*10*). Collin and Zucker's two-volume work (*9*) also includes two other extensively documented chapters that are quite germane to the analyses of optical waveguides: "General Characteristics of Travelling Wave Antennas" by Hessel and "Leaky-Wave Antennas" by Tamir. We defer to these comprehensive and well-written works for both the history and content of the important and relevant microwave literature of the last two decades.

The first dielectric waveguide to be studied at optical frequencies was the glass-coated, glass fiber developed originally for fiber optics imaging applications (*11*). By the late 1950's, several laboratories had developed the technology required to produce close-packed assemblies of glass fibers, each a few microns in diameter, with adjacent fibers less than 1 μm apart. When one of these fibers was illuminated at its end with white light, multicolored intensity distributions could be observed at the output ends of adjacent fibers. These early observations of waveguide-mode patterns and the coupling between parallel guides were a subject of only academic interest at the time. They were, in fact, a phenomenon that had to be designed out of fiber optics faceplates (*11*) by judicious choice of refractive indices, fiber spacing, and plate thickness. Thus, the authors originally (1961) approached the study of waveguide effects in fiber optics (*12*) with a view to substantially eliminating them in devices designed for incoherent systems. In the same year, Snitzer and Osterberg (*13*) reported detailed experimental studies of several low-order modes that appeared alone and in combinations on fiber waveguides that they had fabricated. The theory of the higher-order hybrid modes of dielectric guides was considerably advanced at this time by Schlesinger and King (*14*), Schlesinger *et al.* (*15*), Snitzer (*16*), and Kapany and Burke (*12*). We have extended and generalized these earlier studies of the modes of the dielectric rod in preparing the comprehensive theoretical treatment given in Chapter 4. Biernson and Kinsley (*17*) have presented considerable numerical and graphical data on several of the low-order modes in connection with their studies of retinal rods and cones. A volume of numerical data was prepared by Luneberg and Snitzer (*18*).

A theoretical study of the launching of $HE_{1,m}$ modes by a plane wave with the wave vector parallel to the axis of a circular rod was reported by Snyder (*19*) in 1966. A general launching technique appropriate to optical waveguides was described by Kapany *et al.* (*20*) in 1970; it is described in Chapter 5. Studies of the radiation characteristics of fiber optical waveguides have been reported by Snitzer (*21*) in 1961 and Kapany *et al.* (*22*) in 1965. This work is described in Chapter 6.

The discovery of neodymium-glass lasers, described by Snitzer (*23*) in 1964, led to several interesting investigations of coupling between parallel fiber optical waveguides and laser resonators. Koester (*24*) described experiments with coupled fiber laser resonators in 1963. In 1968, Kapany *et al.* (*25*) reported the quantitative comparison between theory and experiment with both coupled guides and coupled resonators. Theoretical predictions of the coupling strengths between these elements had been provided by Bracey *et al.* (*26*) in 1959 and Jones (*27*) in 1965. Their theoretical results were compared and generalized by Burke (*28*) in 1967. This work is discussed in some detail in Chapter 7. In 1971, the Corning Glass Works succeeded in fabricating single-mode glass fibers with loss rates of 20 dB/km. Techniques for measuring the absorption and scattering in these waveguides have been described by Tynes *et al.* (*29*). These guides are thus very promising candidates for optical transmission lines in the communications systems of the future. It is ironic that such transmission systems should be out of the question at microwave frequencies, but the most promising optically. Such is the difference in the properties of materials in the two ranges and in the ratio of practical bending radii to the wavelength.

The dielectric-sheet waveguide at optical frequencies is at least as old as the development of optical thin-film technology. We have been unable to determine when the dielectric-sheet guide was first studied. There can be little doubt that everyone who has studied the modes of the dielectric rod has also investigated those of the slab, which are so much simpler mathematically that they provide a means for obtaining considerable insight into more complicated geometries. It was not until the late 1960's, however, that thin-film guides became candidates for potential communications applications. In 1969, the Bell Telephone Laboratories (*30*) launched a broad range of theoretical and experimental studies in "integrated optics." This new field of endeavor, which today commands the efforts of many scientists and engineers in both industrial and university laboratories, promises to bring the techniques of electronic integrated circuits to optical waveguides, so that planar arrays of passive and active signal processing elements can be fabricated. Work is just beginning today toward the development of both the materials and techniques that will be needed. Two review papers have already appeared: "Integrated Optical Circuits" by Goell and Standley (*31*) and "Optical Waveguide Transmission" by Gloge (*32*). Extensive references to the most recent literature may be found in these papers.

There does exist a small amount of literature on thin-film guides that preceded the current avalanche. We might mention, in particular, the extensive theoretical and experimental work of Anderson (*33*), who was using integrated circuit techniques to fabricate waveguides for the near-infrared in the early 1960's. Osterberg and Smith (*34*) did experimental studies of the launching of surface waves from prisms in 1964. Schineller *et al.* (*35*) studied guides formed by proton irradiation of fused silica in 1968. In 1969, Harris and Shubert (*36*) and Tien *et al.* (*37*) reported the first experimental studies of the widely publicized prism-film coupler. This system is simply the classical frustrated total reflection (FTR) filter of Leurgans and Turner (*38*), with its output prism removed. The latter was described in waveguide terms by the authors in 1961 (*11*). Detailed theoretical descriptions of both the FTR filter and the prism-film coupler were reported by Iogansen in 1962, 1964, and 1967 (*39*). Its rediscovery and wide publicity by Bell Telephone Laboratories have now precluded the possibility of further rediscovery. It has now been analyzed in at least six different ways, two of which are considered in Chapter 3, while the others are referenced there. The launching of guided waves on thin films by gratings was reported in 1970 by Kuhn *et al.* (*40*). This technique appears to have very considerable advantages for integrated optics applications.

The semiconductor laser has provided another avenue for research into the properties of planar dielectric waveguides and resonators. Yariv and Leite (*41*) described a theory of dielectric waveguide modes in *p–n* junctions in 1963. Ashkin and Gershenzon (*42*) and Bond *et al.* (*43*) reported experimental observations of these modes in 1963. Anderson (*44*) studied mode confinement and gain in junction lasers in 1965. Nelson and McKenna (*45*) reported a theoretical analysis of anisotropic film guides in 1967, while Oldham and Bahraman (*46*) studied *p–n* and *p–i–n* junctions as light modulators in the same year. This is only a small sampling of the very extensive literature on semiconductor junctions as waveguides and lasers. We will not treat this subject explicitly in this work. The references here are thus intended to serve simply as an introduction to the literature. There can be no doubt that semiconductor junctions will play a major role in optical signal processing devices of the future. Precisely what form they will take and which materials will be involved are uncertain today. New materials are being produced regularly. Tomorrow's communications and signal processing systems are only weak visions today.

REFERENCES

1. D. Hondros and P. Debye, *Ann. Phys.* **32**, 465 (1910).
2. H. Zahn, *Ann. Phys.* **49**, 907 (1916).
3. O. Schriever, *Ann. Phys.* **64**, 645 (1920).
4. W. L. Barrow, *Proc. IRE* **24**, 1298 (1936).
5. G. C. Southworth, *Bell Syst. Tech. J.* **15**, 284 (1936).
6. J. R. Carson, S. P. Mead, and S. A. Schelkunoff, *Bell Syst. Tech. J.* **15**, 310 (1936).
7. D. G. Kiely, "Dielectric Aerials." Methuen, London, 1953.
8. F. J. Zucker, *in* "Antenna Theory Part 2" (R. E. Collin and F. J. Zucker, eds.), p. 298. McGraw-Hill, New York, 1969.
9. R. E. Collin and F. J. Zucker, "Antenna Theory," Vols. I and II. McGraw-Hill, New York, 1969.
10. R. E. Collin, "Field Theory of Guided Waves," Chapter 11. McGraw-Hill, New York, 1960.
11. N. S. Kapany, "Fiber Optics Principles and Applications." Academic Press, New York, 1967.
12. N. S. Kapany and J. J. Burke, *J. Opt. Soc. Amer.* **51**, 1067 (1961).
13. E. Snitzer and H. Osterberg, *J. Opt. Soc. Amer.* **51**, 499 (1961).
14. S. P. Schlesinger and D. D. King, *IRE Trans.* **MTT-6**, 291 (1958).
15. S. P. Schlesinger, P. Diament, A. Vigants, *IRE Trans.* **MTT-8**, 252 (1960).
16. E. Snitzer, *J. Opt. Soc. Amer.* **51**, 491 (1961).
17. G. Biernson and D. J. Kinsley, *IEEE Trans.* **MTT-13**, 354 (1965).
18. I. Luneberg and E. Snitzer, Final Rept. Cont. No. AF 19(604)7207 "Characteristics of Dielectric Waveguide Modes," July, 1963.
19. A. W. Snyder, *J. Opt. Soc. Amer.* **56**, 601 (1966).
20. N. S. Kapany, J. J. Burke, and T. Sawatari, *J. Opt. Soc. Amer.* **60**, 1178 (1970).
21. E. Snitzer, *in* "Advances in Quantum Electronics" (J. R. Singer, ed.), p. 348. Columbia Univ. Press, New York, 1961.
22. N. S. Kapany, J. J. Burke, and K. L. Frame, *Appl. Opt.* **4**, 1534 (1965).

23. E. Snitzer, *in* "Quantum Electronics" (P. Grivet and N. Bloembergen, eds.), p. 999. Columbia Univ. Press, New York, 1964.
24. C. J. Koester, *in* "Optical Processing of Information" (D. K. Pollock, C. J. Koester, and J. T. Tippett, eds.), p. 74. Spartan Books, Baltimore, Maryland, 1963.
25. N. S. Kapany, J. J. Burke, K. L. Frame, and R. E. Wilcox, *J. Opt. Soc. Amer.* **58**, 1176 (1968).
26. M. F. Bracey, A. L. Cullen, E. F. F. Gillespie, and J. A. Staniforth, *IRE Trans.* **AP-7**, S219 (1959).
27. A. L. Jones, *J. Opt. Soc. Amer.* **55**, 261 (1965).
28. J. J. Burke, *J. Opt. Soc. Amer.* **57**, 1056 (1967).
29. A. R. Tynes, A. D. Pearson, and D. L. Bisbee, *J. Opt. Soc. Amer.* **61**, 143 (1971).
30. *Bell Syst. Tech. J.* **48** (September 1969).
31. J. E. Goell and R. D. Standley, *Proc. IEEE* **58**, 1504 (1970).
32. D. Gloge, *Proc. IEEE* **58**, 1513 (1970).
33. D. B. Anderson, *in* "Optical and Electro-Optical Information Processing" (J. T. Tippett, D. A. Berkowitz, L. C. Clapp, C. J. Koester, and A. Vanderburgh, Jr., eds.), p. 221. MIT Press, Cambridge, Massachusetts, 1965.
34. H. Osterberg and L. W. Smith, *J. Opt. Soc. Amer.* **54**, 1078 (1964).
35. E. R. Schineller, R. Flam, and D. Wilmot, *J. Opt. Soc. Amer.* **58**, 1171 (1968).
36. J. H. Harris and R. Shubert, Conf. Abstracts, Int. Sci. Radio Union, Spring Meeting, Washington, D.C., p. 71, April 1969.
37. P. K. Tien, R. Ulrich, and R. J. Martin, *Appl. Phys. Lett.* **14**, 291 (1969).
38. P. Leurgans and A. F. Turner, *J. Opt. Soc. Amer.* **37**, 983 (1947).
39. L. V. Iogansen, *Sov. Phys.—Tech. Phys.* **7**, 295 (1962); **8**, 985 (1964); **11**, 1529 (1967).
40. L. Kuhn, M. L. Dakss, P. F. Heidrich, and B. A. Scott, *Appl. Phys. Lett.* **16**, 523 (1970); **17**, 268 (1970).
41. A. Yariv and R. C. C. Leite, *Appl. Phys. Lett.* **2**, 55 (1963).
42. A. Ashkin and M. Gershenzon, *J. Appl. Phys.* **34**, 2116 (1963).
43. W. L. Bond, B. C. Cohen, R. C. C. Leite, and A. Yariv, *Appl. Phys. Lett.* **2**, 57 (1963).
44. W. W. Anderson, *IEEE J. Quant. Electron.* **QE-1**, 228 (1965).
45. D. F. Nelson and J. McKenna, *J. Appl. Phys.* **38**, 4057 (1967).
46. W. G. Oldham and A. Bahraman, *IEEE J. Quant. Electron.* **QE-3**, 278 (1967).

CHAPTER 2

Slab Waveguides

In this and the next chapter, we will provide an introductory theoretical treatment of guided waves in planar dielectric waveguides in terms of the characteristic waves (modes) of these structures. During the development of the mathematical analysis, we will relate the modal descriptions of the waves to familiar optical models of propagation involving homogeneous plane waves, collimated beams, rays, and interference effects. These concepts occupy a minor role in the classical literature of microwave guides, where the fields themselves are observable quantities. At optical frequencies, however, it is possible to observe only intensity distributions and polarization characteristics, and these only outside the guides. An interferometric description of the modes will accordingly suggest how the experimental tools and techniques of optics can be fruitfully applied to the study of optical waveguides. It also will provide a useful alternative point of view from which to introduce the entire subject.

A. WAVEGUIDE CONCEPTS AND TERMINOLOGY

A waveguide can be defined as any structure (usually cylindrical) capable of guiding the flow of electromagnetic energy in a direction parallel to its axis, while substantially confining it to a region either within or adjacent to its surfaces. The well-known fiber optics light guide clearly fits this definition. In fiber optics light guides, however, large-diameter fibers are usually used simply to conduct polychromatic radiation provided by an incoherent light source. It is generally unnecessary to analyze this form of wave propagation as a superposition of numerous elementary electromagnetic waves of different frequencies and spatial characteristics, because they would have randomly related amplitudes and phases and would produce no observable interference effects.

If, on the other hand, the radiation incident at the entrance end of the guide is spatially and temporally coherent, we would need a nongeometric analysis to describe the interference effects at the exit end of the guide and in the radiation field beyond it. Physical optics considerations suggest that this analysis could take the form of a superposition of uniform plane waves propagating down the guide, undergoing successive internal reflections at various angles with its axis. Except for guides with plane walls, however, such an analysis would become very complicated. We therefore look for a simpler set of elementary waves that may be used to describe the interference patterns and polarization conditions to be expected. These elementary waves, which are characteristic of the particular waveguide under consideration, are called the modes of that waveguide.

A waveguide mode is an electromagnetic wave that propagates along a waveguide with a well-defined phase velocity, group velocity, cross-sectional intensity distribution, and polarization. Each component of its electric and magnetic field is thus of the form $f(x, y)e^{i\omega t - ihz}$, where the z axis of the coordinate system coincides with the axis of the guide. The modes are referred to as the "characteristic waves" of the structures because their field vectors satisfy the homogeneous wave equation in all the media that make up the guide, as well as the boundary conditions at the interfaces. Modes thus characterize the structure, in terms of its electromagnetic resonances, and depend in no way on the sources of radiation that may feed it. Each propagating mode in the interior of a dielectric waveguide is itself describable as a superposition of uniform plane waves propagating at a fixed angle with the guide axis. This superposition is so simple in the case of planar waveguides that all the basic equations governing their elementary waves may be easily derived from a direct analysis of plane waves incident and reflected from the plane interfaces that are the guiding surfaces. To introduce the most common waveguide concepts and terminology, we will examine the simplest of all such guides, that formed by two parallel, perfectly reflecting mirrors.

1. Propagating Modes of Plane Parallel Mirror Waveguides

Figure 2-1 illustrates a guide consisting of two parallel plane mirrors. Let us assume that the mirrors are infinite in extent and perfect reflectors (infinite conductivity). We wish to obtain a mathematical description of waves that can propagate between the mirrors with a well-defined phase velocity in the z direction. Let us begin with the case of plane-wave reflection from a conducting interface by removing the lower conductor to $x = -\infty$. Consider a uniform plane wave incident on the upper conducting surface with its wave normal (Poynting vector) at an angle θ with the normal

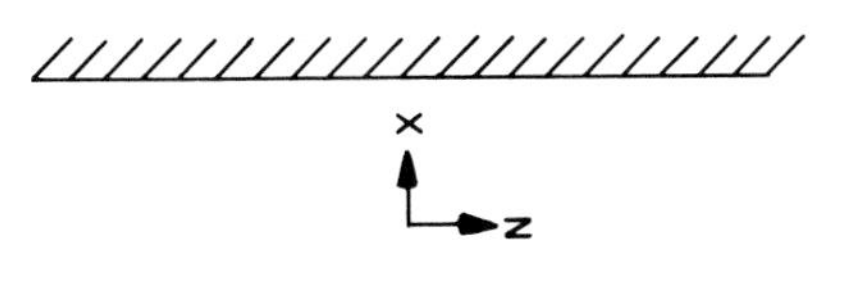

Fig. 2-1. Schematic illustration of a waveguide formed by parallel plane mirrors.

to the conductor. We will assume that the incident wave is plane polarized with the electric vector oscillating perpendicular to the plane of incidence, i.e., the plane containing the wave normal and the surface normal. Then, the electric vector for the incident wave has only one nonvanishing Cartesian component, given by

$$E_y(\text{incident}) = E_0 e^{i\omega t - ikx\cos\theta - ikz\sin\theta}, \tag{2.1}$$

where $k = 2\pi n/\lambda_0$ is the wavenumber, n is the refractive index, and λ_0 is the free-space wavelength. In order to satisfy the boundary condition, the reflected wave must be such that the tangential electric field vanishes at the perfect conductor. Therefore,

$$E_y(\text{reflected}) = -E_0 e^{i\omega t + ikx\cos\theta - ikz\sin\theta}. \tag{2.2}$$

The total electric field in the semiinfinite region below the conductor thus has the form

$$E_y(\text{total}) = E_0[\sin(kx\cos\theta)]e^{i\omega t - ikz\sin\theta}, \tag{2.3}$$

where all constants have been absorbed into the single complex-valued constant E_0. From a waveguide point of view, E_y(total) represents the electric field of a nonuniform plane wave propagating in the z direction with a phase velocity $\omega/(k \sin\theta)$. The wave is of the transverse electric (TE) type, because the electric field is transverse to the assumed direction of propagation. The wave is also called an H wave because it has a nonvanishing component of the magnetic field parallel to the conductor and thus also to the assumed direction of propagation. In the microwave literature, reference to the angle at which the component uniform plane waves propagate is hidden in the representation of the total field

$$E_y = E_0(\sin \beta x)e^{i\omega t - ihz}. \tag{2.4}$$

The quantity h is called the propagation constant or longitudinal wavenumber of the mode. Through it, a guide wavelength $\lambda_g = 2\pi/h$ is defined. In an obvious way, an equivalent refractive index may also be defined as $n_e = \lambda_0/\lambda_g = c/v_{\text{phase}}$. The quantity β is called the transverse wavenumber. We note that, as long as only one conductor is involved, there is no restriction on k and θ; that is, all angles of component plane waves are allowed for all values of k. In waveguide terminology, this means that all values of h and β are allowed, provided only that

$$h^2 + \beta^2 = k^2. \tag{2.5}$$

Suppose now, however, that we wish to confine the energy to a finite region by introducing the second conducting surface. We can clearly maintain the same electric field between the two conductors as we had for one, provided that the separation d between them satisfies the condition

$$\sin \beta d = 0. \tag{2.6}$$

For other values of d, the two plane waves (incident and reflected at the angle θ from two perfectly conducting interfaces) cannot exist, because they cannot satisfy the appropriate boundary conditions. We have thus obtained, for this waveguide, a very simple relationship between the allowed angles of propagation and the thickness of the guide. In waveguide terminology, we have found the "characteristic equation" for this structure, i.e., Eq. (2.6). For a guide of given thickness d, this equation determines the discrete set of allowed values of the transverse wavenumber β_N, and thus also, by Eq. (2.5), the propagation constant h_N of the modes. Each root β_N of the characteristic equation defines the parameters of the corresponding field distributions. Thus, the successive roots $\beta_0, \beta_1, \beta_2, \ldots,$ together with the general expressions for the corresponding field vectors, define the TE_0, TE_1, TE_2, ... modes.

So far, we have considered only the superposition of two uniform plane waves with an electric vector perpendicular to the plane of incidence. We can also have the electric vector parallel to this plane, in which case the magnetic field will be purely transverse to the axis of the waveguide. Such waves are called transverse magnetic (TM) modes in the waveguide literature. They are also called E waves, because they have a component of the electric field along the axis of the guide. It is easy to see that TE waves between perfectly conducting planes will have propagation constants satisfying the same equation [Eq. (2.6)] as those of TE waves. [The longitudinal component of the electric field now varies as $\sin(\beta_N x)$.] This means that we can have at least two entirely different field distributions, that is, modes with the same phase velocity. This condition is referred to as mode degeneracy. The planar waveguide is clearly infinitely degenerate in this sense, because our assumed direction of propagation z was arbitrarily selected.

2. Observable Characteristics of the Propagating Modes

In the analysis thus far, we assumed that the mirrors were infinite in extent. In order to discuss the optically observable manifestations of the modes, let us suppose, now, that the waveguide is made up of a thin sheet of glass a few centimeters long and 1 cm wide, say, coated on both its faces with perfectly conducting material and that the two ends of the guide are polished flat, perpendicular to the axis. The radiation is coupled into one end of this guide from a monochromatic source in such a way as to launch only one mode. The finite width of the guide may be ignored if the coupler collimates the launching radiation in one dimension, so that its lateral divergence is insignificant over the length of the guide. We can then view the intensity distribution on the end of the guide with the aid of a microscope. If we ignore the diffraction and reflection effects at the end, we would expect to see the familiar intensity distribution associated with the interference of two plane waves traveling at angles $\pm(\frac{1}{2}\pi - \theta)$ with the midplane, i.e., a distribution varying as $\sin^2(\beta_N x)$ or $\cos^2(\beta_N x)$, where $\beta_N = N\pi/d = k \cos \theta_N$. These mode patterns are proportional to the time average of the longitudinal component of the Poynting vector for the mode. The higher the mode number N, the smaller is the angle θ_N and the more nulls are observed in the intensity distribution. The highest-order mode pattern manifests the largest number of maxima, corresponding to the number of half-wavelengths required to span the separation between the mirrors. However, it is not generally possible to observe patterns corresponding to small angles of incidence, because most of the mode power is reflected back into the guide.

If we defocus the microscope by moving it away from the output end of the guide, we can observe the intensity distributions in the radiation field, first in the Fresnel region, then into the far-field. For the dielectric waveguides to be studied in this book, we will see that a vector formulation of Huygens's principle gives a good description of the observed radiation patterns.

Our description of the propagating modes of the planar conducting guide through the use of plane waves also suggests how they might be launched on a guide in the laboratory. At optical wavelengths, we would obtain efficient launching of a desired mode by causing two collimated beams at appropriate angles to be incident on the entrance face of the guide. Obviously, the launch angles must satisfy Snell's law at the entrant interface. From another point of view, we would expect to obtain efficient launching of a given mode if we could generate, at the entrance face of the guide, a flux distribution (with appropriate polarization) that matches that of the mode pattern. The two points of view are clearly equivalent, because the interference pattern formed where the beams intersect at the entrance face is just the desired flux distribution.

3. Nonpropagating Modes and Improper Modes on Planar Waveguides

The plane-wave approach to modal analysis is sufficient for studying those characteristic waves of the planar, perfectly conducting guide that can propagate unattenuated down its length. Additional modes are needed, however, to represent a general electromagnetic disturbance in the guide. These may be inferred from Fourier analysis. Thus, on a guide of finite thickness, there is only a finite number of propagating modes at each frequency. Each has an electric field E_y varying as $\sin(N\pi x/d)$ for $0 < x < d$ in any plane $z = \text{const}$. From the theory of Fourier series, we know that we need an infinite number of such functions in order to characterize a general TE field distribution in any transverse plane. What we have considered so far, then, is just those modes that have propagation constants h_N that are real valued, with $\beta_N < k$. Fourier arguments suggest that there are also an infinite number of nonpropagating TE modes for which $\beta_N = N\pi/d > k$, and h_N has imaginary values. Some of these modes will be excited by any source of radiation or by discontinuities in the waveguide, but, because they decay as $e^{-h_N z}$ away from such sources or discontinuities, they will not be observable at large distances from them. Together with the propagating modes, however, they do provide a complete set of orthogonal functions in terms of which the general field between the conductors may be represented. Modes having imaginary values of h_N are said to be cut off, i.e., they cannot propagate.

We shall shortly be treating the dielectric-slab (-sheet, -film) waveguide, wherein the unattenuated modes may be thought of as the superposition of two plane waves confined to the interior of the guide by total internal reflection. A mode is then said to be cut off when the corresponding angle of incidence is equal to or less than the critical angle. For a guide of finite thickness, we will again have only a finite number of such unattenuated propagating modes, referred to variously as surface wave modes, bound modes, proper modes, true modes, and spectral modes. To describe a general field, we will have to add to these a continuous spatial spectrum of modes corresponding to all real angles less than the critical angle in the guide, as well as a continuously distributed set of nonpropagating waves, that is, waves that decay in the z direction away from sources or discontinuities. These last modes are not familiar optical disturbances. They are evanescent in the z direction but propagating (or standing) in the x direction with phase velocities greater than those of uniform plane waves in either medium. We will discuss them further when we treat the dielectric slab in detail.

The representation of a general field in terms of a set of orthogonal functions does not always provide the most informative description of its properties. We shall see in the discussions which follow that a representation involving characteristic waves that do not, in themselves, satisfy the radiation condition can be more fruitful in certain cases. These waves describe resonant responses of the structure to sources of radiation within or near it and, therefore, provide useful descriptions of the radiation field of such structures over a limited region of space. Characteristic waves that do not satisfy the radiation condition fall into a broad class of modes called "improper." The most important of these, originally called "leaky waves" in the microwave literature, have many of the characteristics of the true, bound modes. They differ from them primarily in that, as their name implies, some of their power is continuously radiated through the walls of the guide. The power in the guide thus decays exponentially with distance along the guide. Until recently, these leaky waves were not called modes, the term mode being restricted to those waves that, in lossless structures, have propagation constants that are either purely real or purely imaginary valued. Because of their radiation losses, the leaky waves have complex-valued propagation constants. In the earlier literature, they were referred to as "nonmodal" characteristic waves, or as nonmodal resonances, improper modes, quasimodes, or pseudomodes. More recently, however, some writers have referred to them as "leaky wave modes" or simply "leaky modes." *

Additionally, in the laser literature of the past decade, the term mode has been applied to the resonances of the laser interferometers. These modes

* On certain structures, such as gratings and plasma slabs (1), leaky waves exist that are proper, spectral modes. It is thus not strictly correct to equate "leaky" and "improper."

have more in common with the improper leaky waves of dielectric waveguides than with the true proper modes of conventional metallic guides. As indicated in the beginning of this chapter, we have adopted the broader definition of the term mode, applying it to all the characteristic resonances of the structures to be studied. In Chapter 3, we will illustrate the occurrence of leaky waves in a familiar optical structure (viz., the frustrated total internal reflection filter).

B. PLANAR WAVEGUIDES: A CLASSICAL APPROACH TO THE DETERMINATION OF THE MODES

Consider the slab of thickness $2a$ illustrated schematically in Fig. 2-2. It is of infinite extent in the y and z directions and has permitivity ε_1, conductivity σ_1, and permeability μ_0 equal to that of free space. It is embedded in an infinite medium characterized by the constants ε_2, σ_2, and μ_0. When $\sigma_1 = \sigma_2 = 0$, the normal modes of this structure are inhomogeneous plane

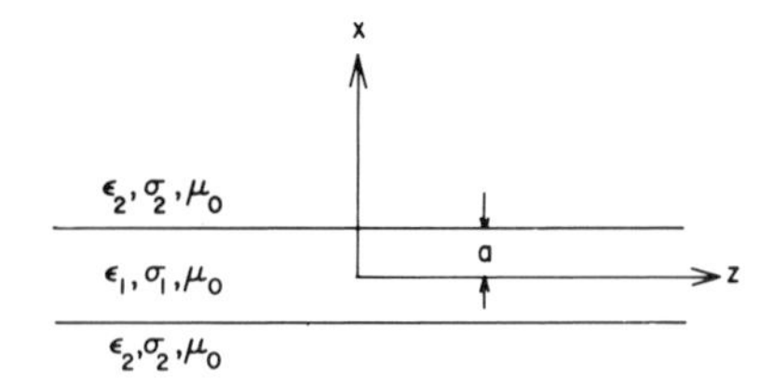

Fig. 2-2. Schematic diagram of thin-film waveguide of thickness $2a$ embedded in a homogeneous medium. The permitivity, conductivity, and permeability of the media are as indicated.

waves traveling in any direction perpendicular to the x axis. (They thus exhibit an infinite degeneracy; that is, an infinite number of modes have the same propagation constant.) Without loss of generality, we may choose the z axis to coincide with the direction of propagation and assume the mode fields to be independent of y.

1. TE Modes

We seek the simplest solutions of Maxwell's equations in each medium having the form of monochromatic waves propagating with a well-defined phase velocity in the guide. Thus, by definition, every component of the electric and magnetic field must be proportional to $e^{i\omega t - ihz}$. It is clear that plane interfaces will not mix waves polarized either in or perpendicular to the plane of incidence. Thus, TE and TM waves must each be capable of satisfying the boundary conditions independently. For the TE waves,

$E_x = E_z = 0$, and the general wave equation in each medium,

$$\nabla^2 \mathbf{E} - \mu\varepsilon(1 - i\sigma/\omega\varepsilon)\frac{\partial^2 \mathbf{E}}{\partial t^2} = 0, \tag{2.7}$$

reduces to

$$(d^2E_y/dx^2) + \beta^2 E_y = 0, \tag{2.8}$$

where

$$\beta^2 = k^2 - h^2 \qquad \text{and} \qquad k^2 = \omega^2 \varepsilon\mu(1 - i\sigma/\omega\varepsilon). \tag{2.9}$$

(The quantities σ, ε, k, and β must, of course, be appropriately subscripted when these equations are specifically applied to media 1 and 2.)

If the media were perfect dielectrics and we were primarily interested in waves that are completely bound to the guide, we would want oscillating solutions of Eq. (2.8) in medium 1 with evanescent waves in medium 2. Thus, $k_2 < h < k_1$, so that $\beta_1{}^2$ is positive and $\beta_2{}^2$ is negative (β_2 is imaginary). A general solution of Eq. (2.8) inside the guide is then given by

$$E_y(\text{inside}) = [A\cos(ux/a) + B\sin(ux/a)]e^{i\omega t - ihz}. \tag{2.10}$$

Outside the guide, we have the solutions

$$E_y(\text{outside}) = \begin{cases} Ce^{-qx/a}e^{i\omega t - ihz}. \\ De^{qx/a}e^{i\omega t - ihz} \end{cases} \tag{2.11}$$

The parameters $u = \beta_1 a$ and $q = i\beta_2 a$ are the dimensionless quantities that arise naturally in applying the boundary conditions. The constants A, B, C, and D are to be determined from the boundary conditions. Because we are looking for the simplest possible solutions, however, we may use the symmetry of the structure to determine the values of some of these. Thus, it must be possible to have purely symmetric fields with $B = 0$ and $C = D$, and purely antisymmetric fields with $A = 0$ and $C = -D$.

The magnetic fields for either the symmetric or antisymmetric TE modes are now determined from Eqs. (2.10) and (2.11) through the Maxwell equation

$$\nabla \times \mathbf{E} = -\mu_0 \frac{\partial \mathbf{H}}{\partial t}, \tag{2.12}$$

which reduces, in this case, to

$$H_z = -(1/i\omega\mu_0)\frac{\partial E_y}{\partial x}; \qquad H_x = (1/i\omega\mu_0)\frac{\partial E_y}{\partial z}. \tag{2.13}$$

We now apply the boundary conditions, which require that both E_y and H_z

be continuous at $x = \pm a$. It is straightforward to determine the value of C in terms of A for symmetric modes, or in terms of B for antisymmetric modes, as well as the characteristic equations, i.e., the relations between u and q that must be satisfied in order to meet the boundary conditions. From Eq. (2.9), we also have an additional relation between u and q that, together with the characteristic equations, determines all the allowed values of these parameters on any particular physical waveguide. We obtain this relation by writing out Eq. (2.9) for each medium, recognizing that h must be the same in each (Snell's law). Thus,

$$k_1^2 = \beta_1^2 + h^2; \qquad k_2^2 = \beta_2^2 + h^2. \tag{2.14}$$

We now multiply each of these equations by a^2 and subtract them, remembering that $\beta_2^2 = -q^2/a^2$. The result defines the important physical parameter R^2 given by

$$R^2 = (k_1^2 - k_2^2)a^2 = u^2 + q^2. \tag{2.15}$$

This quantity R depends only on the physical constants of the two media and the wavelength. Because u and q are also related by the characteristic equations, it determines all the parameters of the modes. The characteristic equation for the symmetric TE modes, found by applying the boundary conditions as described above, is given by

$$q = u \tan u. \tag{2.16}$$

For the antisymmetric TE modes, it is given by

$$q = -u \cot u. \tag{2.17}$$

We shall discuss the general solution of Eqs. (2.15)–(2.17) in Section C.

2. TM Modes

For these modes, the electric vector is in the plane of incidence, while the magnetic field is perpendicular to it. Thus, $H_x = H_z = 0$. We can solve the wave equation for either H_y or E_z and then generate from their general solution the other field components. It is customary, in the microwave literature, to determine the transverse field components from solutions of the wave equation for the longitudinal component. The observable optical manifestations of the modes depend on the transverse components, however, so that it would seem advisable to concentrate on these.

We again assume that the fields are invariant with y, so that the magnetic field satisfies the equation

$$(d^2H_y/dx^2) + \beta^2 H_y = 0, \tag{2.18}$$

where β is again given by Eqs. (2.9) and (2.14). As before, we take oscillating solutions in the interior with evanescent ones in the exterior. Availing ourselves of the symmetries of the structure, we find that the magnetic field of TM modes may have either the symmetric form

$$H_y(\text{inside}) = A[\cos(ux/a)]e^{i\omega t - ihz} \tag{2.19}$$

and

$$H_y(\text{outside}) = Ce^{\pm(qx/a)+i\omega t-ihz} \tag{2.20}$$

or the antisymmetric form

$$H_y(\text{inside}) = A[\sin(ux/a)]e^{i\omega t - ihz} \tag{2.21}$$

and

$$H_y(\text{outside}) = \pm Ce^{\pm(qx/a)+i\omega t-ihz}, \tag{2.22}$$

where the plus sign on the exponential applies in the region $x < -a$.

The corresponding components of the electric field are derived from Eqs. (2.19) and (2.20) or Eqs. (2.21) and (2.22) through the application of the Maxwell relation

$$\nabla \times \mathbf{H} = \tilde{\varepsilon}\, \partial\mathbf{E}/\partial t = i\omega\tilde{\varepsilon}\mathbf{E} = i\omega\varepsilon(1 - i\sigma/\omega\varepsilon)\mathbf{E}. \tag{2.23}$$

The tilde over the ε indicates that it may be complex-valued, as shown. This simplifies to

$$E_z = (1/i\omega\tilde{\varepsilon})\, \partial H_y/\partial x; \qquad E_x = -(1/i\omega\tilde{\varepsilon})\, \partial H_y/\partial z. \tag{2.24}$$

We now apply the boundary conditions on E_z and H_y, thereby determining the value of C in terms of A and the characteristic equations relating u and q. For the symmetric TM modes, the characteristic equation is

$$q = (\tilde{\varepsilon}_2/\tilde{\varepsilon}_1)u \tan u. \tag{2.25}$$

For the antisymmetric TM modes, it is

$$q = -(\tilde{\varepsilon}_2/\tilde{\varepsilon}_1)u \cot u. \tag{2.26}$$

Equations (2.25) and (2.26), together with Eq. (2.15), determine all the parameters of the TM modes. When the complex dielectric constants $\tilde{\varepsilon}$ have only real values, the solutions of Eqs. (2.15), (2.25), and (2.26) for real values of u and q are easily obtained by the methods described in Section C. When all these quantities are allowed to take on complex values, however, it is not possible to get a general graphical picture of the root loci. It is necessary to fix the value of the ratio of the complex dielectric constants, for example, in order to present a graphical picture of the appropriate solutions. Some interesting cases have been treated in the literature (*1*, *2*).

C. CHARACTERISTIC TE MODES OF THE PLANAR SLAB WAVEGUIDE

In the last section, we described how the fields of the slab waveguide are derived. We now wish to solve the characteristic equations so as to determine the properties of all possible TE modes on a simple symmetric guide. This analysis will introduce the many kinds of modes that a thin-film guide can support. We will repeat the most important field expressions here, so that they will be grouped together for easy reference. The geometry of the guide is as shown in Fig. 2-2.

1. General Expressions for the Mode Field

By carrying through the analysis of the last section, we find that the *symmetric TE modes* have fields given by

$$\left.\begin{aligned} E_y &= A[\cos(ux/a)]e^{i\omega t - ihz} \\ H_x &= (-h/\omega\mu_0)E_y \\ H_z &= (-iu/\omega\mu_0 a)[\tan(ux/a)]E_y \\ E_x &= E_z = H_y = 0 \end{aligned}\right\} \quad |x| \leq a \tag{2.27}$$

and

$$\left.\begin{aligned} E_y &= Ae^{\pm q}(\cos u)e^{\mp(qx/a)+i\omega t - ihz} \\ H_x &= (-h/\omega\mu_0)E_y \\ H_z &= (\mp iq/\omega\mu_0 a)E_y \\ E_x &= E_z = H_y = 0, \end{aligned}\right\} \quad a \leq |x| \tag{2.28}$$

where the upper sign applies for $a < x$ and the lower sign for $x < -a$. The various parameters are related to one another, to the free-space wavelength λ_0, and to the complex refractive indices n_1 and n_2 by the equations

$$\begin{aligned} \omega^2\varepsilon_1\mu_0(1 - i\sigma_1/\omega\varepsilon_1) = k_1^2 = k_0^2n_1^2 = (2\pi n_1/\lambda_0)^2, \\ \omega^2\varepsilon_2\mu_0(1 - i\sigma_2/\omega\varepsilon_2) = k_2^2 = k_0^2n_2^2 = (2\pi n_2/\lambda_0)^2, \end{aligned} \tag{2.29}$$

$$h^2 = k_1^2 - (u^2/a^2) = k_2^2 + (q^2/a^2), \tag{2.30}$$

$$q = u \tan u, \tag{2.31}$$

and

$$u^2 + q^2 = k_0^2a^2(n_1^2 - n_2^2) = R^2. \tag{2.32}$$

Any particular solution $u(R)$ and $q(R)$ of Eqs. (2.31) and (2.32), for a given value of the physical parameter R, completely characterizes the corresponding mode through Eqs. (2.27), (2.28), and (2 30). We list also the general defining relations between a complex refractive index $n = n' - in''$ (with n' and n'' real) and the corresponding dielectric constant $K = \varepsilon/\varepsilon_0$ and conductivity σ:

$$K = (n')^2 - (n'')^2; \qquad \sigma/\omega\varepsilon_0 = 2n'n''$$

and

$$\left.\begin{matrix} n' \\ n'' \end{matrix}\right\} = (K/2)^{1/2}[(1 + \sigma^2/\omega^2\varepsilon^2)^{1/2} \pm 1]^{1/2}.$$

For antisymmetric modes, the principal field component E_y is antisymmetric in the transverse coordinate. The fields are

$$\left.\begin{aligned} E_y &= A[\sin(ux/a)]e^{i\omega t - ihz} \\ H_x &= (-h/\omega\mu_0)E_y \\ H_z &= (iu/\omega\mu_0 a)[\cot(ux/a)]E_y \\ E_x &= E_z = H_y = 0 \end{aligned}\right\} \quad |x| \le a \tag{2.33}$$

and

$$\left.\begin{aligned} E_y &= \pm Ae^{\pm q}(\sin u)e^{\mp(qx/a) + i\omega t - ihz} \\ H_x &= (-h/\omega\mu_0)E_y \\ H_z &= (\mp iq/\omega\mu_0 a)E_y \\ E_x &= E_z = H_y = 0, \end{aligned}\right\} \quad a \le |x| \tag{2.34}$$

where, as before, the upper sign applies for $a < x$ and the lower for $x < -a$. The defining relations given by Eqs. (2.29), (2.30), and (2.32) again apply. The characteristic Eq. (2.31), however, is replaced by

$$q = -u \cot u. \tag{2.35}$$

When the conductivities σ_1 and σ_2 are vanishingly small, Eqs. (2.31) and (2.35) can be analyzed together. We will begin the analysis of the characteristic equations in the next section with this special case.

2. Solutions of the Characteristic Equation

a. Proper Surface Waves on Lossless Guides

To ensure that the energy is confined to the vicinity of the waveguide, we impose the additional boundary condition that the fields vanish as $x \to \pm\infty$. (Modes are referred to as "proper" when this condition is satisfied.) If, in addition, the conductivities vanish, then there exist solutions of Eqs. (2.31), (2.32), and (2.35) satisfying this condition for real, positive values of u and q, provided that $n_2 < n_1$. From the field expressions (2.27), (2.28), (2.33), and (2.34), we see that such roots provide for lossless propagation in the z direction. The waves outside the slab are evanescent in nature, so there is no radiation away from the interface. These evanescent waves are the well-known surface waves of the microwave literature.

A physical appreciation of the characteristic equations for this case may be obtained by combining Eq. (2.32) with Eqs. (2.31) and (2.35) to obtain

$$\cot u = \pm u/(R^2 - u^2)^{1/2} \tag{2.31a}$$

and

$$-\tan u = \pm u/(R^2 - u^2)^{1/2}. \tag{2.35a}$$

These equations, with the plus sign on the right-hand side, are solved graphically in the upper half of Fig. 2-3 for $R = \pi$ and 4π. The solid curves represent the right-hand side of both equations. The long-dashed curves give the successive branches of the left-hand side of Eq. (2.31a); the short-dashed curves show the left-hand side of Eq. (2.35a). The intersections of the solid and dashed curves locate the eigenvalues $u(R)$. Barone (*2*) numbers the successive roots of Eq. (2.31a) with the even integers, those of Eq. (2.35a) with odd integers. Each mode and its parameters can then be identified by the use of a subscript N. The root $u_N(R)$ then determines all the parameters of the TE_N mode. The various values of N are labeled in Fig. 2-3. Even values give symmetric modes; odd values, antisymmetric modes. We note that R must be greater than $(N - 1)\pi/2$ to ensure that the TE_N surface wave mode is defined on the corresponding waveguide.

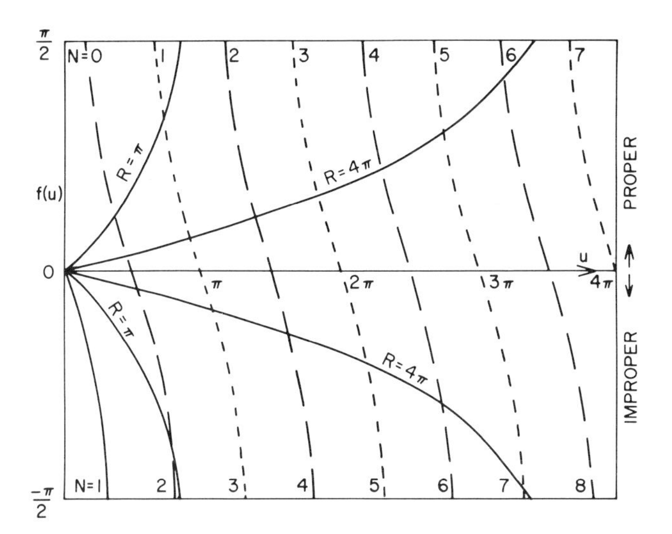

Fig. 2-3. Illustrating a graphical solution of Eqs. (2.31a) and (2.35a) for real values of u and $R^2 = (2\pi a/\lambda_0)^2(n_1^2 - n_2^2)$. The intersections of the solid and dashed curves above the u axis yield the eigenvalues of the surface wave modes; those below the axis give negative values of q, so that the exterior fields increase with distance from the slab. The integers identify the eigenvalues of the TE_N mode (2).

The optical interpretation of these TE_N modes is straightforward. If, in Eq. (2.27), the transverse dependence of the fields is rewritten in terms of exponential functions, it becomes apparent that a TE_N mode is the superposition of two plane waves whose normals make an angle θ_N with the normal to the interface such that

$$u/a = k_1 \cos\theta_N; \qquad h = k_1 \sin\theta_N.$$

We note that, because u_N is less than R for all possible surface waves, θ_N is less than θ_C, the critical angle, for these waves. The waves outside the slab are the familiar evanescent waves associated with the total internal reflection of plane waves incident on an interface between two dielectric media at angles greater than θ_C. [For an elaborate treatment of the optical interpretation of the modes of slab waveguides, the reader is referred to Lotsch (*3*).]

b. Improper Surface Waves on Lossless Guides

If we do not impose the condition that the fields must vanish as x approaches infinity, then Eqs. (2.31), (2.32), and (2.35) clearly have a second set of solutions for real q and u found by letting q take on negative values. The corresponding fields increase exponentially with distance from the slab. For this reason, they have been called improper modes. The roots of the characteristic equation for this case are shown in the lower half of Fig. 2-3. Clearly, the number of such roots is again finite for finite R. Be cause the solid and dashed curves are now both monotonically decreasing, however, it is not obvious from the figure precisely how many roots exist or how they should be ordered. By employing a different graphical representation of the characteristic equations, used by Barone (*2*) and Collin (*4*), we can show that two roots generally exist on all but the first intersected branch of the tangent and cotangent functions. In this representation, we graph Eqs. (2.31), (2.32), and (2.35) separately, plotting q as a function of u in each case. The result is shown in Fig. 2-4 for $R = 1$, π, and $31\pi/20$. The

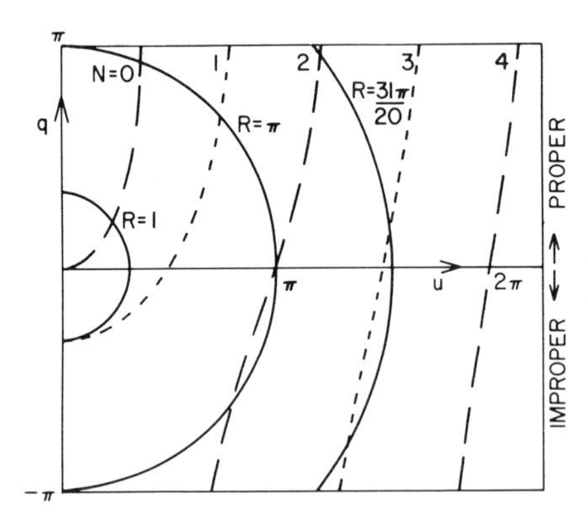

Fig. 2-4. An alternate graphical solution of Eqs. (2.31), (2.32), and (2.35) for three values of R^2, with u and q real.

long-dashed curves represent Eq. (2.31); the solid curves, Eq. (2.32); the short-dashed curves, Eq. (2.35). This representation indicates very lucidly that the number of roots for real q and u is finite and dependent on R. The first root, which represents a surface wave, exists for all values of R. The second root yields a proper surface wave for $R > \pi/2$ and an improper surface wave for $1 < R < \pi/2$. For modes of order greater than two, the roots appear in pairs if they exist at all. The highest pair may be a double root. This occurs when the circle of radius R is tangent to the corresponding dashed curve. As R increases above the value yielding a double root, the corresponding pair of roots first yields two improper modes, then one improper and one proper surface wave mode.

The complete behavior of the roots corresponding to the first two dashed curves for real values of q becomes apparent only when u is allowed to become imaginary. Accordingly, we set $u = iu''$, so Eqs. (2.31), (2.32), and

(2.35) become

$$q = u'' \tanh u'', \tag{2.31b}$$

$$q^2 - (u'')^2 = R^2, \tag{2.32a}$$

and

$$q = -u'' \coth u''. \tag{2.35b}$$

Apparently, q must be negative when u is imaginary, so the corresponding roots again yield improper modes. By squaring Eq. (2.35b) and substituting for q^2 from Eq. (2.32a), we obtain

$$R^2 = (u'')^2/\sinh^2 u''. \tag{2.35c}$$

The roots u'' for all permissible values of R^2 are easily found by plotting R^2 as a function of u'', as shown in the short-dashed curve of Fig. 2-5. The corresponding values of q are found from Eq. (2.35b). From Eq. (2.30), we then have

$$h^2 = k_1{}^2 + [(u'')^2/a^2],$$

so that the propagation constant in the z direction is greater than that of a plane wave in a slab of infinite thickness and index n_1. The electric field in the slab, Eq. (2.33), now has the transverse dependence $\sinh(u''x/a)$.

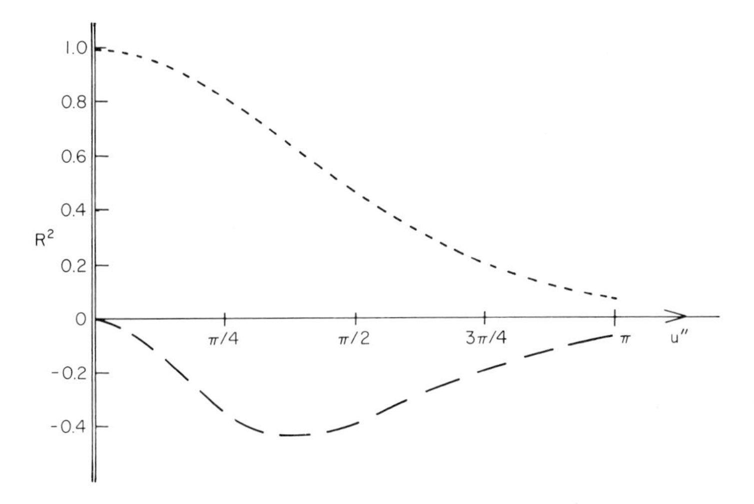

Fig. 2-5. Solutions of Eqs. (2.31) and (2.32) (short dashes) and Eqs. (2.32) and (2.35) (long dashes) for imaginary values of $u = iu''$ and real values of q. These are plots of Eqs. (2.31c) and (2.35c) of text.

This rather unfamiliar TE wave appears to be of physical consequence only in an asymptotic description of the field strengths in the vicinity of sources.*

By squaring Eq. (2.31b) and substituting from Eq. (2.32a), we obtain

$$R^2 = -(u'')^2/\cosh^2 u''. \tag{2.31c}$$

There are clearly no roots of this equation for $R^2 > 0$. Roots do exist, however, for $-0.44 < R^2 < 0$, as shown by the long-dashed curve of Fig. 2-5.

The introduction of negative values of R^2 has led to complications; for example, Eq. (2.31c) yields two roots u'' for each value of R^2 in the range specified. It will be necessary to consider complex values of R^2 before we can appreciate how these two roots should be treated. It may also be apparent to the reader that Eqs. (2.31), (2.32), and (2.35) have an infinite number of complex roots, even for real values of R^2 (lossless media). These provide the parameters of the so-called leaky waves, whose importance in predicting the radiation field of dielectric antennas is well known to microwave scientists (*1*, *2*, *4–9*). Rather than treat these separately, we will proceed directly to the general case—the complex roots for complex values of R^2.

c. The General Case: Complex Waves in Lossy Media

Before proceeding to the detailed analysis, we first consider some of the general properties of the characteristic Eqs. (2.31) and (2.32) [or Eqs. (2.35) and (2.32)]. Note that these equations are quadratic in u, so the values of u^2, not u, are restricted by the boundary conditions. (This is because the boundary conditions are the same at $x = \pm a$.) We see also that R^2, not R, is the physical parameter which arises naturally from the wave equations, and that it is linear in the electromagnetic constants of the media, i.e., ε_1, ε_2, σ_1, σ_2. From Eq. (2.30), we observe that the characteristic equation is also quadratic in h, the direction of propagation (toward positive or negative z) having no effect on the permissible values of the propagation constant. The only linear characteristic parameter, then, is q. With it, we may impose appropriate boundary conditions on the radiation field far from the slab. If we eliminate q from the characteristic equations, as in the transformations leading to Eqs. (2.31c) and (2.35c), we obtain equations in the quadratic parameters R^2 and u^2.

* On certain structures, such as plasma slabs in air (2), proper TM modes of this type may be supported. In this case, R^2 is less than zero, and the ratio of the dielectric constants, $\varepsilon_1/\varepsilon_2$, is also less than zero. Equations (2.25) and (2.26) then yield proper modes that are evanescent in both directions away from the interfaces.

We begin by squaring Eq. (2.31) and substituting for q^2 from Eq. (2.32). Then,

$$R^2 = u^2/\cos^2 u. \tag{2.31d}$$

We now set $u = u' + iu''$ (where u' and u'' are real) and rewrite Eq. (2.31d) in terms of its magnitude and argument (phase). For clarity, we introduce the symbols M_s and M_a for the magnitude of R^2 for symmetric and antisymmetric modes, respectively, and ϕ_s and ϕ_a for the corresponding arguments of R^2. Then, Eq. (2.31d) can be rewritten in the form

$$M_s = 2[(u')^2 + (u'')^2]/(\cosh 2u'' + \cos 2u') \tag{2.31e}$$

and

$$\phi_s = 2\tan^{-1}(u''/u') + 2\tan^{-1}(\tan u' \tanh u''). \tag{2.31f}$$

Similarly, from Eqs. (2.32) and (2.35), we obtain

$$R^2 = u^2/\sin^2 u, \tag{2.35d}$$

$$M_a = 2[(u')^2 + (u'')^2]/(\cosh 2u'' - \cos 2u'), \tag{2.35e}$$

and

$$\phi_a = 2\tan^{-1}(u''/u') - 2\tan^{-1}(\cot u' \tanh u''). \tag{2.35f}$$

These equations may be represented graphically by contour plots on the complex u plane. We note that M_s and M_a are invariant with changes of sign of either u' or u'', so their contours are symmetric about both axes and only those in the first quadrant of the u plane need be plotted. The same is true for the contours of equal phase ϕ. However, the value of the phase on a particular contour curve in the first quadrant changes sign when the mirror image of that contour in either axis is drawn. This is apparent from Eqs. (2.31f) and (2.35f), which show that $\phi(-u', u'') = \phi(u', -u'') = -\phi(u', u'') = -\phi(-u', -u'')$. As will be evident from the graphs, the contours of M_s (or M_a) are orthogonal to those of ϕ_s (or ϕ_a) except at isolated (branch) points. A contour map of Eqs. (2.31e) and (2.31f) is shown in Fig. 2-6 in the range $0 < u' < 3\pi/2$, $-\pi < u'' < \pi$. Contours illustrating Eqs. (2.35e) and (2.35f) appear in Fig. 2-7 for the same range of u'' with $0 < u' < 19\pi/10$. These ranges are sufficient to illustrate the character of all the roots of these equations for all values of R^2.

Let us first discuss the relatively uncomplicated root loci shown in Fig. 2-7. The double lines, corresponding to the phase contours $\phi_a = \pm 0$ or $\pm\pi$, separate the right half of the u plane into branches in each of which

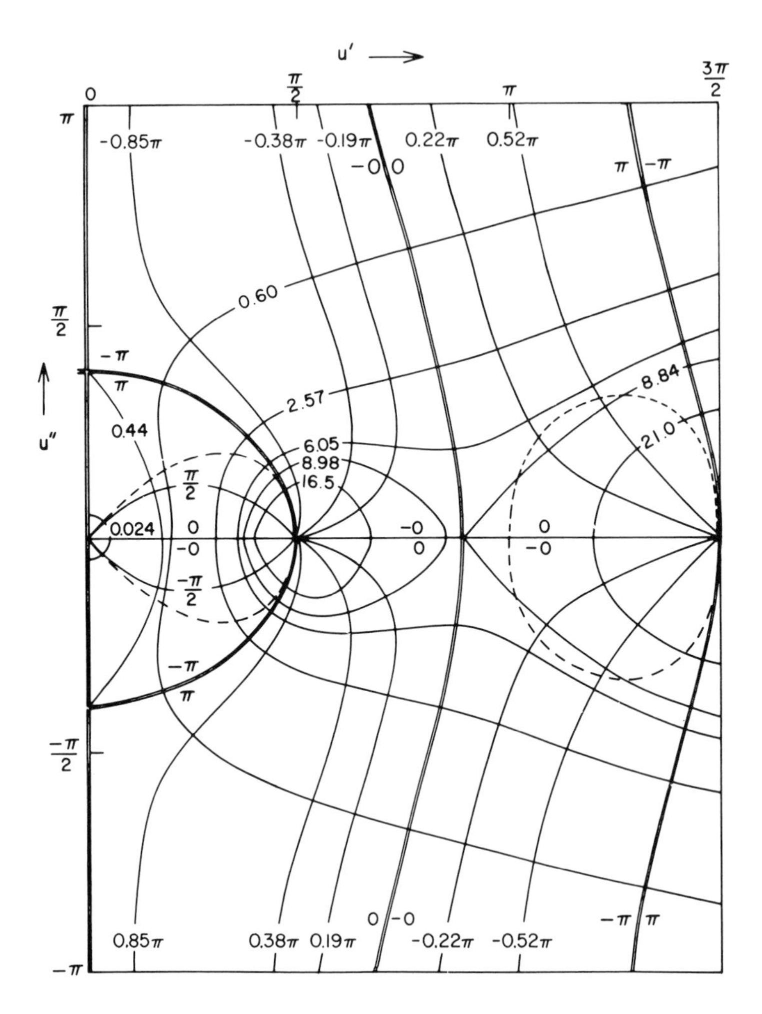

Fig. 2-6. Root loci of Eqs. (2.31) and (2.32) on the right half of the complex $u \equiv (u' + iu'')$ plane for complex values of $R^2 \equiv M_s e^{1\phi_s}$[Eqs. (2.31d–f)]. $\phi_s = 0$ on the u' axis. The u' and u'' axes are scaled above and to the left of the figure, respectively. The double lines divide the half-plane into domains (branches) in each of which R^2 takes on all possible physical values. On the first branch, lines of equal ϕ_s run from $(u', u'') = (0, 0)$ (where $M_s = 0$) to $(\pi/2, 0)$, where $M_s = \infty$. On the other branches, they originate at $u'' = \infty$, where $M_s = 0$, and terminate at $u' = (2N + 1)\pi/2$, where $M_s = \infty$. Roots lying inside the dashed cutoff curves, where $\text{Re}\, q = 0$, correspond to complex surface waves which are localized in and around the guide. Roots lying outside these curves yield leaky waves that radiate away from the guide.

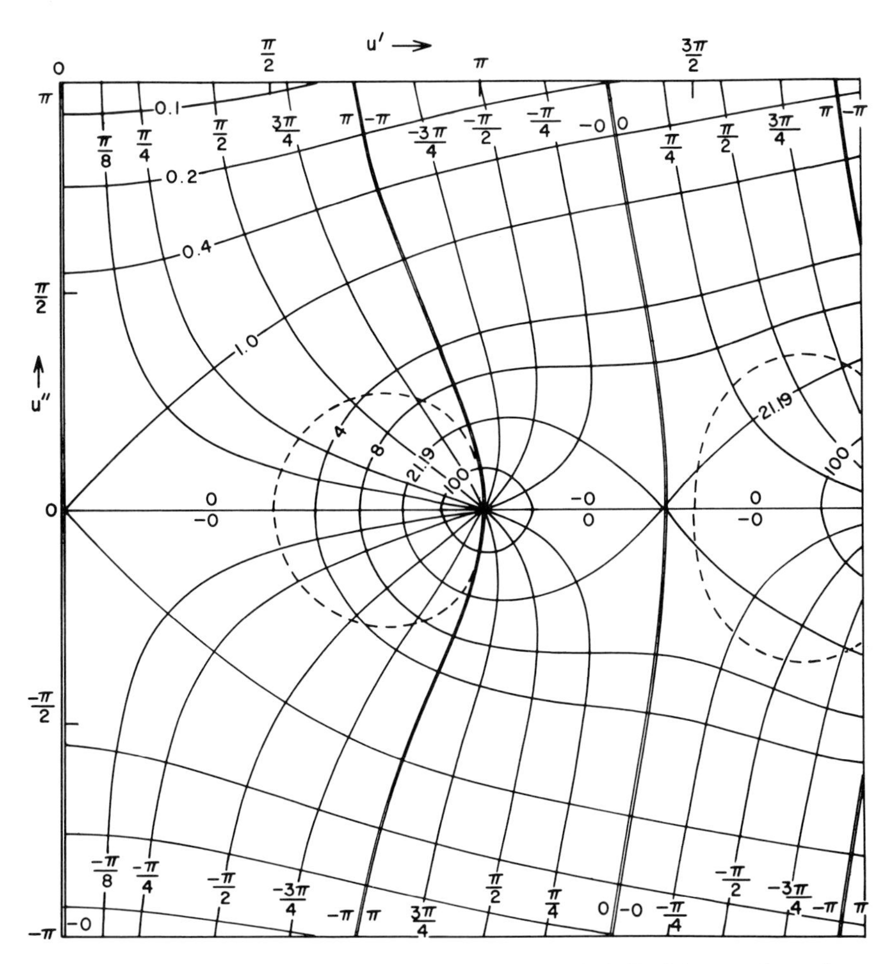

Fig. 2-7. Root loci of Eqs. (2.32) and (2.35) on the right half of the complex u plane for complex values of $R^2 \equiv M_a e^{i\phi_a}$[Eqs. (2.35d-f)]. Lines of equal ϕ_a originate at $u'' = \infty$, where $M_a = 0$, and terminate at $u = N\pi$, where $M_a = \infty$ ($N = 1, 2, 3, \ldots$). As in Fig. 2-6, double lines separate successive branches, while the dashed cutoff curves, where Re $q = 0$, separate the complex surface wave roots from those of leaky waves.

R^2 takes on all possible physical values; i.e., $0 \leq M_a \leq \infty$; $-\pi \leq \phi_a \leq \pi$. This becomes evident on examining Eq. (2.35f) together with Fig. 2-7.

As the phase contours of Fig. 2-7 suggest, the lines of equal ϕ_a run parallel to the imaginary axis for very small values of M_a. From Eq. (2.35f), we obtain

$$\phi_a \approx 2u' \qquad \text{when} \quad M_a \approx 0. \tag{2.35g}$$

[We have chosen to label the phase contours in Figs. 2-6 and 2-7 according to their equivalents in the range $-\pi < \phi < \pi$ rather than as specified in Eq. (2.35g). This choice will facilitate the discussion of the solutions for negative values of u'' and the resulting separation of the various branches.] From Eq. (2.35g), we see that all the phase contour lines in the first quadrant originate at $u'' = \infty$, where $M_a = 0$, and have infinite slope there. Lines representing successive equal increments $\Delta\phi_a$ are displaced from each other by equal increments $\Delta u'$, with $\Delta\phi_a = 2\,\Delta u'$. When $M_a \approx 0$ and $u'' \gg u'$, we note that as u' increases from 0 to π, ϕ_a covers the full range of 2π radians. As u' increases beyond π, the values of ϕ_a repeat periodically. Thus, there are an infinite number of root loci in the first quadrant for every value of ϕ_a.

The phase contours $\phi_a = 0$ in Fig. 2-7 give the root loci for the important case of vanishing conductivity in both media. The first contour with $\phi = 0$ in the first quadrant coincides with the positive u'' axis for $0 < M_a < 1$ and with the u' axis in the range $0 < u' < \pi$ for $1 < M_a < \infty$. (Compare the short-dashed curve of Fig. 2-5 and the first such curve of Fig. 2-4.) All other contours with $\phi = 0$ in the first quadrant of the u plane start from $(u', u'') = (N\pi, \infty; N = 1, 2, 3, \ldots)$, when $M_a = 0$, and proceed toward the real axis as M_a increases, following curves described by

$$u'' \coth u'' = u' \cot u' . \tag{2.35h}$$

(As described by Barone (*2*) and Collin (*4*) and discussed in the next section, this equation defines the root loci of the so-called leaky wave modes.) They meet the u' axis perpendicularly at point u'_N given by

$$\tan u'_N = u'_N . \tag{2.35i}$$

They then turn 90° and follow the real axis to the point $(u', u'') = (N\pi, 0)$, where $M_a = \infty$.

It is apparent that the successive $\phi_a = 0$ contours, described by Eq. (2.35h), divide the first quadrant into domains in each of which R^2 takes on all possible physical values. If we now consider the reflections in the u' axis of these contours, which are also described by Eq. (2.35h), we see that they similarly divide the fourth quadrant. Thus, considering the entire right half of the u plane, we find that there are two roots u for any value of R^2 in each domain bounded by the nonreal parts of the $\phi_a = 0$ contours. The branch lines $\phi_a = \pm\pi$, described by Eq. (2.35j), then separate each of these domains into two branches, so that there is only one root locus on each branch of Eq. (2.35d) so defined:

$$u'' \tanh u'' = -u' \tan u' . \tag{2.35j}$$

This choice of branch cuts, although not unique, conforms best to the analysis of Eq. (2.35c) for real values of u and/or q. In particular, Fig. 2-8 illustrates the successive root loci of Eq. (2.35d) for a very small positive value of ϕ_a.

The situation for the root loci of symmetric modes shown in Fig. 2-6 is similar. Again, the $\phi_s = 0$ contours (defined by $u' \tan u' = -u'' \coth u''$) and the $\phi_s = \pm\pi$ contours (defined by $u' \cot u' = u'' \tanh u''$) divide the right half of the u plane into an infinite number of branches in each of which R^2 takes on all possible physical values.

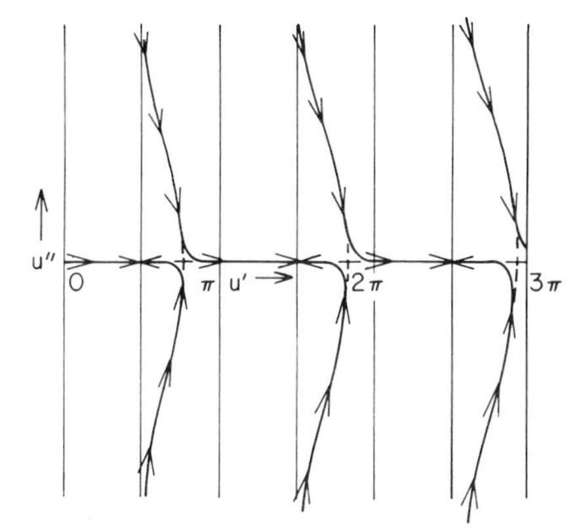

Fig. 2-8. Successive root loci of Eqs. (2.32) and (2.35) for a small value of arg R^2.

The first two branches are somewhat peculiar and require special attention. On these branches, all the phase contours converge on the point $(u', u'') = (\frac{1}{2}\pi, 0)$ when $M_s = \infty$. As M_s goes to 0, however, the root loci on one branch approach the origin of the u plane, while those on the other approach $(\frac{1}{2}\pi - \frac{1}{2}\phi_s, \ -\infty)$ for $0 < \phi_s < \pi$, or $(\frac{1}{2}\pi + \frac{1}{2}\phi_s, \ \infty)$ for $-\pi < \phi_s < 0$. The first $\phi_s = \pi$ contour in the first quadrant, as shown in Fig. 2-6, together with the real axis for $u' < \frac{1}{2}\pi$, forms the boundary of the subdomain within which one root is defined for all values of M_s with $0 < \phi_s < \pi$. This boundary coincides with the imaginary axis in the range $0 < u'' < 1.2$, where $u'' = \coth u''$ from Eq. (2.31f). Over this range of u'', M_s increases from 0 to 0.44. As M_s increases beyond 0.44, the contour turns 90° to the right, then bends down toward the real axis, which it meets perpendicularly at $u' = \frac{1}{2}\pi$ when M_s becomes infinite. The second root locus for a given value of $\phi_s > 0$ is below and to the right of the reflection of this subdomain in the real axis. To illustrate this, the successive root loci for a positive value of ϕ_s just less than π are sketched in Fig. 2-9. For the corresponding negative value of ϕ_s near $-\pi$, the root loci are simply the

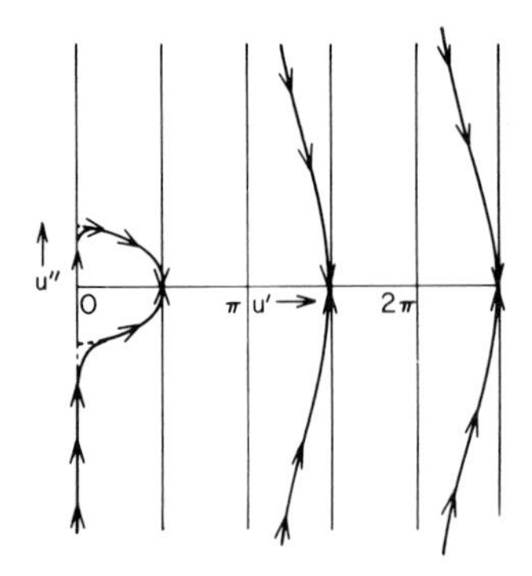

Fig. 2-9. Successive root loci of Eqs. (2.31) and (2.32) for arg R^2 slightly less than π.

reflections of the curves of Fig. 2-9 at the real axis, so that we again find, for symmetric modes, that $u(R^{2*}) = u^*(R^2)$.

For a small positive value of ϕ_s, the root loci are sketched in Fig. 2-10. Note that one of the roots in the first domain is just above the real axis for the full range of M_s. All other root loci behave in a manner very similar to those for antisymmetric modes, as in Fig. 2-7, the only major difference being the shift of $\pi/2$ in the origins of the curves and their terminations,

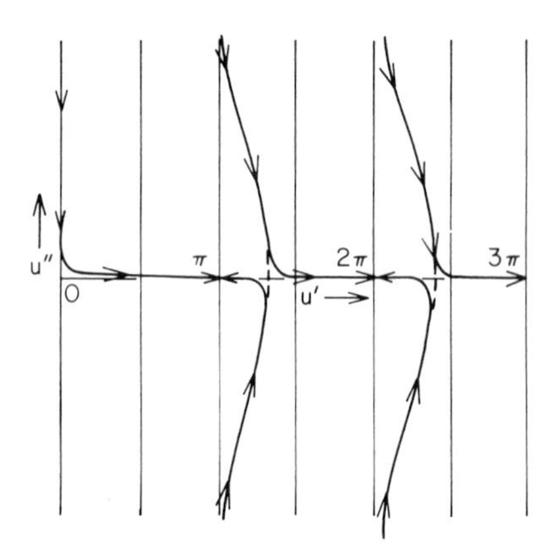

Fig. 2-10. Successive root loci of Eqs. (2.31) and (2.32) for a small value of arg R^2.

corresponding to $M_s = 0$ and ∞, respectively. For $M_s \approx 0$ and $u'' \gg u'$, we find from Eq. (2.31f), analogous to Eq. (2.35g), that

$$\phi_s \approx -\pi + 2u' \qquad \text{when} \quad M_s \approx 0. \tag{2.31g}$$

We also note from Eqs. (2.31f) and (2.35f) that each phase contour meets its termination (where $M_s = \infty$) at an angle with the u' axis given by $\theta = \tan^{-1}(u''/-\Delta) = \phi_s/2$, where Δ is a small increment in u' from the point of intersection of the contour with the u' axis.

3. Physical Characteristics of the Modes

From the root loci $u(R^2)$ found in the last section, we can interpret and classify the modes. Let us first rewrite E_y in Eq. (2.27) in the form of the superposition of two nonuniform plane waves. The complex propagation constant h is first rewritten in terms of its real and imaginary parts h' and h''. Then,

$$E_y = A \exp[-(u''x/a) + h''z] \exp[i\omega t + (iu'x/a) - ih'z] \\ + A \exp[(u''x/a) + h''z] \exp[i\omega t - (iu'x/a) - ih'z].$$

The normals to the planes of constant phase make angles $\theta_{\text{ph}} = \pm\tan^{-1}(h'a/u')$ with the slab walls; the planes of constant amplitude make angles $\theta_{\text{am}} = \pm \tan^{-1}(h''a/u'')$ with the walls. It can be shown that these planes are perpendicular to each other if the conductivity is zero.

Outside the guide $(a < x)$, we have, with $q = q' + iq''$,

$$E_y = C \exp[-(q'x/a) + h''z] \exp[i\omega t - (iq''x/a) - ih'z].$$

a. Complex Surface Waves

One means of differentiating between the roots is by imposing the condition $q' = \text{Re}\, q > 0$. This condition is appropriate for all waves that are bound to the slab and its immediate surround. A bound mode is said to be cut off when $q' = 0$. From Eq. (2.31), we find that, in the u plane, this cutoff condition is described by

$$u' \sin 2u' = u'' \sinh 2u''. \tag{2.31h}$$

The two dashed curves in Fig. 2-6 represent the first two loci of the cutoff condition (2.31h). There are an infinite number of them, originating at $u = (N\pi, 0)$, $N = 0, 1, 2, \ldots$, and terminating at $u = ([2N + 1]\pi/2, 0)$. In regions of the u plane bounded from above and below by curves satisfying Eq. (2.31h), the corresponding fields outside the guide have $q' > 0$, and thus decay away from the slab. These may be called complex surface waves. They are akin to surface waves in that they are evanescent in a direction away from the slab. They differ from surface waves in lossless media, however, in that the corresponding wave normals are not perpendicular to the slab walls, but are tilted at an angle to it to allow for a net transfer of power from one medium to the other to compensate for losses (or to distribute gains, as in a laser amplifier). The normal to the phase fronts outside the slab has a component toward the slab if the interior medium is the more lossy. Note that in each successive domain defining a pair of roots of the characteristic equation, only one root can yield a proper complex surface wave.

The dotted cutoff curves in Fig. 2-7, which satisfy

$$u' \sin 2u' = -u'' \sinh 2u'', \tag{2.35k}$$

divide the modes in the same way. Roots inside these curves also give complex surface waves, since $q' > 0$. Note also that, according to Eqs. (2.31) and (2.35), q depends on u^2 rather than u, so that $\operatorname{Im} q = q''$ takes its sign from the product $u'u''$. It is thus positive in the first quadrant and negative in the fourth. This is just what is required to give the appropriate power flow from the surface wave when the slab medium is more lossy than the exterior medium (or into the surface wave when the opposite is true). Note that only a finite number of complex surface waves can exist on a given guide (a given value of R^2), since successive domains of definition of values of u yielding surface waves correspond to larger values of R^2, i.e., larger values of the slab thickness $2a$.

From Eq. (2.30), we see that the propagation constants in the z direction, h, are defined not only in terms of R^2 and $u(R^2)$, but also in terms of the refractive indices. We may express this explicitly by combining Eqs. (2.30) and (2.32) to indicate the dependence of h^2 on R^2:

$$\begin{aligned} h^2 = k_0^2 n_e^2 &= (n_1^2/\Delta n^2)(R^2/a^2) - (u^2/a^2) \\ &= (n_2^2/\Delta n^2)(R^2/a^2) + (q^2/a^2) \\ &= (k_2^2 u^2 + k_1^2 q^2)/R^2, \end{aligned}$$

where n_e defines the effective refractive index of the mode and $\Delta n^2 = n_1^2 - n_2^2$. In the usual passive optical waveguide, one would select h to yield waves having decreasing amplitudes with increasing z. For waveguide media with gain (e.g., laser amplifiers), however, the opposite choice may be appropriate. It should be noted that the phase velocity need not, and sometimes cannot, be selected to yield waves progressing away from a source at $z = 0$. This is because of the possibility, as with plasma slabs in air (1), of backward waves, i.e., waves whose phase and group velocities have opposite sign.

b. Leaky Waves

All the complex roots u that yield values of $q' < 0$ (those outside the dashed cutoff curves in Figs. 2-6 and 2-7) determine the parameters of the so-called improper modes or leaky modes.* Other names, such as quasi-modes and pseudomodes, are also encountered in the literature. Because the corresponding fields do not satisfy the radiation condition at $x = \pm\infty$, these are not proper modes. Nevertheless, some of these roots can play an important part in the description of the field at large distances (relative to the wavelength) from sources above or in the slab. When so used, the corresponding fields apply to a restricted angular region in the space outside the slab, not to the entire space as implied by Eqs. (2.28) and (2.34). For passive waveguides, they then describe nonuniform plane waves that contribute a net power away from the slab at some angle θ to the interface. The amplitude of these waves decreases exponentially along any radius vector within their angle of definition in a direction away from the region of the source or sources. Such waves have been studied extensively only for cases where the refractive indices are real; the imaginary part of h then describes a loss rate due to radiation only. In general, only a small number of these leaky modes will be of importance in a given guide, since the loss rates for most of them are quite high. The larger the thickness of the slab, however, the greater will be the number of observable leaky waves. The guiding mechanism is simply Fresnel reflection at the interface, the reflection coefficients being very high for almost-axial propagation. The radiation losses are due to refraction.

For the hollow dielectric waveguide studied theoretically by Marcatili and Schmeltzer (*10*) and experimentally by Sawatari and Kapany (*11*), R^2 is negative, since $n_1^2 < n_2^2$. The corresponding roots u of the characteristic equation lie on the $\phi = \pi$ contours in Figs. 2-6 and 2-7, outside the cutoff curves. Thus, all the modes of this system are leaky waves or improper modes. That they are physically significant was demonstrated by Sawatari and Kapany (*11*). A rigorous field-theoretic formulation of problems involving these leaky modes should include explicit consideration of the impressed field. Such formulations have been given by Collin (*4*), Barone (*2*), and others (*7*, *9*).

Certain roots of Eqs. (2.31) and (2.35) never contribute to the radiation field, through they may be important at distances less than a few wavelengths from the source, where higher-order approximations are required (*5*). Among these are those roots lying along the imaginary u axis in Figs. 2-6

* Certain TM modes on plasma slabs (1) and modes of more complicated lossless structures (7), such as gratings, may have complex-valued wavenumbers yielding proper, spectral waves. The term leaky mode is often applied to these waves also, so that it is not generally correct to equate leaky with improper, though we can do so here.

and 2-7 and along the real axis outside the dashed cutoff curves.* Thus, the roots corresponding to $q' < 0$ in Fig. 2-4 and those of Fig. 2-5 are generally of no consequence in optical problems. These are sometimes called improper modes, as distinguished from leaky modes on waveguides where the refractive indices are real and $n_2^2 < n_1^2$. For general complex indices, however, it does not seem possible to give a general classification according to some scheme of relative propriety. An interesting microwave experiment in which all these wave types came into play was reported by Norton (*5*). In Chapter 3, we will use a leaky wave representation to describe the launching of modes in thin-film waveguides by the mechanism of frustrated total reflection.

REFERENCES

1. T. Tamir and A. A. Oliner, *Proc. IEEE* **51**, 317 (1963).
2. S. Barone, Rep. No. R-532-56, Microwave Res. Inst., Polytechnic Inst. of Brooklyn, New York (November 1956); S. Barone and A. Hessel, Rep. No. R-698-58, Microwave Res. Inst., Polytechnic Inst. of Brooklyn, New York (December 1958).
3. H. K. V. Lotsch, *Optik* **27**, 239 (1968).
4. R. E. Collin, "Field Theory of Guided Waves," Chapter 11. McGraw-Hill, New York, 1960.
5. D. E. Norton, 1965 *IEEE Int. Conv. Record* Part 5, p. 200 (1965).
6. E. S. Cassedy and M. Cohn, *IEEE Trans. Microwave Theory Tech.* **MTT-9**, 243 (1961).
7. T. Tamir and A. A. Oliner, *Proc. IEE* (*London*) **110**, 310 (1963).
8. H. M. Barlow and A. L. Cullen, *Proc. IEE* (*London*) **100**, 329 (1953).
9. R. E. Collin and F. J. Zucker, eds., "Antenna Theory Part II." McGraw-Hill, New York, 1969.
10. E. A. J. Marcatili and R. A. Schmeltzer, *Bell Syst. Tech. J.* **43**, 1783 (1964).
11. T. Sawatari and N. S. Kapany, *J. Opt. Soc. Amer.* **60**, 132 (1970).

* See footnote preceding Eq. (2.31c).

CHAPTER 3

Distributed Coupling in Slab Waveguides

In the last chapter, we derived the general properties of all the characteristic TE waves (modes) of a symmetric slab guide by direct solution of the homogeneous Maxwell equations. This is the only waveguide problem for which an explicit general graphical solution can be obtained for all possible values of the material parameters. It is by no means, however, the only case of interest. TM modes are at least equally important, as are the bound and leaky modes of guides formed of several thin films. Nevertheless, the symmetric TE mode analysis provides the necessary foundation from which to study modes on more complicated guides. In this chapter, we will treat questions involving the coherent exchange of energy between waves on two parallel slab guides and between an optical beam and a leaky thin-film mode.

We will begin (Section A) with an interferometric (geometric) description of the coupling of a uniform beam of light into a thin film through the mechanism of frustrated total reflection. In the initial treatment, we assume that transmitted and reflected rays are not shifted with respect to incident rays at totally (or nearly so) reflecting surfaces. In Section B, we present an analysis of the same problem in terms of leaky waves. In Section C, we consider the coupling between parallel guides. By comparing wave and ray descriptions of this process, we show that the Goos–Haenchen shift (Section D) provides the necessary modifications to bring geometric and wave theories into agreement. In Section D, we treat the Goos–Haenchen shift directly from the viewpoint of waveguide theory. By comparing the loss rates of leaky waves with those found geometrically, we are able to obtain an essentially correct theory of prism-film couplers from an interferometric theory of waveguide mode propagation.

A. FRUSTRATED TOTAL REFLECTION FILTER AND PLANAR WAVEGUIDES

We mentioned in Chapter 2 that the characteristic waves of slab waveguides could be launched by generating, at the entrance face of the guide, an intensity distribution of the form $\sin^2(\beta_N x)$ or $\cos^2(\beta_N x)$, where $\beta_N = k_1 \cos \theta_N$ is determined from the characteristic equation. As readers familiar with optical thin films may recognize, however, there are better ways to accomplish this launching.

Consider, for example, the well-known frustrated total reflection (FTR) filter first described by Leurgans and Turner (*1*) in 1947. It is illustrated schematically in Fig. 3-1. It consists of two prisms of high refractive index separated by three thin films: two of low refractive index in contact with the prisms, and a third, central layer called the spacer with a high refractive index. If a collimated beam of white light in one prism is incident on the films at an angle greater than the critical angle, most of the beam is reflected. For one wavelength (actually a narrow band) and polarization, however, a strong field builds up in the spacer layer, and most of the incident light is transmitted into the second prism and beyond. This filtering action may be most readily understood in terms of the distributed coupling of a beam to a leaky waveguide mode.

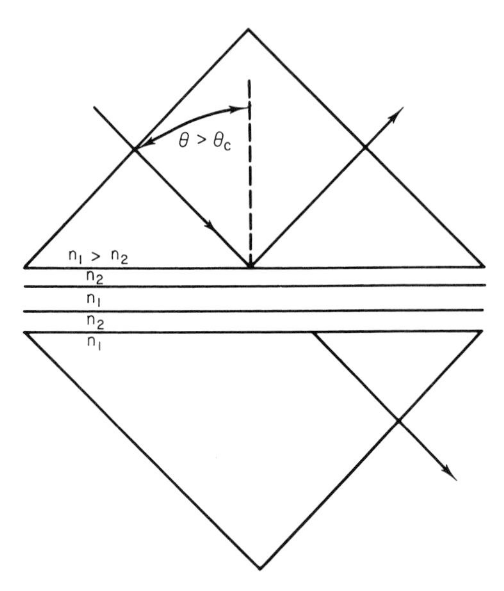

Fig. 3-1. Schematic illustration of a symmetric, frustrated total reflection (FTR) filter.

1. FTR Coupling: A Qualitative Description

In the system shown in Fig. 3-1, the incident beam is viewed as a source that efficiently launches a characteristic leaky wave of the waveguide whose interior is the spacer layer. Thus, when the angle of incidence, the wavelength of the incident beam, and the thickness of the spacer layer are such as to satisfy a transverse resonance condition, a strong wave builds up with distance along the waveguide from one side of the beam. For a wide beam of uniform amplitude, an equilibrium power density is reached at a distance down the guide governed by the reflectivity of the low-index films. The guided wave maintains this density over the remainder of the beam's width, while transmitting part of its radiation into the second prism. At the same time, it cancels, by destructive interference, the primary reflected wave in the

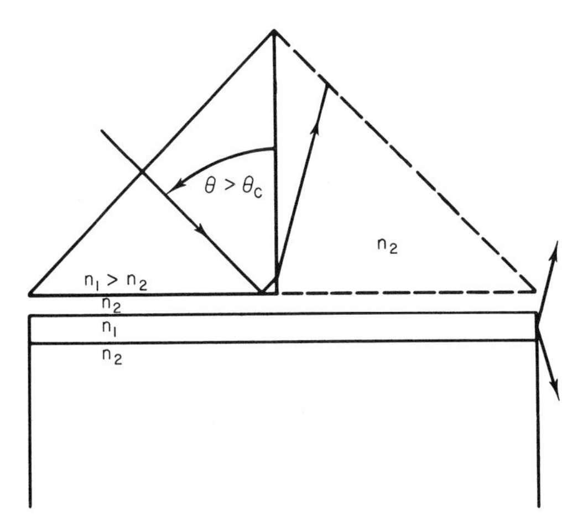

Fig. 3-2. Schematic illustration of a one-sided, truncated FTR system used to couple a beam to a mode of a thin-film guide.

first. At that distance down the slab where the incident beam terminates, the characteristic wave in the spacer begins to decay exponentially, because of radiation into both prisms as it progresses further down the spacer. It is probably now obvious to the general reader that the second prism is not necessary if one is interested in the FTR principle as a means of mode launching rather than filtering. Though the incident beam will be entirely reflected rather than transmitted when the second prism is removed, an essentially identical situation will be obtained in the waveguide spacer. The power density there will build up exponentially from one side of the beam until it reaches an equilibrium, after which the reflectivity of the system will be unity. From that point along the guide where the end of the incident beam is reached, the energy in the spacer decays exponentially through radiation back into the prism. Obviously, if one truncates the prism at the end of the incident beam, as indicated in Fig. 3-2, so that only the spacer surrounded by lower-index material remains, the power stays in the mode.

The first detailed theoretical description of prism couplers was published by Ioganseu (*2*). Harris and Shubert (*3*) and Tien *et al.* (*4*) demonstrated them experimentally in 1969. Several different theories of prism-film couplers subsequently appeared in the literature (*5*, *8*). The theoretical description presented in this section is similar to that of Tien and Ulrich (*6*) and is a precise statement of the analysis we outlined in 1961. The altenate description in Section B is new.

2. FTR Coupling: A Quantitative Description

The FTR filter provides an excellent means of demonstrating some of the characteristic waves discussed in a general way in the last section, specifically, the leaky waves. Consider again the frustrated total reflection arrangement shown schematically in Fig. 3-2. A simple optical analysis of this case follows the theoretical approach to multiple-beam interferometry introduced by Airy in 1833. It is called the "ray summation method" by Sommerfeld (*9*) who compares it with the "boundary value method" now in more prevalent use in optical thin-film theory. It appears that its power in the analysis of waveguide problems has only rarely been put to use (*6*, *10*).

Let us idealize the situation somewhat by treating the prism as an infinite medium of index n_1. The spacer is an infinite thin film, also of index n_1, and rests on a semiinfinite medium of index n_2. The spacer is separated from the upper medium by a thin film, also of index n_2. A collimated beam of finite cross section and uniform amplitude is incident from above at an angle $\theta > \theta_C = \sin^{-1}(n_2/n_1)$.

A rigorous field-theoretic approach to the determination of the fields resulting in each region from the impressed field (the finite beam) would involve finding those solutions of Maxwell's equations that match all the boundary conditions at the interfaces. This might be done, for example, by expanding the incident beam in an angular spectrum of uniform plane waves (*7*, *8*). The solution obtained is in the form of a complicated integral that cannot be evaluated analytically. In contrast, the ray summation method is both mathematically simple and physically visualizable. It quickly predicts the most important observable phenomena, missing only relatively minor effects, such as those of diffraction at the edges of the beam.

In the ray summation method, we consider the field along a ray at some plane $z = z_0$ in the spacer. It is assumed that this field is the resultant of the superposition of all rays transmitted through the reflecting layer along the width of the incident beam up to this plane and multiply reflected down the guide. Similarly, the reflected ray at any point along the structure is the superposition of a ray transmitted through the layer from the spacer guide and a directly reflected ray.

The well-known results of the theory of frustrated total reflection (*11*) give us formulas for the reflectance and transmittance of a plane wave incident at $\theta > \theta_C$ on a thin, low-index film separating two infinite regions of index n_1. Assuming these to apply locally along thin pencils of rays, we examine the consequences of superposing the amplitudes of many such pencils.

Let $A(z_0)$ represent the amplitude of the ray in the spacer, as illustrated in Fig. 3-3. We obtain it by adding up the contributions from all incident ray pencils that effect it. Thus, from the figure, it is clear that incident ray num-

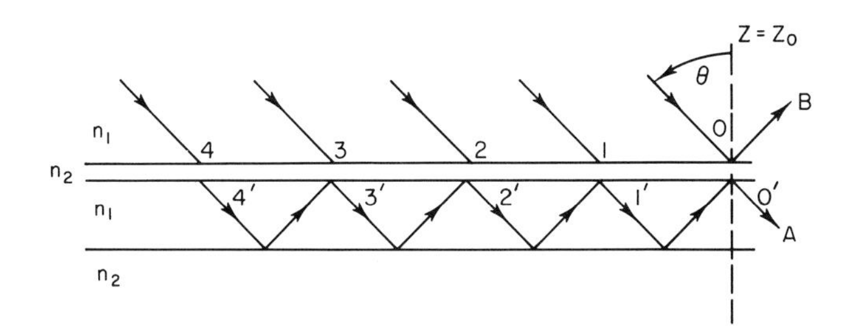

Fig. 3-3. The geometric interpretation of the buildup of the mode amplitude in a thin film under the influence of a collimated beam incident on an FTR low-index film above it.

bered 0 contributes the pencil 0′ by way of direct transmission through the low-index layer. The incident pencil numbered 1 contributes indirectly the pencil 1′ after two internal reflections in the spacer. Incident pencil 2 contributes the pencil 2′ after four internal reflections, etc. We add up all such contributions until we reach the edge of the incident beam. We must sum these contributions with the appropriate relative phases, because it is their interference that provides the effect we wish to study. Let r and r' represent the magnitude of the wave amplitude reflectance at the top and bottom of the spacer, respectively, and t represent the magnitude of the amplitude transmittance of the low-index layer. The corresponding phase changes on reflection and transmission are given by ϕ_r, $\phi_{r'}$, and ϕ_t. Successive contributions also have a phase difference $4\pi n_1 d(\cos\theta)/\lambda_0$ arising from their optical path difference in reaching the plane z_0. Thus, successive contributions have the total phase difference δ given by

$$\delta = \phi_r + \phi_{r'} - 4\pi n_1 d(\cos\theta)/\lambda_0. \tag{3.1}$$

If we assume that the incident beam has unit amplitude, we then readily calculate that the sum of the contributions of N incident pencils yields a wave amplitude in the spacer given by

$$A(z_0) = te^{i\phi_t}[1 + r'r e^{i\delta} + (r')^2 r^2 e^{2i\delta} + \cdots + (r')^{N-1} r^{N-1} e^{i(N-1)\delta}]. \tag{3.2}$$

This geometric progression reduces to

$$A(z_0) = te^{i\phi_t}[1 - r^N (r')^N e^{iN\delta}]/(1 - rr' e^{i\delta}). \tag{3.3}$$

Consequently, the amplitude of the reflected wave is given by

$$B(z_0) = re^{i\phi_r} + (t/r)e^{i(\phi_t - \phi_r)}[A(z_0) - te^{i\phi_t}]. \tag{3.4}$$

Now, because the film is almost totally reflecting, $r \approx 1$, while $t \ll 1$. Even for large r, however, r^N becomes negligibly small as N increases. The power inside the guide thus quickly reaches its maximum value given by

$$|A|^2(N = \infty) = t^2/[1 + r^2(r')^2 - 2rr'\cos\delta]. \tag{3.5}$$

When the incident wavelength, angle of incidence, and spacer thickness are such that $\cos\delta = 1$, we then have

$$|A|^2 = t^2/(1 - r'r)^2 = t^2/[1 - (1 - t^2)^{1/2}]^2 \approx 4/t^2. \tag{3.6}$$

This result assumes that $r' = 1$, as in Fig. 3-2. If $r' = r$, as in the FTR filter (Fig. 3-1), we find that the equilibrium power density in the spacer is approximately $1/t^2$ times the incident power density. Because t can be quite small, it is obvious that one can build up very high power densities in the film through the use of the FTR principle.

Let us return again to Eq. (3.3) and Fig. 3-3 to estimate the distance required to reach the maximum power density given by Eq. (3.6). The distance corresponds to $2N$ internal reflections in the guide, where r^N is negligible. As shown in Fig. 3-3, this distance would be $2Nd\tan\theta$. The figure assumes, however, that a ray incident at a point z on the interface is reflected or transmitted precisely at that point. We know, from the experiments of Goos and Haenchen (*12*), that this is not the case. Thus, the actual distance over which the characteristic wave builds up is longer than that implied by the figure.

In Section D, we will show that it is given by $2Nd_e \tan\theta$, where d_e is the effective geometric thickness of the film. In the weak coupling approximation appropriate to the analysis of distributed couplers, $d_e - d$ is proportional to the Goos–Haenchen shift.

We said earlier that the power buildup in the mode was exponential with distance from the edge of the incident beam. Our ray treatment, however, illustrates a buildup in discrete steps over this distance. To bring these apparently different descriptions into harmony, we simply rewrite our equations for A on resonance ($\cos\delta = 1$) in another form, and then make use of the fact that $t \ll 1$. Thus,

$$\begin{aligned} A(N) &= [t \exp i\phi_t/(1-r)](1-r^N) \\ &= A_\infty\{1 - \exp[N \log_e(1-t^2)^{1/2}]\} \\ &\approx A_\infty[1 - \exp(-Nt^2/2)], \end{aligned} \tag{3.7}$$

where A_∞ gives the equilibrium value of A for an infinitely wide beam. A representative value of t^2 might be 10^{-3}, so that 2000 incident ray pencils would be needed to bring A to within $1/e$ of its final value. (It must be remembered that N represents the number of incident pencils contributing to A, rather than the number of internal reflections, which is $2N$). The power density in the spacer is then 4000 times that in the incident beam. The buildup occurs over a length given approximately by $2Nd_e \tan\theta$. When d_e is of the order of λ_0, this length is a few millimeters.

There are some interesting differences between the one-sided (Fig. 3-2) and the two-sided FTR arrangements (Fig. 3-1). Because $r' = r < 1$ in the two-sided case, the equilibrium power density in the spacer considered in our example would be only $1/t^2 = 1000$ times that of the incident wave for the FTR filter on resonance. The *transmitted power*, with the assumed ideal films, is then

$$|C_\infty|^2 = t^2(1/t^2) = 1, \tag{3.8}$$

where C_∞ is the equilibrium amplitude of the transmitted wave in the FTR filter.

In the one-sided case on resonance, we have, from Eqs. (3.4) and (3.3),

$$B(z_0) = e^{i\phi_r}[r + (1+r)(1-r^{N-1})e^{2i(\phi_t-\phi_r)}]. \tag{3.9}$$

Now, because $r \approx 1$, we can write this as

$$B(z_0) = e^{i\phi_r}[1 + 2(1-r^{N-1})e^{2i(\phi_t-\phi_r)}]. \tag{3.10}$$

Then, because the reflected power cannot be greater than the incident power, it can be readily shown that $(\phi_t - \phi_r) = \pm\pi/2$. Thus, $|B| \approx 1$ for the one-sided case, whereas $|B| \approx 0$ for the two-sided case. Thus, the primary reflected field is canceled in the FTR filter after equilibrium is reached.

We have now determined that the equilibrium power in the spacer waveguide is four times greater in the one-prism FTR arrangement than in the classical FTR filter and that it requires twice as great a distance along the waveguide to reach this equilibrium. We note also, from Eq. (3.10), that, in the single-prism arrangement, the reflected field in the prism first decays in strength until $r^{N-1} = 1/2$, where $B(N) \approx 0$. It then increases approximately exponentially until $|B| \approx 1$. A pictorial comparison of these

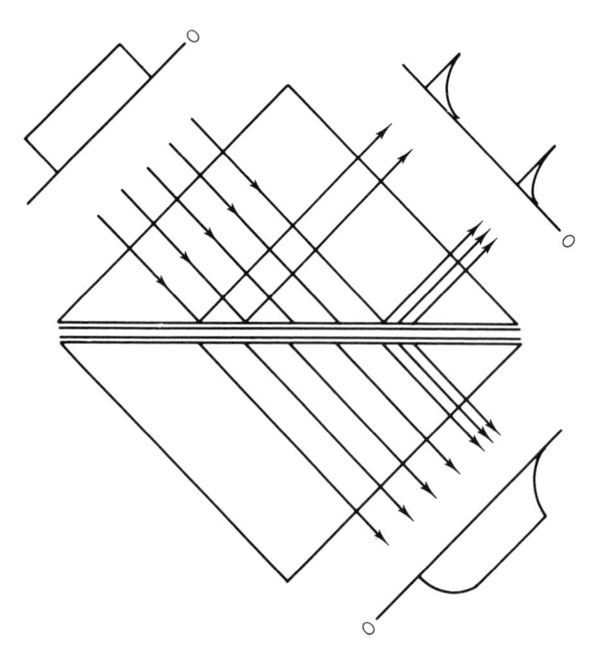

Fig. 3-4. The amplitude profile of the reflected field in a one-sided FTR coupler driven on-resonance by a wide, collimated beam of uniform amplitude.

two FTR arrangements thus has the properties shown schematically in Figs. 3-4 and 3-5, which indicate the amplitude (square root of the power density) profiles of the incident, reflected, and transmitted beams. (The sharp discontinuities would not be present, of course, if we included the effects of diffraction.)

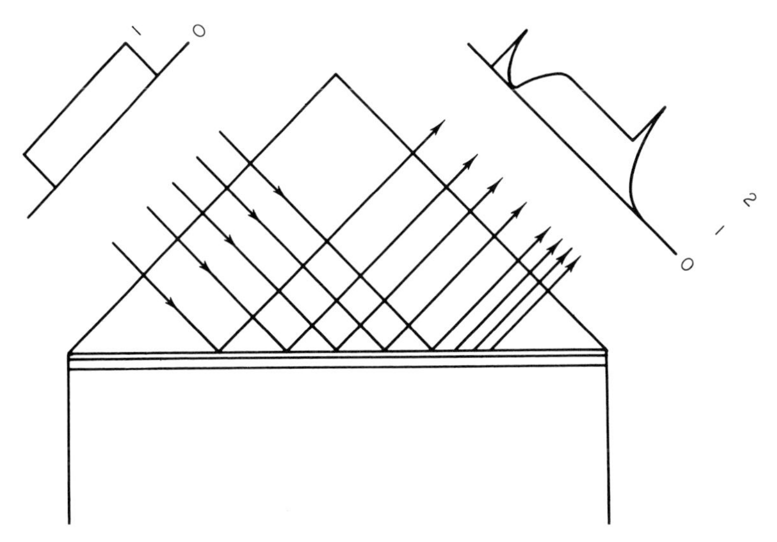

Fig. 3-5. Schematic illustration of the amplitude profiles of the reflected and transmitted beams when an FTR filter is driven on-resonance by a wide, collimated beam of uniform amplitude.

If we now assume that the beam terminates in the plane $z = z_1$, then, in planes $z > z_1$, we clearly have

$$\begin{aligned} |A(z - z_1)| &= |A(z_1)|\, r^M \approx |A(z_1)| \exp(-Mt^2/2) \\ |B(z - z_1)| &= t\,|A(z_1)|\, r^{M-1} \approx t\,|A(z_1)| \exp(-Mt^2/2) \end{aligned} \tag{3.11}$$

for one prism, and

$$\begin{aligned} |A(z - z_1)| &\approx |A(z_1)| \exp(-Mt^2) \\ |C(z - z_1)| &= |B(z - z_1)| \approx t\,|A(z_1)| \exp(-Mt^2) \end{aligned} \tag{3.12}$$

for two prisms, where M is the number of internal reflections beyond the plane $z = z_1$.

The collections of ray pencils described by B and C in these equations have the physical characteristics of the leaky wave modes described in Chapter 2. These rays are parallel to the primary reflected or transmitted rays. Their planes of constant phase (wavefronts) are perpendicular to these rays, and the amplitude decays exponentially along any wavefront in the direction away from the primary reflected or transmitted rays. Notice that in the upper prism to the right of the plane where the incident beam terminates the wave amplitude increases exponentially in the direction away from the waveguide until it reaches the last primary reflected ray. This is precisely the behavior of the leaky wave mode. As pointed out in Chapter 2, the domain of the definition of leaky waves is confined to an angular region above the guide. We see very clearly that this region is defined in our FTR case by the last primary reflected ray.

It should also be noted that the slab waveguide in the FTR arrangement has no "true, bound, spectral" modes, because radiation will always leak out of a slab with a prism over or under it. If the prisms are truncated, however, then in regions beyond the truncation, the analysis given here is obtained, and some of the modes are then "true, proper" modes. At the plane of truncation, we thus have leaky modes coupled to proper modes.

Some of the effects of finite beams considered here have been treated in greater detail by Bergstein and Shulman (*13*), who used a plane-wave modal expansion of the incident beam to determine the transmitted beam in the FTR filter. Ioganseu (*2*) has also considered this problem. The case of Gaussian beams is treated by Midwinter (*8*) and Ulrich (*7*).

B. LEAKY WAVES AND THE CONCEPT OF THE SPATIAL TRANSIENT

In Chapter 2, we found that the characteristic equation defining the parameters of the TE modes on slab waveguides had an infinite number of complex-valued roots. Although most of these roots yielded waves with amplitudes that increased exponentially away from the guide's axis, we nevertheless located them graphically in the complex-wavenumber plane and examined the properties of the corresponding leaky waves. In the last section, we found that in certain regions of space, the reflected and transmitted beams in an FTR filter had properties quite like those of leaky waves. These properties characterize waves that might be called "spatial transients" in analogy with the temporal transient oscillations of lossy electrical and mechanical systems driven by a sinusoidal source over a finite interval of time. This approach will provide an alternate description of the operation of FTR couplers. Let us begin by reviewing the transient analysis of a system that depends on only one dimension, the time, say, and then proceed to cases involving two dimensions, i.e., (x, t) or (x, z).

1. Transient Oscillations of a Point Mass

Consider a mass m on a spring with force constant K whose motion is damped by a velocity-dependent frictional force with proportionality constant γ. We want to know the position of the mass at each instant t when it is driven by a quasimonochromatic source of angular frequency ω_s that is turned on at time $t = 0$ and turned off at $t = T$. If x denotes the displacement of the mass from its rest position, then x satisfies the equations

$$x = 0 \qquad \text{for} \quad t < 0, \tag{3.13}$$

$$(d^2x/dt^2) + \gamma(dx/dt) + \omega_0{}^2 x = e^{i\omega_s t} \qquad \text{for} \quad 0 < t < T, \tag{3.14}$$

$$(d^2x/dt^2) + \gamma(dx/dt) + \omega_0{}^2 x = 0 \qquad \text{for} \quad T < t, \tag{3.15}$$

where $\omega_0 = (K/m)^{1/2}$. The solution to the homogeneous (source-free) equation (3.15) describes what is referred to as the "transient response" of the system. If we assume a solution of the form $e^{i\omega t}$ and substitute it into Eq. (3.15), we find that such a solution is valid providing that ω is complex-valued:

$$\omega = \pm(\omega_0{}^2 - \tfrac{1}{2}\gamma^2)^{1/2} + \tfrac{1}{2}i\gamma. \tag{3.16}$$

The displacement of the system for $T < t$ is therefore given by

$$x(t) = A(T)\exp(\pm i\omega' t - \tfrac{1}{2}\gamma t), \tag{3.17}$$

where ω' is the real part of ω and $A(T)$ is the initial condition for the differential equation (3.15), i.e., the amplitude of the oscillation at time $t = T$. The frequency ω' is called the "natural frequency" of the system. Note that the frequency spectrum of this transient oscillation is Lorentzian, with peak at ω' and half-width γ.

In the domain $0 < t < T$, when Eq. (3.14) applies, we know from the theory of differential equations that the complete solution is a superposition of the particular solution of the inhomogeneous equation (3.14) and the general solution of the homogeneous equation (3.15). Physically, we expect this to describe the transient buildup to the steady-state condition that will be obtained after a time T' satisfying $2/\gamma = \tau \ll T' < T$. If we take as a trial solution to the inhomogeneous equation the particular function $B(\omega_s)e^{i\omega_s t}$, we find that this is appropriate, providing that

$$B(\omega_s) = 1/[(\omega_0{}^2 - \omega_s{}^2) + i\gamma\omega_s]. \tag{3.18}$$

The general solution in the domain $0 < t < T$ is thus given by

$$B(\omega_s)\exp(i\omega_s t) + C\exp(\pm i\omega' t - \tfrac{1}{2}\gamma t) = x(t), \tag{3.19}$$

where the constant C must be chosen to satisfy the initial condition that

$x(t) = 0$ at $t = 0$. Therefore,

$$x(t) = B(\omega_s)[\exp(i\omega_s t) - \exp(i\omega' t - \tfrac{1}{2}\gamma t)] \qquad t < T. \tag{3.20}$$

Under the assumption that $\gamma^{-1} \ll T$, we thus find that $A(T)$ in Eq. (3.17) is given by

$$A(T) \approx B(\omega_s) \exp[i(\omega_s - \omega')T + \tfrac{1}{2}\gamma T]. \tag{3.21}$$

If the source frequency ω_s is equal to the natural frequency ω', then the envelope of the oscillation, as it builds up to its steady-state value, is proportional to the simple exponential function $(1 - e^{-\gamma t/2})$. When these frequencies are not equal, however, the envelope of the oscillation at frequency ω_s is itself a damped oscillation at frequency $\Delta\omega = \omega' - \omega_s$. After steady state is reached, the mass oscillates between $\pm B$ at frequency ω_s. The largest possible value of $|B|$ occurs for a source with angular frequency $\omega_s = \omega_0^2 - \frac{1}{2}\gamma^2$, as can be shown by differentiating $|B|$ with respect to ω_s. Similarly, we can show that the maximum value of the velocity of the mass m results from a source at angular frequency ω_0. At this frequency, called the "resonant frequency" of the system, the mass presents a purely resistive "load" (friction) to the driving source, and power is transferred most efficiently to the system. Note that this frequency is not equal to the natural frequency ω' of the system as given by Eq. (3.16).

2. Transient Analysis of a One-Dimensional Optical Resonator Driven by a Rectangular Wave Packet

Let us now enlarge the problem by one dimension to consider the case of an optical resonator driven for a finite interval of time T by a plane wave incident normally on its semitransparent face located at $x = a$. The geometry is illustrated schematically in Fig. 3-6. To keep the algebra as clear as possible, we assume that the resonator is simply a thin dielectric film of very high refractive index n in contact with air (or vacuum; refractive index $= 1$) at $x = a$ and a perfectly conducting wall at $x = 0$. This provides, at $x =$ a, a reflectivity $r = (n - 1)/(n + 1) \approx 1 - (2/n)$.

Let us suppose that the incident field is a wave packet (wave train) whose electric field is described by

$$E_y(\text{incident}) = U(x + ct)e^{i\omega_s[t+(x/c)]}, \tag{3.22}$$

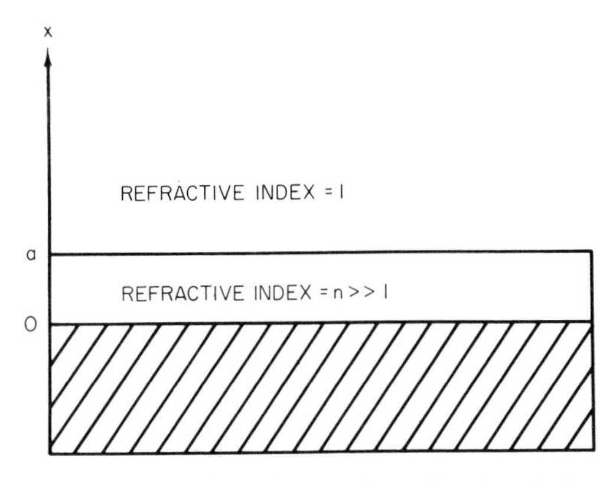

Fig. 3-6. Schematic representation of a one-dimensional optical resonator composed of a single thin film of high refractive index n bounded by air at $x = a$ and a perfect conductor at $x = 0$.

where U is the rectangular function of unit height illustrated in Fig. 3-7(a). For $t < 0$, the leading edge of the packet is located along the plane surface $x = a - ct$ across which the field has a unit discontinuity. The trailing edge, representing a second unit discontinuity in the field, is located at $x = a + cT - ct$, with $t < 0$.

At $t = 0$, the leading edge of the packet strikes the resonator at $x = a$, thereby generating the leading edge of the reflected wave packet, which moves to the right at velocity c. At time T, the trailing edge of the incident packet is at $x = a$, where it generates a second discontinuity in the reflected packet. For $t > T$, the incident packet has vanished from the region $a < x$ and only the reflected packet remains. If the resonator were replaced by a perfectly conducting film of thickness a, the reflected packet would be given by

$$E_y(\text{reflected from perfectly conducting film}) = U(x - ct)e^{i\omega_s[t-(x/c)]}, \qquad (3.23)$$

for $t > T$. The rectangular unit function U is as illustrated by the dashed curve in Fig. 3-7(b). Its leading edge is located at $x = a + ct$, and its trailing edge is at $x = a + c(t - T)$, with $T < t$. We must now investigate the reflected packet when the region $0 < x < a$ is occupied by our thin-film resonator.

A conventional field-theoretic derivation of a solution to this problem might proceed as follows. The incident wave packet is Fourier-analyzed into its spectral components. From Eq. (3.22), we see that these have relative strength $F(\omega) = \text{sinc}[(\omega - \omega_s)T/2]$, where $\text{sinc}\, x = (\sin x)/x$. The in-

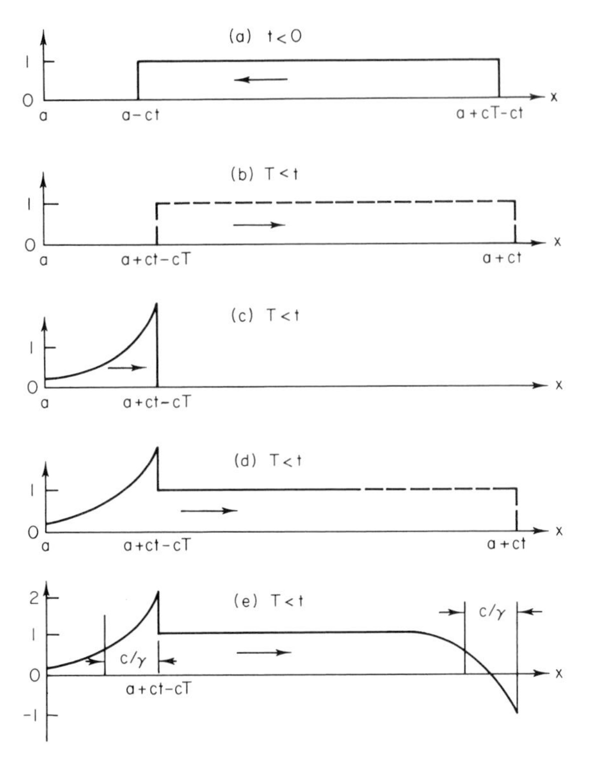

Fig. 3-7. The envelope of the electric or magnetic field outside the thin-film resonator excited by a rectangular wave packet. (a) The incident wave packet of duration T prior to striking the resonator. (b) The reflected packet that would occur if the thin film were a perfect conductor. (c) The leaky wave transient that must be fit to the trailing edge of the steady-state reflected field. (d) Envelope resulting from matching the transient to the steady-state field at $t = T$. (e) The complete field resulting from the incident packet.

ternal and reflected fields are then found for each incident spectral component, i.e., the plane wave $F(\omega)e^{i\omega[t-(x/c)]}$. Each such solution represents a steady state field characterizing the entire space for all time. By integrating this solution over ω, we then obtain the time-dependent field in each region. The resultant integrals constitute what is called a "spectral representation" of the fields. In most cases, the physical interpretation of these integrals is not obvious. It is rare, indeed, that the integrations can be carried out exactly. To obtain useful solutions in closed form, one then deforms the path of integration in the complex-ω plane, thereby obtaining, for example, a saddle-point representation of the reflected field above the resonator. This solution will depend primarily on the location of the poles of the integrand with respect to the path of integration.

What we wish to do in this section is to arrive at a description of the internal and reflected fields which is as accurate as that obtained from the saddle-point technique, but which avoids the associated mathematical complexity by a combination of relatively simple physical and mathematical arguments that are easy to apply to more complicated cases. We accordingly look at our thin-film resonator in terms of its space-time transients.* Physically, we know that the buildup and decay of an oscillation in such a system must be characterized by some time constant τ. Let us assume that the duration T of the incident packet is much greater than τ. Then, after a time T' satisfying $\tau \ll T' < T$, a steady-state condition of forced oscillation at the driving frequency ω_s will obtain in the film, just as it did in the simple mechanical oscillator just considered. Then, after the time T, when the driving field is no longer present, the oscillation will decay by way of radiation to the region outside the film. Let us begin by studying this decay.

We must appreciate that the radiated field associated with the decay of the oscillation applies only to the space–time domain to the left of the trailing edge of the geometrically reflected pulse, as indicated by the dashed line at $x = a + c(t - T)$ in Fig. 3-7(b). This line represents a propagating plane surface along which the fields are discontinuous. It was generated at $x = a$ and $t = T$ by the field discontinuity in the trailing edge of the incident packet.† We may expect that, in the space–time domain $x < a + c(t - T)$, $T < t$, the fields satisfy the homogeneous wave equation subject to the usual boundary conditions at $x = a$ and $x = 0$. We can easily derive these solu-

* Although the ray summation method could clearly be applied to this problem, we are now seeking a wave solution.

† The reader should note that the subject of the reflection, refraction, and propagation of field discontinuities is a nontrivial one. Clearly, the differential forms of Maxwell's equations have no meaning at discontinuities, and an integral formulation is required to obtain appropriate initial conditions. The reader is referred to the comprehensive work of Kline and Kay (*14*) for a detailed analysis of the subject. The principal results of their work that we will need here may be summarized as follows. In homogeneous, isotropic media, the electric and magnetic fields on surfaces of discontinuity are mutually orthogonal and are perpendicular to geometric optic rays whose trajectories satisfy Fermat's principle. The associated family of geometric optic wavefronts is generated by the surface of field discontinuity as it moves through space with velocity c. At any surface of discontinuity in the media, reflected and/or refracted surfaces of field discontinuity are generated. Each satisfies the laws of geometric optics. In addition, the electric and magnetic fields at the interfaces on these new surfaces of field discontinuity have strengths, relative to the incident field discontinuity, given by the familiar Fresnel formulas. This applies locally for each ray and is in no way dependent on that shape of the surfaces of field discontinuity or the refracting surfaces, providing each is continuous with continuous derivatives.

tions by the method described in Chapter 2. Accordingly,

$$E_y(\text{inside}) = A[\sin(ux/a)]e^{i\omega t}, \tag{3.24a}$$

$$E_y(\text{outside}) = A(\sin u)e^{-q[(x/a)-1]+i\omega t}, \tag{3.24b}$$

$$H_z = (i/\omega\mu)\,\partial E_y/\partial x, \tag{3.24c}$$

and

$$u = \omega an/c, \qquad q = i\omega a/c. \tag{3.24d}$$

The characteristic equation for the resonator is given by

$$q = -u \cot u \tag{3.25a}$$

and

$$R^2 = u^2 + q^2 = (\omega^2 a^2/c^2)(n^2 - 1). \tag{3.25b}$$

Note that we have intentionally avoided the use of the customary wave numbers in writing this solution. We are reserving these for the description of the driving, incident fields and the steady-state response of the system for which the wave numbers are real-valued. In the transient fields described by Eqs. (3.24a–d), on the other hand, u, q, and ω are all complex-vauled quantities to be determined from Eqs. (3.25a,b). They define the leaky modes of our resonator.

Evidently, Eqs. (3.25a,b) are special cases of Eqs. (2.35) and (2.32) of Chapter 2. In this case, however, we know that q/u is simply i/n. We will assume that the refractive index n of the film is independent of frequency, so that Eqs. (3.25) are solved quite simply by writing the first of them in terms of its real and imaginary parts, u' and u'', respectively. Thus

$$(\sin 2u')/(\cosh 2u'' - \cos 2u') = 0,$$
$$-(\sinh 2u'')/(\cosh 2u'' - \cos 2u') = -\, i/n. \tag{3.26}$$

The first of Eqs. (3.26) yields

$$u' = N\pi/2, \qquad N = 0, 1, 2, \ldots. \tag{3.27}$$

If N is an odd integer, the second equation then yields

$$\tanh u'' = 1/n \approx u'', \tag{3.28}$$

because $1 \ll n$. If N is an even integer, the second equation yields $\coth u'' = 1/n$. There are no solutions of this equation satisfying the conditions of our problem. We thus have an infinity of roots given by

$$u_N \approx [(2N+1)\pi/2] + (i/n), \qquad N = 1, 2, 3, \ldots,$$
$$q_N = iu_N/n \approx -(1/n^2) + [i(2N+1)\pi/2n], \tag{3.29}$$
$$\omega_N \approx [(2N+1)\pi c/2an] + (ic/n^2 a),$$

and

$$\tau_N \equiv \tau = 1/\text{Im}\ \omega_N = na/(c/n) = 1/\gamma.$$

We see from Eqs. (3.24 a–d) and the expression for q_N in Eqs. (3.29) that the field increases exponentially away from the upper surface of the resonator for these modes. However, we know that these solutions apply only to the space–time region $x < a + c(t - T)$, $t > T$. Note that the time constants for all these resonances are equal and are in accordance with the concept of waves bouncing back and forth in the resonator, losing a fraction $1/n$ of their energy with each round trip. (In equilibrium, the ratio of the Poynting vectors inside and outside the interface at $x = a$ would be n. The time constant τ describes the rate of decay of the amplitude. The energy density decays as $2\tau = n(2an/c)$, illustrating the round-trip time.) Note also that the phase velocity of the radiated field is equal to c. The envelope of one of these transient fields at some time $t > T$ is sketched in Fig. 3-7(c). The constant A is as yet undetermined, so we are not yet ready to fit our solution to the steady-state condition that described the fields in and near the film for times t satisfying $\tau \ll t < T$.

Let us now consider the steady-state field. It corresponds to the familiar solution of the problem posed by an incident monochromatic plane wave at frequency ω_s. We find that the steady-state field is given by

$$E_y(\text{incident}) = e^{i\omega_s t + ik_s(x-a)}, \tag{3.30}$$

$$E_y(\text{reflected}) = [ik_s - k\cot(ka)]/[ik_s + k\cot(ka)]e^{i\omega_s t - ik_s(x-a)}, \tag{3.31}$$

and

$$E_y(\text{inside}) = 2(\sin kx)e^{i\omega_s t}/[\sin(ka) + (k/ik_0)\cos(ka)], \tag{3.32}$$

where $k = n\omega_s/c$ and $k_s = \omega_s/c$ are the usual wave numbers in the film and in free space, respectively. At time $t = T - 0$, the field in the film is assumed to be given, to any measurable accuracy, by Eq. (3.32). At $t = T + 0$, on the other hand, the field should be expressible in terms of the transient solutions given by Eqs. (3.24) and (3.29). These have the form

$$E_y(\text{inside}) = \sum_{N=0}^{\infty} A_N[\sin(u_n x/a)]e^{i\omega_N t}, \qquad T < t. \tag{3.33}$$

If Eqs. (3.33) and (3.32) for the interior fields before and after $t = T$ are set equal at time $t = T$, we obtain the equation determining the values of the constants A_N in the transient field. Thus,

$$\sum_{N=0}^{\infty} A_N[\sin(u_N x/a)]e^{i\omega_N T} = 2(\sin kx)e^{i\omega_s T}/[\sin(ka) + (k/ik_s)\cos(ka)]. \quad (3.34)$$

Because the leaky wave modes are not orthogonal, we must expend a little more effort than usual to determine the coefficients A_N. (The lack of orthogonality of these modes can be viewed as a consequence of the fact that their spectral components, which may be obtained by Fourier transformation, overlap.) The well known Schmidt orthogonalization procedure (*15*) provides the necessary mathematical tools. Accordingly, we form a new set of functions $\phi_N(x)$, which are linear combinations of the leaky mode functions $h_N(x) = \sin(u_N x/a)$. The new set ϕ_N is defined so as to satisfy the orthogonality condition

$$\int_0^a \phi_N^*(x)\phi_M(x)\,dx = \delta_{NM}C_N, \quad (3.35)$$

where δ_{NM} is the Kronecker delta, equaling 0 for $N \neq M$ and 1 for $N = M$, and the * denotes complex conjugate. Each of the $\phi_N(x)$ has the form

$$\phi_N(x) = \sum_{P=0}^{\infty} b_{NP}\sin(u_P x/a). \quad (3.36)$$

After finding the functions ϕ_N, we can express any function $f(x)$ on the interval $0 < x < a$ in terms of them, as in the following equation:

$$f(x) = \sum_{N=0}^{\infty} a_N\phi_N(x), \qquad \text{where} \quad a_N = (1/C_N)\int_0^a \phi_N^*(x)f(x)\,dx. \quad (3.37)$$

But the determination of the coefficients a_N also fully determines the coefficients A_N of Eq. (3.34) through the application of Eqs. (3.36) and (3.37). Accordingly, we define the functions $S_N(k)$ as

$$S_N(k) = 0 \qquad \text{for} \quad N < 0 \quad (3.38a)$$

and

$$S_N(k) = \operatorname{sinc}(u_N^* - ka) - \operatorname{sinc}(u_N^* + ka), \quad (3.38b)$$

where $\operatorname{sinc} x = (\sin x)/x$ and k is the wave number, in the film, of the steady-state oscillation. Then, the coefficient A_N in Eq. (3.34) are given by

$$A_N = 2[\sin(ka) + (k/ik_s)\cos(ka)]^{-1}e^{-i(\omega_N-\omega_s)T} \times [S_N + (2i/n\pi)\Delta + O(1/n^2)]. \quad (3.39a)$$

where $O(1/n^2)$ includes all terms of order $1/n^2$ or smaller, and

$$\Delta = (S_{N+1} - S_{N-1}) - \tfrac{1}{2}(S_{N+2} - S_{N-2}) + \tfrac{1}{3}(S_{N+3} - S_{N-3}) \\ - \tfrac{1}{4}(S_{N+4} - S_{N-4}) + \cdots. \tag{3.39b}$$

As can be seen from Eq. (3.38b), the infinite series Δ is, in general, a slowly converging one.

Let us now specialize these equations to the case where the incident packet drives the system at one of its natural frequencies ω_M'. Then,

$$\omega_s = (c/na) \operatorname{Re} u_M = (c/2an)(2M+1)\pi, \qquad ka = (2M+1)\pi/2. \tag{3.40a}$$

Therefore, from Eq. (3.38), we find that

$$\begin{aligned} S_M &\approx 1 - (i/\pi n)/(2M+1), \\ S_N &\approx \pm(i/n\pi)(2N+1)/[(N-M)(N+M+1)], \qquad N \neq M, \end{aligned} \tag{3.40b}$$

where the plus sign in the second equation applies when $(N - M)$ is an odd integer. Thus, up to terms of order $1/n^2$, we find, from Eqs. (3.39)–(3.40),

$$A_M \approx 2(\exp \omega_M''T)[1 - (i/\pi n)/(2M+1)], \tag{3.41}$$

where $\omega_M'' = \gamma = \operatorname{Im} \omega_M$, from Eq. (3.29), and

$$A_{M+P} \approx 2e^{i(\omega_{M+P}-\omega_s)T}(-1)^P(i/Pn\pi)(2M+1)/(2M+P+1), \quad P \geq -M. \tag{3.42}$$

Equation (3.41) and (3.42) tell us that, for all practical purposes, we can use the approximations

$$A_M \approx 2e^{\gamma T} \quad \text{and} \quad A_N \approx 0 \quad \text{for} \quad N \neq M \quad \text{and} \quad \omega_s = \omega_M'. \tag{3.43}$$

This was intuitively clear from Eq. (3.34). The succeeding equations give us useful estimates of what happens off resonance as well, and provide better approximations for the resonance case, should they be needed. From Eqs. (3.34) and (3.38)–(3.39), we see that many modes participate off resonance, but none of them is strong, because of the term k in the denominator of Eq. (3.39a). Thus, while the decay is much more complicated off resonance, the steady-state condition involves too little power to be interesting.

Now that we have shown that only one mode is important on resonance and that it is described by

$$E_y(\text{inside}) = 2\left(\sin \frac{u_M x}{a}\right) \exp[-\gamma(t-T) + i\omega_s t], \qquad T < t, \tag{3.44a}$$

and

$$E_y(\text{outside}) \approx 2 \exp\left[i\omega_s\left(t - \frac{x-a}{c}\right) - \gamma\left(t - T - \frac{x-a}{c}\right)\right],$$
$$x < a + c(t-T), \quad T < t, \tag{3.44b}$$

we can fit this transient solution for the radiated field to the trailing edge of the steady-state reflected field. The result is shown in Fig. 3-7(d). Because the steady-state reflected field has unit amplitude on resonance [Eq. (3.31) with $\cot(ka) = 0$], while the transient has twice this amplitude, there is a unit discontinuity in the radiated field at the trailing edge of the geometrically reflected field, where $x = a + c(t - T)$, $T < t$.

We will now describe the buildup of the oscillation that begins at $t = 0$. At this time, the field strength inside the film equals zero. This is the initial condition to be satisfied by the superposition of the particular solution of the inhomogeneous wave equation (the steady-state field) and the general solution of the homogeneous equation (the transients). From Eqs. (3.32) and (3.33), at $t = 0$, we thus obtain

$$\sum_{N=0}^{\infty} A_N \sin(u_N x/a) + 2(\sin kx)/[\sin(ka) + (k/ik_s)\cos(ka)] = 0 \tag{3.45}$$

for $0 < x < a$. This is identical, except for the change in sign, with Eq. (3.34). Assuming again that the driving frequency is on the Mth resonance and terms of order $1/n$ can be neglected, we obtain

$$A_M \approx -2; \qquad A_N \approx 0 \quad \text{for} \quad N \neq M. \tag{3.46}$$

Thus, the buildup of the electric field inside the resonator is described by

$$E_y(\text{inside}) \approx 2\{\sin(kx) - [\sin(u_M x/a)]e^{-\gamma t}\}e^{i\omega_s t}. \tag{3.47}$$

The radiated (reflected) field is accordingly given by the superposition of Eqs. (3.24) and (3.31),

$$E_y(\text{radiated}) \approx \exp\left[i\omega_s\left(t - \frac{x-a}{c}\right)\right] - 2(\sin u_M)\exp\left[i\omega_M\left(t - \frac{x-a}{c}\right)\right]. \tag{3.48}$$

We now use the fact that

$$\sin u_M = \sin\left(\frac{2M+1}{2}\pi + \frac{i}{n}\right) = 1 + O\left(\frac{1}{n^2}\right)$$

to reduce this to

$$E_y(\text{radiated}) \approx \left\{\exp\left[i\omega_s\left(t - \frac{x-a}{c}\right)\right]\right\}$$
$$\times \left\{1 - 2\exp\left[-\gamma\left(t - \frac{x-a}{c}\right)\right]\right\}, \qquad x < a + ct, \quad 0 < t. \tag{3.49}$$

We can now combine the results of Eqs. (3.44) and (3.49) to form the envelope of the entire reflected (radiated) wave packet, as illustrated in Fig. 3-7(e).

The analogy between this case and the FTR system treated earlier is obvious. Here, we have treated a space–time (x, t) problem. There, we had an (x, z) problem. The results are essentially the same. As we will show in Section D, the only important difference arises in the calculation of the decay constants. These provide a means of modifying the ray theory so that it is quantitatively correct. Clearly, we could have obtained the principal results of this section much more easily by using the ray summation method. That method also has the advantage of simple physical interpretation. Without it, we would have difficulty understanding why the reflected field actually vanishes at one space–time point during the buildup of the oscillation and increases sharply at the instant the incident packet vanishes. The explanation is that the primary reflected field, which arises from direct reflection at $x = a$, is being opposed by the increasing internal fields in the film. This interpretation is patently obvious in the ray theory, but is well hidden in the mathematical treatment given here and almost totally buried in a spectral or saddle-point representation.

We should note before leaving this case that we have not considered in this analysis the fields on refracted surfaces of field discontinuity and the additional surfaces of field discontinuity that these generate in the radiated field. These fields are of order $1/n^2$, $1/n^4$, $1/n^6$, ..., however, and of no consequence in the approximations taken here. We also note that had we considered a more realistic optical resonator, one with a highly reflecting dielectric stack at $x = a$, the time delays associated with transmission and reflection at the stack would have required study. These time delays are analogous to the undetermined lateral shifts of the reflected and transmitted rays in the FTR case. We normally characterize them in the conventional steady-state solutions simply by phase changes associated with transmission or reflection. In a transient analysis, however, we must treat them according to what they are physically, i.e., time delays in the resonator and lateral shifts associated with similar delays in the FTR problem. Thus, the detailed analysis of transients is intrinsically more complicated than a steady-state analysis.

We note also that the theories of this section and the last apply near the guide, not in its far radiation field (the Fraunhofer region). In this section, we have not taken into account the diffraction effects that arise because of the finite aperture of the resonator. In the last section, we did not consider the spreading of the incident and multiply reflected beams associated with diffraction. These are of no consequence near the structure, but are obviously essential to a description of the far field.

C. DISTRIBUTED COUPLING BETWEEN PARALLEL PLANAR WAVEGUIDES

1. Introduction

The coupling of waves in parallel transmission systems is a subject of both fundamental and practical interest. It provides an excellent illustration of resonance in physical systems as well as a means of reflectionless transfer of signals from one waveguide to another. Our principal objective in this section will be to develop a description of this phenomenon in terms of familiar optical models of propagation. Let us first take a qualitative look at the subject.

Consider first two parallel slab guides, each bounded by perfectly conducting planes. Because the conductors cannot support an electric field, energy cannot be transferred from one guide to the other. The modes of the system (two guides) are thus those of the individual guides. If the guides are identical (same thickness), each mode has a degenerate counterpart corresponding to the added degree of freedom that the TE_N mode, for example, may be propagating on one guide or the other (or with arbitrary relative amplitudes on both guides).

Consider now two parallel films of refractive index n_1 separated by a film of lower index n_2; the three films are embedded in an infinite medium of index n_2. We may ask, for example, what happens along the guides if, at some plane perpendicular to the axes of propagation, we have substantially all the energy localized on one of the guides? Alternately, we may wish to know what waves can propagate on the system such that there is no change in the distribution of power between the guides as the wave advances. The second question is substantially the same as asking for the (normal) modes of the system. The first, on the other hand, implies that each guide has its own modes but that they are somehow modified by the presence of the other. Thus, for example, we might ask how the energy in a TE_1 mode propagating on a given waveguide is transferred to another waveguide brought into its vicinity. This is an important question. To get the principal features of the answer, let us look at a mechanical analogy.

Consider two identical pendula coupled together by a weak spring. We know that if we initially displace one pendulum from its equilibrium position and then release it, it begins to oscillate at its natural frequency. The oscillation is damped, however, as power is transferred to the second pendulum, which was initially at rest. After a time $T/2$, where T is the beat time, the initially displaced pendulum is essentially at rest while the second is oscillating with all the energy originally imparted to the first. After a time T, the energy is again returned to the first pendulum. This process is repeated indefinitely with increasing time. Thus, the phenomenon of temporal beating is observed. The beat frequency depends on the strength of the coupling between the two pendula. The larger the spring constant, the higher the frequency.

The two-pendulum system, we remember, has two normal modes for which there is no net exchange of energy. The symmetric mode, corresponding to parallel displacements of the masses, has a lower frequency than that of the uncoupled pendula. The antisymmetric mode, corresponding to antiparallel displacement of the masses, has a higher frequency. We can describe the phenomenon of temporal beating in terms of the superposition of the symmetric and antisymmetric modes, in which case the beat frequency is expressed in terms of the frequency difference between these two normal modes of the system.

The principal features of mode coupling between identical parallel guides are entirely analogous to the mechanical system. Instead of beat frequencies and beat times, however, we find beat wave numbers and beat lengths. The propagation constant of each mode on the isolated guide is split to provide two new propagation constants, one for a mode that is symmetric about the midplane between the guides, the other for an antisymmetric mode. If all energy is localized on one guide in the transverse plane $z = 0$, it is periodically transferred between the guides along their length. This sinusoidal modulation of the energy on each guide occurs at a spatial frequency determined by the difference between the propagation constants of the symmetric and antisymmetric modes. This is the main result of the theory of weakly coupled guides, where it is assumed that the only major effect of the interaction is the coupling of identical modes on the isolated guides. This assumption is valid providing that the change in the propagation constant of the given mode is small compared with the difference between this propagation constant and that of other modes on the isolated guide. A general mathematical development of the theory of coupled transmission systems (*16*) is essentially identical with textbook developments of the quantum mechanical (QM) perturbation theory. Each involves the solution of a simultaneous set of first-order linear differential equations of the form

$$\frac{\partial a_r}{\partial z} + i \sum_s H_{rs} a_s = 0 ,$$

where a_s is the amplitude of the sth mode, H_{rr} is the uncoupled propagation constant of the rth mode, and H_{rs} is the coupling coefficient between the rth and sth modes. In the QM problem, a_r is the probability amplitude for the rth unperturbed state, which has energy H_{rr}, and H_{rs} determines the transition probability between the rth and sth states. The QM problem concerns the temporal (t) development of the system, whereas the transmission line problem concerns its spatial (z) development. If H_{rs} depends on z in the latter problem, we have parametric coupling. We will treat in this section only nonparametric coupling, where the H_{rs} are constants. This is analogous to QM time-independent perturbation theory. We shall avoid formal techniques for the calculation of the coupling coefficients by using a ray theory to describe the interaction. Fortunately, this can be done for the simple geometry presented by coupled planar waveguides. With more complicated guides, formal techniques, such as variational methods, are needed. As indicated previously, a ray theory of coupling requires an important extension before it can be used to predict the correct beat lengths. This is because a reflected ray is laterally displaced from the corresponding incident ray at total reflection. This is the Goos–Haenchen shift referred to previously. Because of this displacement, the effective thickness d_e of the guide is greater than the actual thickness d, so that the axial distance between reflections, $d_e \tan \theta_1$, is greater than the geometrically calculated distance, $d \tan \theta_1$.

2. Ray Theory of Coupling between Identical Parallel Planar Guides

The conceptual framework for this development is similar to that used in Section A. We consider the arrangement sketched in Fig. 3-8. Suppose we know that in the plane $z = 0$, a ray of amplitude A_0 leaves interface 3 at an angle θ_1, and a ray of amplitude B_0 departs interface 2. For a TE wave, for example, A_0 and B_0 are proportional to the amplitude of the electric field at these interfaces in the plane $z = 0$.

After a distance corresponding to $2N$ internal reflections in either guide (as we shall show later, this distance is not $2Nd \tan \theta_1$), the corresponding amplitudes are A_N and B_N, respectively. We can find A_N and B_N from A_0

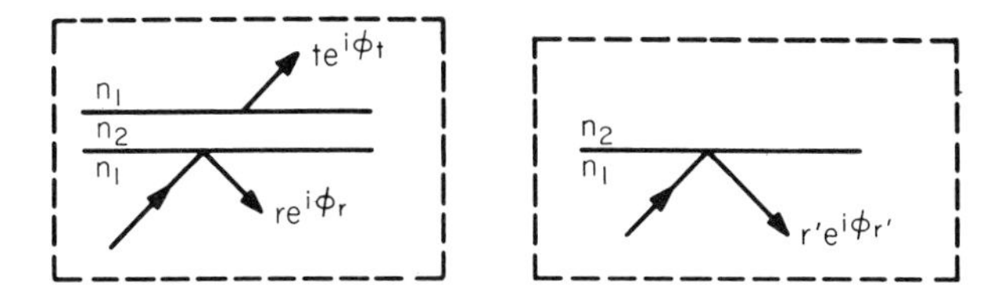

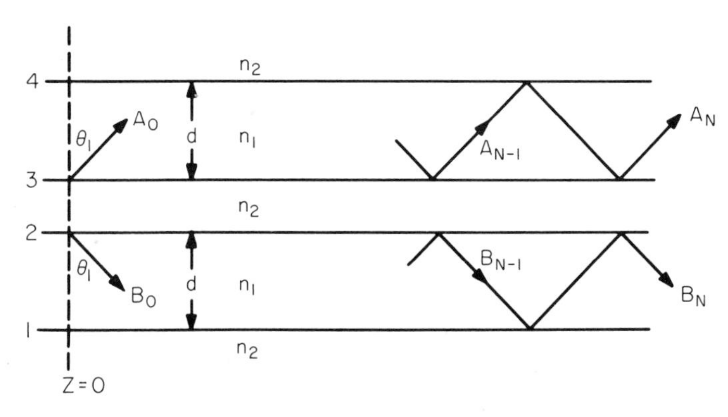

Fig. 3-8. The geometric interpretation of energy transfer between identical parallel thin-film waveguides. Insets define the reflection and transmission coefficients.

and B_0 by iteration, determining them first in terms of A_{N-1} and B_{N-1}. Thus, if $r' \exp i\phi_{r'}$, $r \exp i\phi_r$, and $t \exp i\phi_t$ are the amplitude reflection and transmission coefficients, as sketched in the inserts of Fig. 3-8, we have

$$\begin{aligned} A_N &= rr' \exp i(\phi_r + \phi_{r'} - 2k_1 d \sec \theta_1) A_{N-1} \\ &\quad + r't[\exp i(\phi_t + \phi_{r'} - 2k_1 d \sec \theta_1)] B_{N-1}, \\ B_N &= r't[\exp i(\phi_t + \phi_{r'} - 2k_1 d \sec \theta_1)] A_{N-1} \\ &\quad + r'r[\exp i(\phi_r + \phi_{r'} - 2k_1 d \sec \theta_1)] B_{N-1}. \end{aligned} \tag{3.50}$$

Now, $r' = 1$ and $\phi_t = \phi_r - \frac{1}{2}\pi$ (see the appendix to this chapter), so that this reduces to

$$\begin{aligned} A_N &= (rA_{N-1} - itB_{N-1}) \exp i(\phi_r + \phi_{r'} - 2k_0 n_1 d \sec \theta_1), \\ B_N &= -(itA_{N-1} - rB_{N-1}) \exp i(\phi_r + \phi_{r'} - 2k_0 n_1 d \sec \theta_1). \end{aligned} \tag{3.51}$$

This may be written in matrix form as

$$\begin{pmatrix} A_N \\ B_N \end{pmatrix} = \mathbf{M} \begin{pmatrix} A_{N-1} \\ B_{N-1} \end{pmatrix}, \tag{3.52}$$

where

$$\mathbf{M} = \begin{pmatrix} r & -it \\ -it & r \end{pmatrix} \exp i(\phi_r + \phi_{r'} - 2kn_1 d \sec \theta_1). \tag{3.53}$$

Clearly, we can now write

$$\begin{pmatrix} A_N \\ B_N \end{pmatrix} = \mathbf{M}^N \begin{pmatrix} A_0 \\ B_0 \end{pmatrix}, \tag{3.54}$$

so that our problem is solved. The solution is not convenient, however, because of the N matrix multiplications. Obviously, we could simplify the calculation considerably by writing the initial condition in terms of the eigenvectors of the matrix $\mathbf{M}$. This is equivalent to determining the relative values of A_{N-1} and B_{N-1} such that A_N is proportional to A_{N-1} and B_N is proportional to B_{N-1}. These will describe the steady-state conditions (normal modes), for which there is no net change in the power distribution. Rather than developing the formal matrix algebra, let us go back to Eq. (3.51) to solve for these steady-state amplitudes. Thus, let $\psi = \phi_r + \phi_{r'} - 2k_0 n_1 d \sec \theta_1$. Then, for the steady state

$$\begin{aligned} A_N &= e^{i\psi}(rA_{N-1} - itB_{N-1}) = \gamma A_{N-1} \\ B_N &= e^{i\psi}(-itA_{N-1} + rB_{N-1}) = \gamma B_{N-1}, \end{aligned} \tag{3.55}$$

where γ is a complex constant to be determined. Equations (3.55) have a nontrivial solution only if the determinant of the coefficients of A_{N-1} and B_{N-1} vanishes. Accordingly,

$$\begin{vmatrix} re^{i\psi} - \gamma & -ite^{i\psi} \\ -ite^{i\psi} & re^{i\psi} - \gamma \end{vmatrix} = 0 \tag{3.56}$$

or

$$\gamma_\pm = (r \pm it) \exp i\psi = \exp[i\psi \pm i \tan^{-1}(t/r)], \tag{3.57}$$

because $r^2 + t^2 = 1$. Thus, we apparently have two steady-state conditions for the assumed angle θ. Inserting γ_- into Eq. (3.55), we find that the ray amplitudes A_{N-1} and B_{N-1} are equal in magnitude and phase, so that γ_- characterizes the symmetric mode. Substituting γ_+ into Eq. (3.55), we find that $B_{N-1} = -A_{N-1}$, so that γ_+ characterizes the antisymmetric mode.

Suppose now that all the power is in the upper guide in the plane $z = 0$. Then $B_0 = 0$. This condition can be described as the equal amplitude superposition of the symmetric and antisymmetric steady-state conditions. In the upper slab, the two rays add in the plane $z = 0$, while in the lower slab they cancel. The power distribution changes, however, as the rays move down the guide. Thus, in the upper slab we obtain

$$A_N = \tfrac{1}{2}A_0 \exp[iN\psi + iN\tan^{-1}(t/r)] + \tfrac{1}{2}A_0 \exp[iN\psi - iN\tan^{-1}(t/r)], \quad (3.58)$$

while in the lower slab, we find

$$B_N = \tfrac{1}{2}A_0 \exp[iN\psi + iN\tan^{-1}(t/r)] - \tfrac{1}{2}A_0 \exp[iN\psi - iN\tan^{-1}(t/r)]. \quad (3.59)$$

If we now assume that $t/r \ll 1$, so that $\tan^{-1}(t/r) \approx t/r$, we then find that the power in the upper guide varies as

$$|A_N|^2 \approx |A_0|^2 \cos^2(Nt/r), \quad (3.60)$$

while that in the lower guide is

$$|B_N|^2 \approx |A_0|^2 \sin^2(Nt/r). \quad (3.61)$$

The energy clearly oscillates back and forth between the guides along their length. The beat length L_B corresponds to that axial distance for which $Nt/r = \pi$, where $N = L_B/(2d_e \tan\theta_1)$ and d_e is the effective thickness of the guide. We shall show later that in the weak coupling approximation that we are using here ($t/r \ll 1$),

$$d_e = d + \lambda_0\pi^{-1}(n_1^2 \sin^2\theta_1 - n_2^2)^{-1/2}$$

for TE modes. In this approximation, the ray displacement on reflection is thus independent of the thickness of the guides, and it is the same for isolated or coupled guides. Thus, the layer of index n_2 separating the guides acts like an infinitely thick medium with respect to ray displacements.

The foregoing analysis implies that all angles $\theta > \theta_C$ are allowed. We know, however, that this cannot be so for waves bound to the system. To obtain the characteristic angles, we could solve the complete boundary value electromagnetically. Alternatively, we can use a ray argument. To do this, we consider the rays in each guide separately. We make use of the fact that the eigenvalues (propagation constants) of the normal modes are real, so that the fields in each guide can be viewed as arising from the superposition of two uniform plane waves. The plane surfaces of constant phase can then be viewed as geometric wavefronts satisfying the Eikonal equation. Ray A_{N-1} in Fig. 3-9 reaches the wavefront illustrated by the

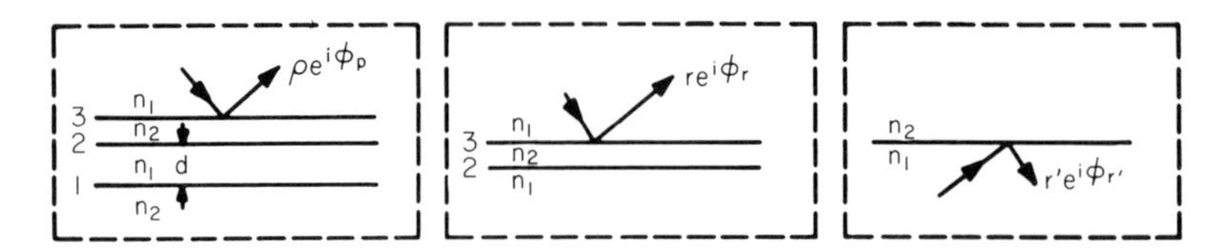

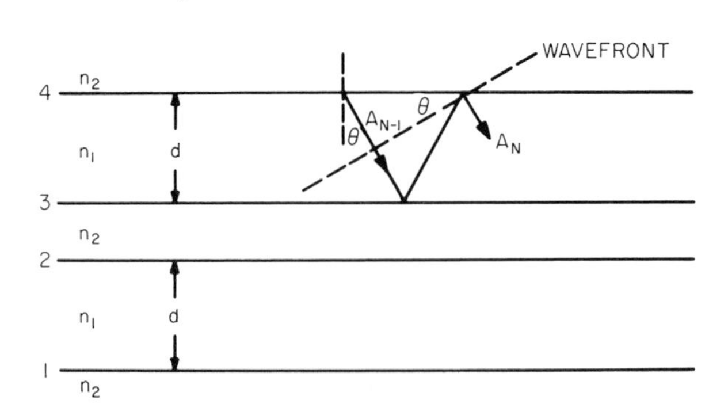

Fig. 3-9. A geometric technique for determining the characteristic angles of dielectric waveguides. Insets define the reflection and transmission coefficients.

dotted line by two paths from point P; one path is direct, the other involves reflections at interfaces 3 and 4. Because rays A_{N-1} and A_N have equal magnitudes, the optical path difference between these two paths to the wavefront must be an integral multiple of wavelengths. We thus have

$$-2k_1 d \sec\theta + \phi_\varrho + \phi_{r'} + 2k_1 d \tan\theta \sin\theta = 2N\pi .$$

This can be reduced to

$$-2k_1 d \cos\theta + \phi_\varrho + \phi_{r'} = 2N\pi . \tag{3.62}$$

The same path-difference conditions must also be met in the lower guide at the characteristic angle θ. Because the guides are identical in the case we are now considering, the application of these conditions again leads to Eq. (3.62).

The reflection coefficients defining the phase changes ϕ_ϱ and $\phi_{r'}$ in Eq. (3.62) are illustrated in the insets in Fig. 3-9. From classical optical thin-film theory (*11*) or from Eqs. (3.3) and (3.4), we can calculate that

$$\phi_\varrho = \phi_r - \tan^{-1} \frac{(1 - r^2) \sin \delta}{2r - (1 + r^2) \cos \delta}, \tag{3.63}$$

where $\delta = \phi_r + \phi_{r'} - 2k_1 d \cos \theta$, as defined in Section *A*. By substituting Eq. (3.63) into Eq. (3.62), we obtain in the condition $\tan \delta = \pm t/r$. Therefore, the characteristic equation (3.62) becomes

$$\delta_\pm \pm \tan^{-1}(t/r) = 2N\pi. \tag{3.64}$$

Comparing this equation for the characteristic angles of the normal modes with Eq. (3.57), we see that the geometrically calculated eigenvalue $\gamma_\pm$ are accordingly equal to $\exp(2ik_1 d \tan \theta_\pm \sin \theta_\pm)$. Note that $2d \tan \theta_\pm$ is just the axial distance between rays A_{N-1} and A_N and that $k_1 \sin \theta_\pm$ is the axial wavenumber. Because Eq. (3.64) defines the characteristic angles in pairs, we should modify the statement following Eq. (3.57) concerning the steady-state conditions for the rays. We do not, in fact, have two such conditions for each assumed angle θ. Rather, the characteristic angles θ_+ and θ_- arise in pairs centered around the discrete angles θ_i that define the modes of the isolated guide. We may use the geometric technique described earlier to show that the angles θ_i satisfy the characteristic equation

$$2\phi_{r'} - 2k_1 d \cos \theta_i = 2N\pi. \tag{3.65}$$

Note further that the characteristic angles $\theta_\pm$ of the coupled guides also bracket those angles θ_p that satisfy the leaky mode resonance condition, i.e.,

$$\delta \equiv \phi_r + \phi_{r'} - 2k_1 d \cos \theta_p = 2N\pi. \tag{3.66}$$

(Note that this is not the characteristic equation of a leaky mode, because it does not account for losses.) The difference between the angles θ_i and θ_p can be neglected in obtaining a first-order approximation of the beat phenomenon. To show this, we examine the phases ϕ_r and $\phi_{r'}$. From the appendix to this chapter, we find that

$$\phi_{r'} = \tan^{-1} \frac{2uq}{u^2 - q^2}; \qquad \phi_r = \tan^{-1} \frac{2uq \coth(qs/a)}{u^2 - q^2}, \tag{3.67}$$

where

$$u = k_0 n_1 a \cos\theta; \qquad q = k_0(n_1^2 \sin^2\theta_1 - n_2^2)^{1/2};$$

$2a$ is the thickness of each guide and s is the thickness of the FTR layer. To first order in the quantity $e^{-qs/a}$, which is assumed much less than unity, the function $\coth(qs/a)$ can be set equal to unity, so that $\phi_r \approx \phi_{r'}$. On the other hand, the phase angle,

$$\tan^{-1}(t/r) = \tan^{-1}\frac{2uq}{(u^2+q^2)\sinh(qs/a)} \approx \tan^{-1}\frac{4uqe^{-qs/a}}{u^2+q^2}, \qquad (3.68)$$

is of first order in $e^{-qs/a}$ and must be retained in the geometric derivation of the beat phenomenon, Eqs. (3.58)–(3.61). The angles θ to be associated with the phase ψ in these four equations should satisfy either Eq. (3.65) or Eq. (3.66). They are average angles lying between the successive pairs of angles $\theta_\pm$ that satisfy the normal mode equation (3.64). Because θ satisfies Eq. (3.66), we see that the average phase ψ that appears in Eqs. (3.58) and (3.59) is equal to $k_1(\sin\theta_p)(2d\tan\theta_p)$. The quantity $k_1 \sin\theta_p$ is clearly the average axial wavenumber of the normal modes. Finally, it can be noted that, to first order in $e^{-qs/a}$, we can evaluate the quantity (t/r) at any of the angles θ_+, θ_i, θ_p, or θ_-.

While the geometric derivation of characteristic equations is direct, it is not generally to be recommended for numerical work; this is because the resulting equations are often unnecessarily complicated algebraically. The correct characteristic equation can always be generated by solving the homogeneous Maxwell equations in each medium and satisfying the boundary conditions. It is instructive to do so for the simple structure we are studying here. This is because, as one of the few problems in coupled mode theory that can be solved exactly, it provides a means of verifying the propriety and limitations of perturbation techniques used to solve more complicated problems. Let us therefore study the structure again as a boundary value problem.

3. Wave Theory of Coupling between Identical Parallel Guides

In Fig. 3-10, we again illustrate the system shown in Fig. 3-8. For the present analysis, we must superpose a coordinate system as shown. The central layer of index n_2 and thickness s is bisected by the plane $x = 0$. The two guiding layers of index $n_1 > n_2$ and thickness $d = 2a$ are centered on the planes $x = \pm c$. The space $|x| > c + a$ is filled with a medium of index n_2.

To derive the TE modes ($E_x = E_z = 0$) of this system, we proceed as in Chapter 2. We seek the elementary solutions of Maxwell's equations that

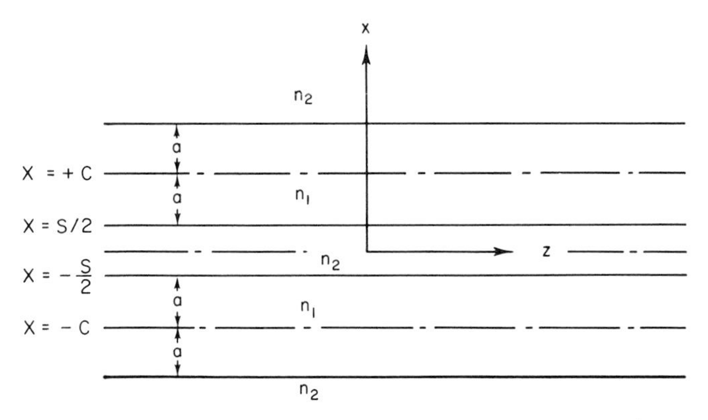

Fig. 3-10. Coordinate system used in analyzing a thin-film directional coupler.

describe waves that do not radiate away from the structure. These have the form $E_y = f(x)e^{i\omega t - ihz}$ in each medium. Exponential or hyperbolic functions of real arguments are appropriate for $f(x)$ in regions of index n_2, sine and cosine functions in regions of index n_1. Both E_y and H_z must be continuous at the interfaces between the media. Taking advantage of the symmetry of the problem, we divide the solutions into two classes: those for which E_y is symmetric about the plane $x = 0$, and those for which E_y is antisymmetric. This reduces the algebra considerably by substituting two relatively simple problems, each involving four equations in four unknown constants, for one problem involving eight equations in eight unknowns. The algebra is straightforward, so that we will only present and discuss the results. For the *symmetric modes*, we have

$$E_y = C_S e^{i\omega t - ih_S z} \begin{cases} \dfrac{e^{-q_S(x-c-a)/a}}{\cos u_S + \Delta \sin u_S}; \quad x > c + a \\[2ex] \left[\cos \dfrac{u_S(x-c)}{a}\right] + \Delta \sin \dfrac{u_S(x-c)}{a}; \\ \qquad\qquad c - a < x < c + a \\[2ex] \left(\cosh \dfrac{q_S x}{a}\right) \dfrac{\cos u_S - \Delta \sin u_S}{\cosh q_S[(c/a) - 1]}; \\ \qquad\qquad 0 < |x| < c - a \\[2ex] \cos\left(\dfrac{u_S x}{a} + \dfrac{u_S c}{a}\right) - \Delta \sin\left(\dfrac{u_S x}{a} + \dfrac{u_S c}{a}\right); \\ \qquad\qquad -c - a < x < -c + a \\[2ex] \dfrac{e^{(q_S x/a) + (q_S c/a) + q_S}}{\cos u_S + \Delta \sin u_S}; \quad x < -c - a, \end{cases} \tag{3.69}$$

where C_S is an arbitrary amplitude constant and

$$\Delta = (u_S \sin u_S - q_S \cos u_S)/(u_S \cos u_S + q_S \sin u_S), \tag{3.70}$$

$$u_S^2 + q_S^2 = R^2 = (\pi d/\lambda_0)^2(n_1^2 - n_2^2), \tag{3.71}$$

and

$$u_S^2 + h_S^2 a^2 = k_1^2 a^2; \qquad -q_S^2 + h_S^2 a^2 = k_2^2 a^2. \tag{3.72}$$

The characteristic equation for these *symmetric modes* is

$$2 \cot 2u_S = \frac{u_S}{q_S} - \frac{q_S}{u_S} + \left(\frac{u_S}{q_S} + \frac{q_S}{u_S}\right) e^{-q_S s/a}, \tag{3.73}$$

where s is the thickness of the low-index layer separating the guides.

The electric fields of the *antisymmetric modes* are quite similar. They are described by

$$E_y = C_A e^{i\omega t - ih_A z} \begin{cases} \dfrac{e^{-q_A(x-c-a)/a}}{\cos u_A + \Delta \sin u_A}; & x > c + a \\ \left[\cos \dfrac{u_A(x-c)}{a}\right] + \Delta \sin \dfrac{u_A(x-c)}{a}; & c - a < x < c + a \\ \left(\sinh \dfrac{q_A x}{a}\right) \dfrac{\cos u_A - \Delta \sin u_A}{\sinh q_A[(c/a) - 1]}; & 0 < |x| < c - a \\ -\left[\cos \dfrac{u_A(x+c)}{a}\right] + \Delta \sin \dfrac{u_A(x+c)}{a}; & -c - a < x < -c + a \\ -\dfrac{e^{q_A(x+c+a)/a}}{\cos u_A + \Delta \sin u_A}; & -c - a < x, \end{cases} \tag{3.74}$$

where C_A is an arbitrary amplitude constant for the antisymmetric mode and Δ, u, q, and h again satisfy Eqs. (3.70)–(3.72). The characteristic equation for the *antisymmetric modes* is

$$2 \cot 2u_A = \frac{u_A}{q_A} - \frac{q_A}{u_A} - \left(\frac{u_A}{q_A} + \frac{q_A}{u_A}\right) e^{-q_A s/a}. \tag{3.75}$$

Equations (3.73) and (3.75) can be considerably simplified by assuming that

$e^{-qs/a} \ll 1$. Accordingly, we obtain

$$2 \cot 2u \approx \frac{u}{q}(1 \pm e^{-qs/a}) - \frac{q}{u(1 \pm e^{-qs/a})}. \tag{3.76}$$

These equations can be reduced to

$$\cot u = (u/q)(1 \pm e^{-qs/a}); \qquad \tan u = -(u/q)(1 \pm e^{-qs/a}), \tag{3.77}$$

where the positive sign applies to symmetric modes and the negative sign to antisymmetric modes. Comparing Eqs. (3.77) with Eqs. (2.31) and (2.35) of Chapter 2, we see that for each mode on the isolated guide, we obtain both a symmetric and an antisymmetric mode on the coupled system. When $e^{-qs/a} \ll 1$, the eigenvalues of these modes are slightly greater and slightly less than that of the corresponding mode on the isolated guide. We further note, by comparing Eqs. (3.77) with Eq. (3.70), that the constant Δ, which specifies the relative contributions of the cosine and sine functions for the fields inside the guides, is either much greater or much less than one. Accordingly, the mode fields inside the guides can be well approximated by the single mode fields of the isolated guides. These properties of the eigenfunctions and eigenvalues of weakly coupled waveguides are quite general. They do not depend on the details of the geometry. It is only in very simple cases, such as this one, however, that these properties can be observed by a straightforward solution of Maxwell's equations.

We will now derive explicit approximate expressions for the eigenvalues u and q of the coupled system. To obtain them, let us rewrite the first of Eqs. (3.77) as

$$u \tan u - q = \mp q e^{-qs/a}. \tag{3.78}$$

We now expand both sides of Eq. (3.78) about the zeros (u_0, q_0) of the left-hand side and neglect second- and higher-order terms. Then,

$$u_0 \tan u_0 - q_0 + (1/u_0)(u_0^2 \sec^2 u_0 + u_0 \tan u_0)\, \Delta u - \Delta q = \mp q_0 e^{-q_0 s/a}. \tag{3.79}$$

We now use the fact that $R^2 = u^2 + q^2 = u_0^2 + q_0^2$ is a constant of the system, so that

$$\Delta q = -(u_0/q_0)\, \Delta u. \tag{3.80}$$

We also know that $R^2 = u_0^2 \sec^2 u_0$. Thus, Eq. (3.79) reduces to

$$(1/u_0)(R^2 + q_0)\, \Delta u + (u_0/q_0)\, \Delta u = \mp q_0 e^{-q_0 s/a}$$

or

$$\Delta u = \mp u_0 q_0{}^2 e^{-q_0 s/a}/[q_0(R^2 + q_0) + u_0{}^2]$$
$$= \mp u_0 q_0{}^2 e^{-q_0 s/a}/[R^2(1 + q_0)]. \qquad (3.81)$$

The propagation constant h satifies the equation $h^2 + (u^2/a^2) = k_1{}^2$, so that

$$\Delta h_{\text{wave}} = \frac{u_0}{h_0}\frac{\Delta u}{a^2} = \pm\frac{u_0{}^2 q_0{}^2 e^{-q_0 s/a}}{h_0 \mathrm{R}^2(1 + q_0)a^2}, \qquad (3.82)$$

where the positive sign applies to the symmetric modes.

4. Comparison of Ray and Wave Theories

Let us now consider the case where the symmetric and antisymmetric modes are excited with equal amplitude in the plane $z = 0$. Then, $C_\mathrm{S} = C_\mathrm{A}$ $(= C/2)$ in Eqs. (3.74) and (3.60). By summing these expressions, we see that the electric field in the upper guide is given by

$$E_y \approx Ce^{i\omega t - ih_0 z}\cos(\Delta hz)\left[\cos\left(\bar{u}\,\frac{x - c}{a}\right) + \Delta\,\sin\left(\bar{u}\,\frac{x - c}{a}\right)\right], \qquad (3.83)$$

where $\bar{u} = (u_\mathrm{S} + u_\mathrm{A})/2 \approx u_0$, while in the lower slab it is

$$E_y \approx iCe^{i\omega t - ih_0 z}\sin(\Delta hz)\left[\cos\left(\bar{u}\,\frac{x - c}{a}\right) + \Delta\,\sin\left(\bar{u}\,\frac{x - c}{a}\right)\right]. \qquad (3.84)$$

The power thus oscillates with distance along the guide at the spatial frequency Δh. In the weak coupling approximation, Δh is given by Eq. (3.81).

We are now in a position to determine the effective thickness of the guides d_e to be used in determining the beat length from the ray theory. From Eqs. (3.58) and (3.59) and Fig. 3-8, we see that a purely geometric analysis suggests that $N = z/2d(\tan\theta_1)$, so that the geometric beat wavenumber is given by

$$\Delta h_{\text{geom}} = [\tan^{-1}(t/r)]/(2d\tan\theta). \qquad (3.85)$$

From the Airy formula (appendix to this chapter), we have

$$t/r \approx (4uq/R^2)e^{-qs/a}, \qquad (3.86)$$

so that the geometric beat frequency, in the approximation $\tan^{-1}(t/r) \approx t/r$, is given by

$$\Delta h_{\text{geom}} = u_0 q_0 e^{-q_0 s/a}/(R^2 a\tan\theta_1). \qquad (3.87)$$

Comparing this with Eq. (3.82), after substituting $h_0 a/u_0 = \tan\theta_1$, we find that

$$\Delta h_{\text{geom}}/\Delta h_{\text{wave}} = 1 + (1/q_0). \tag{3.88}$$

This result shows that the geometrically calculated coupling is stronger than the actual coupling by the factor $1 + (1/q_0)$. We can obtain the correct (in the weak coupling approximation) result geometrically, nevertheless, simply by using the effective width

$$d_e = d[1 + (1/q_0)] \tag{3.89}$$

in place of the actual width of the guide. The increment in width obtained from the wave theory is just

$$\Delta d = d/(R^2 - u_0^2)^{1/2} = \lambda_0/\pi(n_1^2 \sin^2\theta_1 - n_2^2)^{1/2}. \tag{3.90}$$

This applies to TE waves. A different form is obtained for TM waves. It should be noted that this increment is entirely independent of d and, therefore, independent of the particular resonances of the structure. It is, in fact, in agreement with theoretical predictions of the Goos–Haenchen shift—the lateral shift of a finite collimated beam of light incident on a totally reflecting interface at an angle greater than the critical angle.

5. Ray Theory of Distributed Coupling between Parallel Guides of Arbitrary Width

Consider the guides sketched in Fig. 3-11. We assume that we have two rays A_N and B_N departing the interior surfaces of the upper and lower

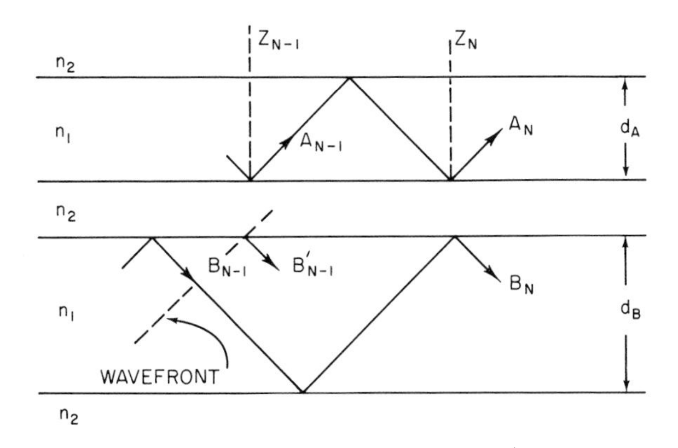

Fig. 3-11. Geometric interpretation of energy transfer in guides of different widths.

guides in the plane $z = z_N$. We first write two equations relating A_N and B_N to A_{N-1} and B_{N-1}. After solving these for the steady-state ray amplitudes, we describe energy transfer between the guides in terms of the superposition of two steady-state normal modes of appropriate relative amplitude and phase. Accordingly, we have

$$\begin{aligned} A_N &= re^{i\phi_A}A_{N-1} - ite^{i\phi_B}B_{N-1} = \gamma A_{N-1} \\ B_N &= -ite^{i\phi_A}A_{N-1} + re^{i\phi_B}B_{N-1} = \gamma B_{N-1}, \end{aligned} \tag{3.91}$$

where we have used the fact that $e^{i(\phi_t - \phi_r)} = -i$ and the definitions

$$\begin{aligned} \phi_A &= \phi_{r'} + \phi_r - 2k_0 n_1 d_A \sec\theta_1 \\ \phi_B &= \phi_{r'} + \phi_r - 2k_0 n_1 d_B \sec\theta_1 . \end{aligned} \tag{3.92}$$

The condition that the determinant of the coefficient of A_{N-1} and B_{N-1} in Eq. (3.91) must vanish gives the steady-state characteristic values

$$\begin{aligned} \gamma_\pm &= (\exp i\bar{\phi})[r(\cos\Delta\phi) \pm i(1 - r^2\cos^2\Delta\phi)^{1/2}] \\ &\equiv \exp i\{\bar{\phi} \pm \tan^{-1}[(\tan^2\Delta\phi)/r^2 + t^2/r^2]^{1/2}\}, \end{aligned} \tag{3.93}$$

where

$$\bar{\phi} = (\phi_A + \phi_B)/2; \qquad \Delta\phi = (\phi_A - \phi_B)/2. \tag{3.94}$$

We note from the second form of Eq. (3.93) that, when $t^2 \ll \tan^2\Delta\phi$ and $r^2 \approx 1$, the phase change between A_N and A_{N-1} or B_N and B_{N-1} is substantially independent of the other guide; $\gamma \approx e^{i\phi_A}$ or $e^{i\phi_B}$. When $\Delta\phi = M\pi$, on the other hand, we have phase matching and maximum interaction.

Substituting $\gamma_\pm$ from Eq. (3.93) back into Eq. (3.91) we find that the relative amplitudes of the steady-state rays in the two guides are given by

$$\begin{aligned} \frac{B_{N-1}}{A_{N-1}} &= -e^{i\Delta\phi}\left\{\pm\left[\frac{r^2}{t^2}(\sin^2\Delta\phi) + 1\right]^{1/2} - \frac{r}{t}\sin\Delta\phi\right\} \\ &\equiv K_\pm e^{i\Delta\phi}. \end{aligned} \tag{3.95}$$

Note that $K_- = -1/K_+$. When $\Delta\phi = 0$, as for the identical guides, we have $B_{N-1} = \pm A_{N-1}$. When $t^2/r^2 \ll \sin^2\Delta\phi$, we see that the ray amplitudes in the two guides are in the ratio $(t/2r\sin\Delta\phi)^{\pm 1}$. The coupling between the guides is thus generally quite small unless resonance conditions are satisfied.

In order to describe in geometric terms a length-dependent modulation of the power, we must be able to translate the two rays A_{N-1} and B_{N-1}

into the same plane. Thus, for example, we must find, in terms of B_{N-1}, the ray B'_{N-1}, which departs the interior interface of the lower guide in the plane z_{N-1}. This plane is determined from ray A_{N-1}, as illustrated in Fig. 3-11. In the steady state, rays B_{N-1} and B'_{N-1} must be in phase at the wavefront illustrated by the dotted line in Fig. 3-11. The phase difference between these two rays is

$$-k_1 2(d_B - d_A) \tan \theta_1 \sin \theta_1 = 2\,\Delta p = -i \ln(B'_{N-1}/B_{N-1}). \quad (3.96)$$

We have now determined the amplitudes of the rays that depart the interior surfaces of each guide in a plane z_N in terms of the amplitudes of rays A_{N-1} and B'_{N-1} that depart these surfaces in the plane z_{N-1}. We can repeat this argument to find rays A_{N-1} and B'_{N-1} in terms of rays A_{N-2} and B'_{N-2} that depart the interior surfaces of each guide at distances corresponding to two internal reflections from plane z_{N-1}. Equations (3.91)–(3.95) again apply. In particular, from Eqs. (3.95) and (3.91), we have

$$B_N/A_N = B_{N-1}/A_{N-1} = K_{\pm}e^{i\Delta\phi} = B'_{N-1}/A_{N-1} = B'_{N-2}/A_{N-2}. \quad (3.97)$$

It follows from Eq. (3.97) that $B_{N-1} = B'_{N-1}$. This, in turn, requires that the phase difference in Eq. (3.96) must equal $2M\pi$. By a similar argument, we find that the ray that departs the interior surface of the lower guide in the plane z_{N-2}, defined on ray A_{N-2}, is equal to B'_{N-2}.

The argument can now be repeated $N - 2$ times to determine, finally, the rays A_N and B_N in plane z_N from the rays A_0 and B_0 that depart the interior surfaces in the plane $z_0 = z_N - N(z_N - z_{N-1})$. Accordingly, we find that $A_N = A_0\gamma^N$ and $B_N = B_0\gamma^N$, with $B_0/A_0 = K_{\pm}e^{i\Delta\phi}$. The steady-state rays can accordingly be described by the two ordered pairs

$$\begin{aligned}(A_N,\ B_N) &= C_+[\exp(iN\bar{\phi} + iN\nu)](1,\ K_+ \exp i\Delta\phi)\\ (A_N,\ B_N) &= C_-[\exp(iN\bar{\phi} - iN\nu)](1,\ K_- \exp i\Delta\phi).\end{aligned} \quad (3.98)$$

The first member of each pair gives the ray amplitude in a plane z_N in guide A; the second gives the ray amplitude in the same plane in guide B. The constants C_+ and C_- give the arbitrary amplitudes of the two disturbances. The parameter ν, from Eq. (3.93), is

$$\nu = \tan^{-1}\left[\frac{(\tan^2 \Delta\phi + t^2)}{r^2}\right]^{1/2}. \quad (3.99)$$

Let us now suppose that we know that all the energy is in guide A in the plane z_0. We can describe this initial condition by summing the two steady-state modes, Eqs. (3.98), with $C_+ = C$ and $C_- = K_+^2 C$. Accordingly, we obtain

$$\begin{aligned} A_N &= C(\exp iN\bar{\phi})[(1 + K_+^2)(\cos N\nu) + i(1 - K_+^2)\sin N\nu] \\ B_N &= 2iK_+C(\sin N\nu)\exp(iN\bar{\phi} + i\Delta\phi). \end{aligned} \tag{3.100}$$

We can translate N into z by using the effective width of the guide A, as found in the last subsection. Thus,

$$N = z/(2d_e \tan\theta_1), \tag{3.101}$$

where d_e is given by Eq. (3.89) (TE modes). Note from Eqs. (3.100) that the power in the initially unexcited guide varies sinusoidally with length from zero to a maximum of $4K_+^2$ at the spatial frequency $\nu/(2d_e \tan\theta_1)$. It is apparent from Eqs. (3.95) and (3.99) that this maximum power is small and the frequency large unless the guides are of resonant dimensions.

One of the advantages of the ray approach to the coupling problem is that it separates the description of the coupling process from the determination of the characteristic angles. Because the guides are dissimilar, we must simultaneously satisfy a path difference condition in each of the guides, i.e.,

$$\begin{aligned} \delta_A &= -2k_1 d_A(\cos\theta) + \phi_\varrho + \phi_{r'} = 2N\pi \\ \delta_B &= -2k_1 d_B(\cos\theta) + \phi_\varrho + \phi_{r'} = 2M\pi. \end{aligned} \tag{3.102}$$

Equations (3.102) follow directly from Eq. (3.62) and the arguments leading to it. We can use Eq. (3.63), with $\delta = \delta_B$, to reduce the first of Eqs. (3.102). A similar reduction of the second of these equations is effected by substituting Eq. (3.63) with $\delta = \delta_A$. Equations (3.102) can then be further reduced to yield the equivalent characteristic equations

$$r\cos\Delta\delta = \cos\bar{\delta}; \quad \text{or} \quad \tan(\tfrac{1}{2}\delta_A)\tan(\tfrac{1}{2}\delta_B) = (1 - r)/(1 + r) = t^2/(1 + r)^2; \tag{3.103}$$

where $\Delta\delta = (\delta_A - \delta_B)/2$ and $\bar{\delta} = (\delta_A + \delta_B)/2$. Comparing Eq. (3.103) with Eqs. (2.31) and (2.35) of Chapter 2, we see that the characteristic angles θ for the coupled dissimilar films, defined implicitly by Eq. (3.103), are very near those of the isolated films. When $t = 0$, Eq. (3.99) reduces to the corresponding equations of Chapter 2. The ray angles to be used in Eqs. (3. 100), which describe the coupling process, may be those satisfying either the uncoupled mode characteristic equations, or the corresponding resonances of the leaky guides.

D. LATERAL SHIFTS OF LIGHT BEAMS ON TRANSMISSION AND REFLECTION AT FTR LAYERS

When a beam of light is totally reflected at an interface separating two dielectric media, the reflected beam is physically displaced from the trajectory. We would not expect this on the basis of geometric theory. It appears as though the incident beam enters the medium of lower index of refraction through one region of the interface and then emerges from it through a slightly displaced region. The displacement is in the direction of the component of linear momentum of the incident beam parallel to the interface. If the straight-line trajectories of the centroids of the incident and totally reflected beams in the high-index medium are extended into the low-index medium, their intersection defines the location of a virtual surface, parallel to the actual interface, at which geometric reflection appears to occur. In this section, we wish to determine the location of this virtual surface with a view to completing our ray theories of distributed coupling in planar waveguides. In so doing, we shall find that waveguide concepts themselves contribute significantly to the basic theory of this interesting phenomenon.

Many theoretical studies of the total reflection of finite beams at dielectric interfaces have been reported. A recent treatment by Lotsch (*17*) contains an extensive list of references of this literature. In 1947, Goos and Haenchen (*12*) devised an experiment which yielded a quantitative measure of the displacement of a beam totally reflected from a dielectric interface relative to a beam reflected from an adjacent silvered interface. The effect has thus come to be known as the Goos–Haenchen shift. Specifically, the Goos–Haenchen shift D is the perpendicular distance between the centroids of the actually reflected beam and a beam geometrically reflected at the interface, as sketched in Fig. 3-12.

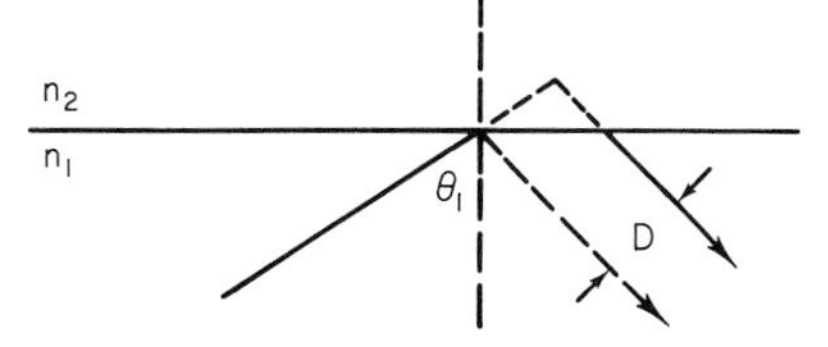

Fig. 3-12. The Goos–Haenchen shift D of a beam of light totally reflected at a dielectric interface.

The measurements of Goos and Haenchen appeared to fit the formula

$$D = Cn_2\lambda_0/(n_1^2 \sin^2 \theta_1 - n_2^2)^{1/2}, \tag{3.104}$$

where C is a constant equal to 0.52 for $n_1 = 1.52$ and $n_2 = 1$. The constant was found to be the same for both polarizations of the incident beams.

Von Fragstein (*18*) thought the experiment to be in error for this constant. He and others (19-21) predicted different shifts for the two polarizations, i.e.,

$$D_{\perp}(\text{TE}) = \lambda_0 n_2 \pi^{-1} n_1^{-1} (n_1^2 \sin^2 \theta_1 - n_2^2)^{-1/2} \tag{3.105}$$

and

$$D_{\parallel}(\text{TM}) = n_1^2 D_{\perp} / n_2^2. \tag{3.106}$$

These formulas, which apply near the critical angle, were verified experimentally in 1949 by Goos and Haenchen (*12*). Several different formulas, applying to all angles of incidence, have been developed in the literature (*17-22*), but, as far as we know, no experiments capable of discriminating between them have been reported. It is thus quite interesting to study this question from a waveguide-mode point of view. To the time of this writing, no such study has been done. In view of the mounting interest in optical guides, however, it seems likely that new theoretical and experimental developments will be forthcoming.

1. Waveguide Modes and Goos–Haenchen Shift

One of the difficulties associated with the theoretical prediction of the Goos–Haenchen shift lies in the formulation of the problem. We should like to be able to think in terms of a single angle of incidence. But this requires an infinite, uniform plane wave. In the classical Goos–Haenchen setup, however, it is not meaningful to assume an infinite plane wave, because this assumption gives us no way to distinguish one part of the reflected wave from another. We must, therefore, have a finite beam and, consequently, a range of angles of incidence. For a bound mode on a lossless, infinitely long waveguide, on the other hand, we not only can, but we must think in terms of a single angle of incidence and reflection, i.e., the characteristic angle of the mode. Though we cannot observe a reflected beam, we can measure the phase velocity along the guide. Let us, therefore, perform the following *gedanken* experiment.

Let us imagine that we have two waveguides. One is a thin film of refractive index n_1 and thickness d' coated above and below with perfectly reflecting (conducting) films. The second is a film of index n_1 and thickness d embedded in a medium of index n_2. These are sketched in Figs. 3-13(a) and 3-13 (b).

Now, we know that the interior fields in both these guides can be viewed as the superposition of two uniform plane waves whose normals make

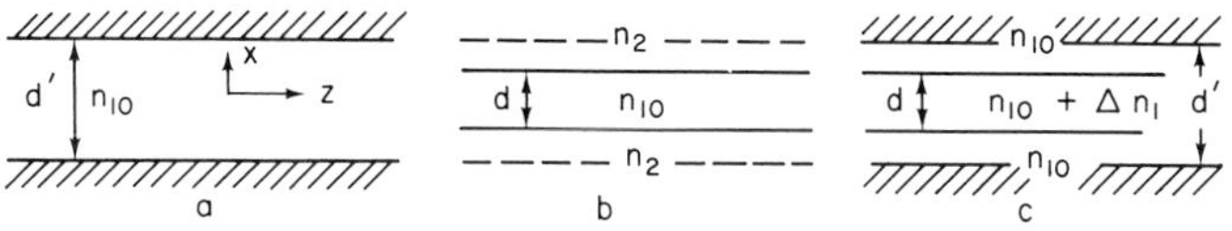

Fig. 3-13. Geometry of three waveguides used to derive the Goos–Haenchen shift.

an angle θ_1 with the bounding interfaces. Let us suppose that we can vary the thicknesses d and d' in such a way that the lowest-order TE mode in each guide has the same phase velocity, $h = k_1 \sin \theta_1$, where

$$k_1 = 2\pi n_1/\lambda_0.$$

Then, the rays in each guide travel at the same angle to the bounding interfaces.

To find the thicknesses d and d' in terms of the characteristic angle θ, we use Eqs. (2.6) and (2.31) of Chapter 2. Thus,

$$d' = \pi/(k_0 n_1 \cos \theta_1) \tag{3.107}$$

and

$$d = 2u/(k_0 n_1 \cos \theta_1) = 2[\tan^{-1}(q/u)]/(k_0 n_1 \cos \theta_1). \tag{3.108}$$

Recall that

$$q^2 = R^2 - u^2 = k_0{}^2 a^2 (n_1{}^2 - n_2{}^2 - n_1{}^2 \cos^2 \theta_1), \tag{3.109}$$

so that Eq. (3.108) indeed gives d in terms of the angle of incidence and the physical parameters of the guide. Now note that the difference between d' and d is equal to the difference between the phase changes at reflection for the two guides, divided by the transverse wave number, $k_1 \cos \theta_1$. This suggests that the effective thickness of the dielectric guide is just that of the conducting one. This suggestion seems quite reasonable, in view of the question posed in the Goos–Haenchen experiment, i.e., what are the relative positions of two beams: one reflected at a good conductor, the other at a dielectric? On the basis of phase considerations, therefore, we would conclude that the virtual surfaces we seek are located at distances $(d' - d)/2$ beyond the actual interfaces.

Let us now study this suggestion on the basis of energy considerations. We may begin by asking whether there is anything fundamental about the point of view that the modes are the superposition of two uniform plane waves at angle θ_1. To answer this question, let us examine how a mode is attenuated by dielectric loss in a perfectly conducting guide. Accordingly, we set $n_1 = n_{1_0} + \Delta n_1$, where n_{1_0} is real and equal to the index of the lossless guide and Δn_1 is a small, imaginary-valued quantity, such that $|\Delta n_1|/n_{1_0} \ll 1$. Then, $k_1 = k_{1_0} + \Delta k_1$, where

$$\Delta k_1 = 2\pi \, \Delta n_1/\lambda_0$$

is also imaginary-valued. Now, the axial propagation constant h will also take on an imaginary part, which will define the rate of loss from the mode. There is no change in the boundary conditions, so that we still have $\sin(\beta_1 d') = 0$, and thus

$$h^2 = k_1^2 - \beta_1^2 = k_1^2 - [\pi^2/(d')^2] \tag{3.110}$$

for the lowest-order TE mode. Therefore, if $h = h_0 + \Delta h$, where h_0 is the axial propagation constant for the lossless case, we have

$$\Delta h \approx k_{1_0} \, \Delta k_1/h_0 = 2\pi \, \Delta n_1/(\lambda_0 \sin \theta_1). \tag{3.111}$$

Thus, over an axial distance L, the loss is proportional to $L \csc \theta_1$. We may conclude that the energy travels down the guide in the way we would expect of multiply reflected ray pencils at angle θ_1, parallel to the normals to the component plane waves that comprise the mode.

Let us now look at the corresponding rays in the dielectric guide. We again assume a small loss in the interior medium, so that $k_1 = k_{1_0} + \Delta k_1$, but we retain the perfect dielectric outside the guiding interfaces. We then have

$$h^2 = k_2^2 + (q^2/a^2), \tag{3.112}$$

where k_2 does not change. Therefore,

$$\Delta h \approx q_0 \, \Delta q/h_0 a^2, \tag{3.113}$$

where the subscript zero represents the lossless case. Now, $q = u \tan u$, $q_0 = u_0 \tan u_0$, and $R_0^2 = u_0^2 + q_0^2 = u_0^2 \sec^2 u_0$, so that

$$\Delta q \approx (q_0 + R_0^2) \, \Delta u/u_0. \tag{3.114}$$

Therefore, from Eqs. (3.113) and (3.114), we obtain

$$\Delta h \approx \frac{q_0^2 + q_0 R_0^2}{h_0 a^2} \frac{\Delta u}{u_0}. \tag{3.115}$$

We now use $R^2 = u^2 \sec^2 u$ to obtain

$$\frac{\Delta u}{u_0} \approx \frac{1}{1 + q_0} \frac{\Delta R}{R_0}, \tag{3.116}$$

and because $R^2 = (k_1^2 - k_2^2)a^2$, we have

$$\Delta R/R_0 \approx k_{1_0}\, \Delta k_1 a^2/R_0^2. \tag{3.117}$$

Combining Eqs. (3.115)–(3.117) yields

$$\Delta h \approx \frac{k_{1_0}\,\Delta k_1}{h_0}\,\frac{q_0^2 + q_0 R_0^2}{R_0^2(1+q_0)} = \frac{\Delta k_1}{\sin\theta_1}\left[1 - \frac{u_0^2}{R_0^2(1+q_0)}\right]. \tag{3.118}$$

Now, it is not difficult to show that the factor multiplying $\Delta k_1/\sin\theta_1$ in Eq. (3.118) is just the fraction of the total TE mode power that is carried in the interior of the lossless guide. To sketch this calculation, recall that, in the interior (subscript 1) and exterior (subscript 2), we had

$$\begin{aligned} E_{y_1} &= [\cos(u_0 x/a)]e^{i\omega t - ih_0 z} \\ E_{y_2} &= (\cos u_0)e^{-q_0[(x/a)-1]}e^{i\omega t - ih_0 z} \\ H_x &= -(h_0/\omega\mu)E_y e^{i\omega t - ih_0 z} \end{aligned} \tag{3.119}$$

so that $S_z = (h/\omega\mu)E_y E_y{}^*$. Therefore, the powers P_1 and P_2 in the interior and exterior are

$$\begin{aligned} P_1 &= \frac{2h_0}{\omega\mu}\int_0^a \cos^2\frac{u_0 x}{a} = \frac{h_0 a}{\omega\mu}\left(1 + \frac{\sin 2u_0}{2u_0}\right) = \frac{h_0 a}{\omega\mu}\left(1 + \frac{q_0}{R_0^2}\right) \\ P_2 &= \frac{2h_0\cos^2 u_0}{\omega\mu}\int_a^\infty e^{-2q_0[(x/a)-1]} = \frac{h_0 a}{\omega\mu}\,\frac{\cos^2 u_0}{q_0} = \frac{h_0 a}{\omega\mu}\,\frac{u_0^2}{q_0 R_0^2}, \end{aligned} \tag{3.120}$$

The final forms of P_1 and P_2 are obtained from those preceding them by using the identities $q_0 = u_0 \tan u_0$, $R_0^2 = u_0^2 \sec^2 u_0$. We thus find that

$$\frac{P_1}{P_1 + P_2} = 1 - \frac{P_2}{P_1 + P_2} = 1 - \frac{u_0^2}{R_0^2(1+q_0)} = \left(1 + \frac{q_0}{R_0^2}\right)\frac{q_0}{1+q_0}. \tag{3.121}$$

Thus, comparing Eqs. (3.111) and (3.118), we conclude that it is incorrect to assume that the ray pencils in the dielectric guide are reflected at the actual interfaces. To find where they are reflected, let us reconsider the guide with perfectly conducting boundaries at $x - \pm a' = \pm d'/2$.

Suppose, as sketched in Fig. 3-13(c), that in region 1 of the conducting guide, defined by $-a < x < a$, the refractive index is $n_{1_0} + \Delta n_1$, while in region 2, defined by $a < |x| < a'$, the index is n_{1_0}. What is the loss rate in this case? It can be shown that, for $|\Delta n_1| < n_{1_0}$, the loss rate is simply

$$\Delta h = \frac{\Delta k_1}{\sin\theta_1}\left(1 - \frac{P_2}{P_1 + P_2}\right) = \frac{\Delta k_1 a}{\sin\theta_1 a'}\left(1 + \frac{\sin 2\beta_1 a}{2\beta_1 a}\right), \quad (3.122)$$

where P_1 and P_2 again represent the power in the two regions for the lossless case. Now if $2a = d$ (i.e., the width of the dielectric guide), and if the thickness of the conducting guide $d' = 2a'$ is adjusted so that the characteristic angles θ_1 are the same in both the conducting and dielectric guides, then $\beta_1 a$ in Eq. (3.122) $\equiv u_0$ in the dielectric guide. Therefore, we can write Eq. (3.122) in the form

$$\Delta h = \frac{\Delta k_1}{\sin\theta_1}\,\frac{a}{a'}\left(1 - \frac{u_0^2}{R^2(1 + q_0)}\right)\frac{1 + q_0}{q_0}. \quad (3.123)$$

Now, the field vectors, as well as the properties of the materials in the regions $|x| < a$, are identical in the two guides of Figs. 3-13(b) and 3-13(c), and there is no mechanism for loss outside these regions. We therefore require that the loss rates be the same in both guides. From Eqs. (3.123) and (3.118), we then have

$$a' = a(q_0 + 1)/q_0$$

or

$$d' - d = d/q_0 = \lambda_0/\pi(n_1^2 \sin^2\theta_1 - n_2^2)^{1/2}. \quad (3.124)$$

This result is based on energy considerations. To compare it with the phase argument, we use Eqs. (3.107) and (3.108) to obtain

$$d' - d = [\pi - 2\tan^{-1}(q/u)]/(k_1 \cos\theta_1). \quad (3.125)$$

For $u \ll q$, this reduces to

$$d' - d \approx 2u/(qk_1 \cos\theta_1). \quad (3.126)$$

Equation (3.126) is identical with Eq. (3.124). As sketched in Fig. 3-14, the lateral (z) shift to be associated with each reflection in the dielectric guide is

$$\Delta z = (d' - d)\tan\theta_1, \quad (3.127)$$

and the Goos–Haenchen shift is

$$D = (d' - d)\sin\theta_1. \quad (3.128)$$

Note that near the critical angle, $\sin\theta_1 \approx n_2/n_1$; therefore, Eqs. (3.124) and (3.126) lead to the classical predictions given by Eq. (3.105).

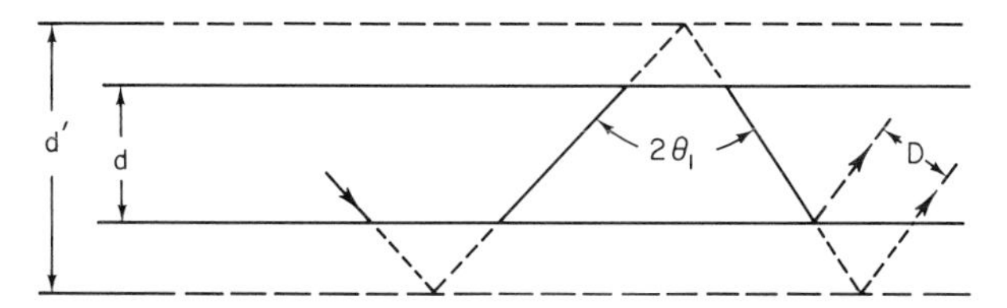

Fig. 3-14. The effective geometrical thickness d' of a dielectric guide of thickness d.

A similar analysis can be carried out for TM modes (parallel polarization). For this case, we find

$$d' - d = [\pi - 2\tan^{-1}(n_1{}^2 q/n_2{}^2 u)]/(k_1 \cos\theta_1).$$

For $u \ll q$, this reduces to

$$d' - d \approx (\lambda_0 n_2{}^2/n_1{}^2)/\pi(n_1{}^2 \sin^2\theta_1 - n_2{}^2)^{1/2}. \tag{3.129}$$

Equation (3.129) results from phase considerations. An energy argument like that for TE modes (power ratios) yields the result

$$d' - d = \frac{\lambda_0}{\pi(n_1{}^2 \sin^2\theta_1 - n_2{}^2)^{1/2}} \quad \frac{n_2{}^2}{n_1{}^2 \sin^2\theta_1 - n_2{}^2 \cos^2\theta_1}. \tag{3.130}$$

This agrees with the classical equation (3.106) near $\theta = \theta_C$ and with the results of Artmann (*19*) for all angles. Near $\theta_1 = \pi/2$, it agrees with the phase result, Eq. (3.129).

From this analysis, it is clear that we must reject the phase argument near $\theta_1 = \theta_C$. This is because the classical predictions, Eqs. (3.105) and (3.106), as well as our energy predictions, Eqs. (3.124) and (3.130), have been verified by the experiments of Goos and Haenchen for angles very close to θ_C. Near grazing incidence, on the other hand, our phase and energy arguments lead to the same results, Eqs. (3.124), (3.126), (3.129), and (3.130). The energy predictions are further corroborated by an independent analysis based on a comparison of geometric and wave-optic estimates of the loss rates of leaky modes in optical waveguides. Taken together, these arguments lead to the following conclusions:

1. Equations (3.124) and (3.130) give the Goos–Haenchen shift for all angles of incidence more than a few minutes greater than θ_C.* For $\theta_1 \approx \pi/2$, they do not predict that the shift vanishes.

* The immediate vicinity of the critical angle requires special treatment, as in Horowitz and Tamir (22).

2. Ray-optics calculations of coupling coefficients for planar waveguides require the use of the effective thickness d_e rather than the actual thickness d of the guide. Whenever the thickness t of the low-index coupling layer is such that $e^{-4qt/d} \ll 1$, this effective geometric thickness is given by d' in Eqs. (3.124) and (3.130).

2. Completion of Geometric Theory of Distributed Coupling

Our objective in this subsection is to complete the geometric theory of prism-film couplers begun in Section A of this chapter by taking into account the displacements which take place when a beam (ray) is reflected from or transmitted through an FTR layer. Accordingly, we will now assume that the effective geometrical thickness W_e of a thin-film guide of index n_1, coated with FTR layers of index n_2, is given by the actual thickness W plus an incremental thickness ΔW given by Eq. (3.124) for TE modes and Eq. (3.130) for TM modes. If the FTR layers are made of dissimilar materials of indices n_2 and n_2', the incremental thickness is given by the sum of two terms having the form of these equations, one for each of the indices, and each multiplied by the factor $\frac{1}{2}$. We have already seen, in Section C, that this procedure leads to the correct prediction of the beat length (coupling strength) between two parallel films forming a symmetric directional coupler. Here, we will show that it is also appropriate for prism-film couplers and directional couplers of arbitrary dielectric materials, provided only that the assumption of weak coupling is met. This is the case in practice.

We noted in Section A of this chapter that the behavior of the FTR filter, when described in terms of the number N of ray pencils contributing to the amplitude of a ray A_N along the guide, was identical to that of a parallel plate interferometer. The important results of such a description are summarized in Fig. 3-15. The quantities R and T in the figure represent the fraction of the power in an incident plane wave that is reflected or transmitted at the interface between the plate and its surround. They are assumed to apply locally along the rays. The amplitude of a ray is proportional to the square root of power density, as sketched in the inset, where the phase changes on reflection and transmission are also shown. The amplitude of the ray A_N in the film depends primarily on the phase difference ψ between the successive contributions to it. When this phase difference equals $2M\pi$, the amplitude can reach a maximum value of $T^{-1/2}$, as indicated in Fig. 3-15.

In Fig. 3-16, the analogous situation for the FTR filter is sketched. The algebraic results shown in Fig. 3-15 apply to this case also, where the meaning of the reflection and transmission coefficients is now as sketched

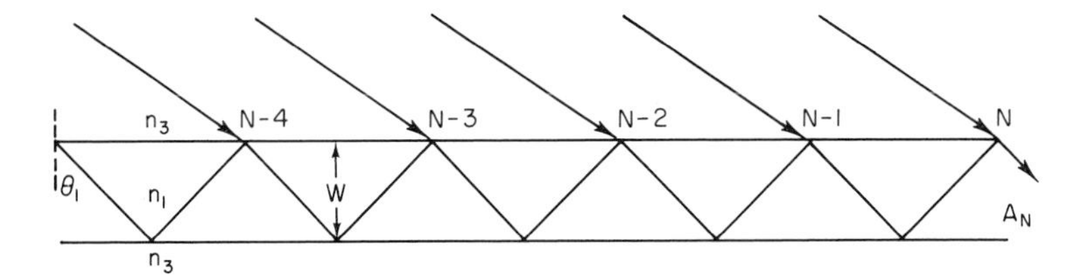

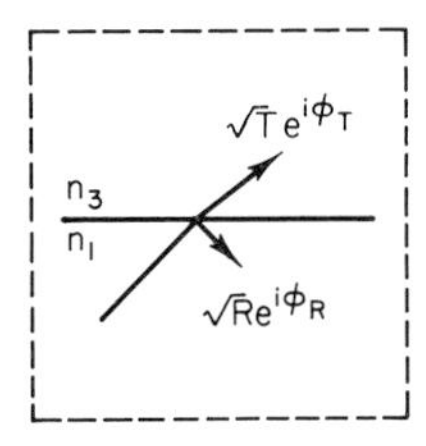

Fig. 3-15. The classical derivation and principal results of a ray analysis of a parallel plate interferometer. Inset defines the reflection and transmission coefficients:

$$\begin{aligned} A_N &= T^{1/2}e^{i\phi_T}(1 + Re^{i\psi} + R^2e^{2i\psi} + \cdots), \\ &= T^{1/2}e^{i\phi_T}(1 - R^Ne^{iN\psi})/(1 - Re^{i\psi}), \\ \psi &= 2\phi_R + 2k_0n_1W\cos\theta_1, \\ A_N &\to T^{-1/2}/T \quad \text{if} \quad R^N \ll 1 \quad \text{and} \quad T \ll 1. \end{aligned}$$

in the inset of Fig. 3-16. In the case of the parallel plate of index $n_1 > n_3$, the phase change at reflection ϕ_R equals zero. For the FTR filter, on the other hand, ϕ_R is a complicated function of the angle of incidence and the indices of refraction n_1, n_2, and n_3 (appendix to this chapter). These complications in no way modify the simple analysis of Fig. 3-15, however. The important phase difference ψ has exactly the same form in both cases; note that it is the actual width W of both the plate and the film that enters into the determination of ψ. It is only in the specification of the relative positions of rays A_N and A_{N-1} that the difference between the parallel plate and the FTR filter becomes apparent. In this section, we will assume that the axial separation between these rays is $2W\tan\theta$ in the plate, but $2W_e\tan\theta$ in the film, where the increment in thickness is found from Eq. (3.124) (TE modes) or Eq. (3.130) (TM modes). We then verify that the geometrically calculated loss rate (attenuation coefficient) resulting from this assumption agrees with that calculated for the corresponding leaky wave, i.e., the (spatial) transient solution to Maxwell's equations.

The ray paths sketched by the dotted lines in Fig. 3-16 illustrate our geometric interpretation of the passage of a ray through an FTR layer, or a ray reflected from it. The actual thickness of the FTR layer is of no

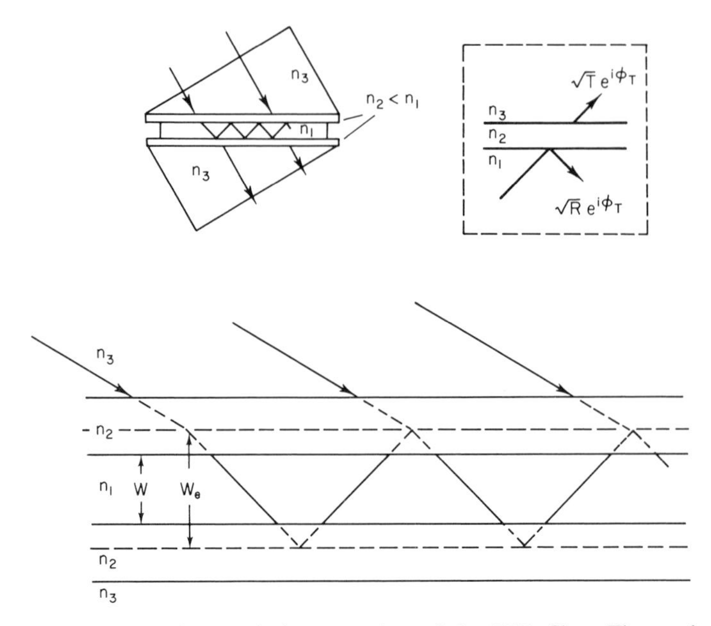

Fig. 3-16. The interferometric interpretation of the FTR filter. The result of Fig. 3-15 applies to this case. The effective thickness of the guide W_e is determined by the Goos–Haenchen shifts at the reflections on its opposite sides. Trasmitted rays are shifted, as illustrated, in two steps, each of which is proportional to one-half on the Goos–Haenchen shift appropriate to the two media on opposite sides of the FTR layer. Inset defines the reflection and transmission coefficients appropriate to the interferometric analysis of this case.

consequence in the geometric theory under the assumption of weak coupling. The layer may be replaced by two thin slabs, one of index n_1, the other of index n_3. The slab of index n_1 has a thickness equal to one-half that given by Eq. (3.124) for TE modes or Eq. (3.130) for TM modes. The slab of index n_3 has a thickness described by these same equations, with n_3 and θ_3 replacing n_1 and θ_1. Because Snell's law applies, it can be seen from these equations that the two slabs that form the geometric equivalent of the FTR layer are of equal thickness for TE modes, but of slightly different thicknesses for TM modes.

We now verify the propriety of the geometric construction described here for the case of a leaky wave in an FTR filter or FTR prism coupler. The basic physical situation is as sketched in Fig. 3-17, which is divided by two vertical dotted lines into three regions. In the region at the left, we illustrate how a wide beam of uniform amplitude is coupled into a leaky mode in the FTR prism-film coupler. If we assume that the angle of incidence satisfies the resonance condition, $\psi = 2M\pi$, then the ray theory leads to a buildup of the amplitude given by the equation (a) in the figure caption. The leaky wave analysis, on the other hand, leads to the equation (b), where $\alpha = \text{Im}\, h$.

In the center of Fig. 3-17, we illustrate a region where the wave is confined to the guiding film. Wave and ray descriptions of the bound mode give the same amplitude A_0. In passing into the region at the right of Fig. 3-17, the wave becomes a leaky mode of a truncated FTR filter. (The dotted lines in the central section of the figure illustrate the parts of the prisms of the FTR filter which have been removed.) Note that the geometric theory describes the decay of the leaky mode according to the formula A_0R^N, this being the amplitude of a ray traveling downward in the film after N reflections at the upper boundary of the film. The wave picture describes the decay of the field as that of a leaky mode with attenuation coefficient $\alpha = \text{Im}\, h$.

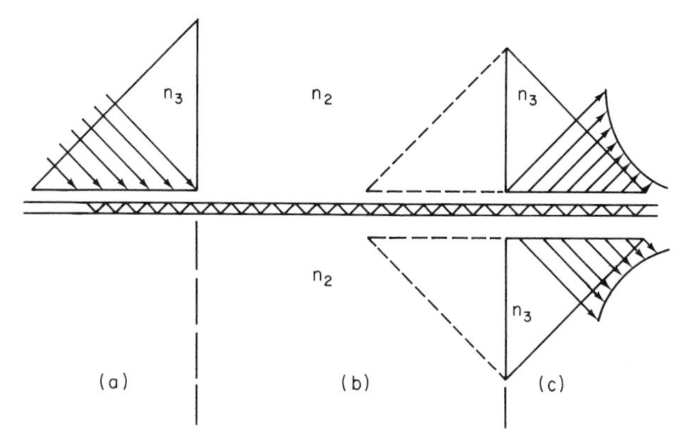

Fig. 3-17. The concepts involved in comparing the ray and wave theories of distributed coupling in (a) prism-film couplers, (b) lossless waveguides, and (c) undriven FTR output coupler: (a) $A_{\text{ray}} = A_\infty(1 - R^{N/2})$, $A_{\text{wave}} = A_\infty(1 - e^{\alpha z})$. (b) $A_{\text{ray}} = A_{\text{wave}} = A_0$. (c) $A_{\text{ray}} = A_0R^N$, $A_{\text{wave}} = A_0e^{-\alpha z}$.

We now assume that $N = z/(2W_e \tan\theta_1)$ and compare the geometric loss rate with that obtained from wave theory. The essentials of this comparison are illustrated in Fig. 3-18. On the left-hand side of the figure, the principal transverse field components in each region of a five-medium symmetric FTR structure are shown. We assume a symmetric mode, so that the fields below the guide are the same as those above it. The constants B, C, and D, as well as the complex eigenvalues u, q, u', and h, are determined by matching the tangential components of $\mathbf{E}$ and $\mathbf{H}$ at the interfaces. In the

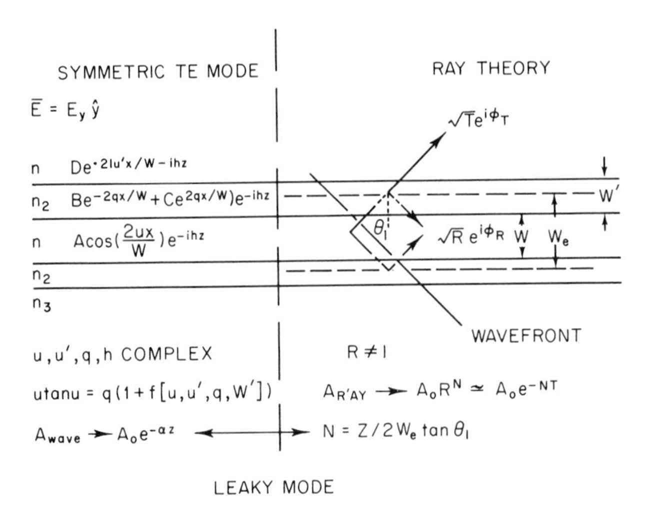

Fig. 3-18. A specific illustration of the comparison of wave and ray theories corresponding to the output coupler at the right of Fig. 3-17.

figure, we have indicated that the field expressions describe the y component of the electric field. The corresponding eigenvalues would be those of the symmetric TE modes, as indicated. Let us assume, however, that the expressions in the figure give the y component of the magnetic field in each region, so that we will seek the characteristics of the symmetric TM modes, which we have not considered in detail before. We use Maxwell's equations to find the corresponding electric field components and then use the boundary conditions to generate the following characteristic equation:

$$\frac{u \tan u}{\varepsilon_1} = \frac{q}{\varepsilon_2} \left\{ \frac{1 + \dfrac{(u'/\varepsilon_3) + (iq/\varepsilon_2)}{(u'/\varepsilon_3) - (iq/\varepsilon_2)} \exp\left(-\dfrac{4qW'}{W}\right)}{1 - \dfrac{(u'/\varepsilon_3) + (iq/\varepsilon_2)}{(u'/\varepsilon_3) - (iq/\varepsilon_2)} \exp\left(-\dfrac{4qW'}{W}\right)} \right\}. \tag{3.131}$$

We now assume that $\exp(-4qW'/W) \ll 1$ and write Eq. (3.131) as

$$\frac{u \tan u}{\varepsilon_1} - \frac{q}{\varepsilon_2} \approx \frac{2q}{\varepsilon_2} \exp\left[2i\left(\tan^{-1}\frac{\varepsilon_3 q}{\varepsilon_2 u'}\right) - \frac{4qW'}{W}\right]. \tag{3.132}$$

As in Section C, we can expand both sides of this equation about a root (u_0, q_0, h_0) of the left-hand side; this root characterizes the corresponding mode of the lossless three-medium structure that obtains when the FTR layers are infinitely thick. Because $[4(u')^2/W^2] + h^2 = k_3^2$ the corresponding value of u' is $u_0' = (k_3 W \cos\theta_3)/2$, where θ_3 is a real angle found from Snell's law, given that $u_0 = (k_1 W \cos\theta_1)/2$. Because of the factor $\exp(-4qW'/W)$ on the right-hand side of Eq. (3.132), its first-order approximation is obtained simply by replacing q and u' by q_0 and u_0'. The left-hand side, to first order, is

$$(\varepsilon_1\varepsilon_2{}^2 q_0 u_0)^{-1}(\varepsilon_1\varepsilon_2 q_0{}^2 + \varepsilon_2{}^2 q_0 u_0{}^2 + \varepsilon_1{}^2 q_0{}^3 + \varepsilon_1\varepsilon_2 u_0{}^2)\,\Delta u, \qquad (3.133)$$

where we have used the fact that $u_0(\tan u_0)/\varepsilon_1 = q_0/\varepsilon_2$.

Equating the first-order approximations of the two sides of Eq. (3.132), we obtain an expression for $\Delta u = u - u_0$, where u is a complex eigenvalue of the five-medium problem and u_0 is the corresponding real eigenvalue of the three-medium problem. We then use the relation $k_1{}^2 = h^2 + (4u^2/W^2)$ to obtain

$$\begin{aligned}\operatorname{Im} h = \operatorname{Im} \Delta h &\approx -\frac{4u_0 \operatorname{Im} \Delta u}{h_0 W^2} = -\frac{2 \operatorname{Im} \Delta u}{W \tan\theta_1}\\ &= -\frac{(8\varepsilon_1\varepsilon_2{}^2\varepsilon_3 u_0 q_0{}^3 u_0'/W\tan\theta_1)\exp(-4q_0W'/W)}{[\varepsilon_1\varepsilon_2 R^2 + q_0(\varepsilon_2{}^2u_0{}^2 + \varepsilon_1{}^2 q_0{}^2)][\varepsilon_2{}^2(u_0')^2 + \varepsilon_3{}^2 q_0{}^2]}.\end{aligned} \qquad (3.134)$$

The geometrically calculated loss, on the other hand, has the form

$$R^N = e^{N\ln R} = e^{N\ln(1-T)} \approx e^{-NT} = e^{-(T/2W_e\tan\theta_1)z}. \qquad (3.135)$$

From the appendix to this chapter, we find that

$$T \approx \frac{u_0'\varepsilon_1}{u_0\varepsilon_3}\,\frac{16\varepsilon_2{}^2u_0{}^2\varepsilon_3{}^2q_0{}^2\exp(-4q_0W'/W)}{(\varepsilon_2{}^2u_0{}^2 + \varepsilon_1{}^2q_0{}^2)[\varepsilon_2{}^2(u_0')^2 + \varepsilon_3{}^2q_0{}^2]}. \qquad (3.136)$$

If Eq. (3.136) is substituted into Eq. (3.135), it is found that the geometric loss rate, i.e., $T/(2W_e\tan\theta_1)$, with W_e calculated from Eq. (3.130), is precisely that given by the wave theory, Eq. (3.134). If, on the other hand, we had used this comparison of ray and wave theories to determine W_e, we would accordingly have developed a new technique for finding the Goos–Haenchen shift.

The problem of the coupling of two dissimilar thin films coated with dissimilar films of lower index of refraction is sketched in Fig. 3-19. The geometric solution to this problem proceeds as indicated by the equations in the legends to that figure, i.e., by finding the eigenstates of the system, characterized by the eigenvalues γ. The analysis parallels that carried out in Section C and thus need not be developed here. The results of the analysis show that the procedure described at the beginning of this section for specifying the shift of a ray transmitted across an FTR layer leads to predictions of the coupling coefficients that agree with first order solutions to Maxwell's equations. Accordingly, we find that a geometric theory of wave propagation in thin films predicts a new result, i.e., the lateral shift of light beams on transmission through an FTR layer. It is remarkable that this shift does not depend on the thickness of the FTR layer, in the weak coupling approximation.

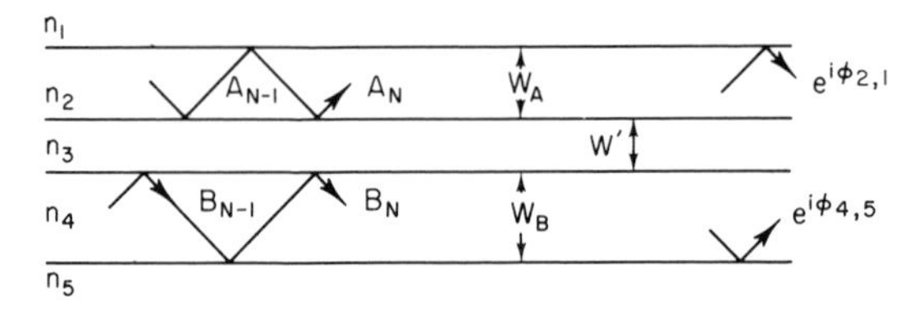

Fig. 3-19. The general analysis of a five-medium directional coupler:

$$A_N = R^{1/2}e^{i\phi_R+i\psi_A}A_{N-1} + T^{1/2}e^{i\phi_T+i\psi_B}B_{N-1} = \gamma A_{N-1},$$
$$B_N = T^{1/2}e^{i\phi_T+i\psi_A}A_{N-1} + R^{1/2}e^{i\phi_R+i\psi_B}B_{N-1} = \gamma B_{N-1},$$
$$\psi_A = \phi_{2,1} + 2k_0n_2(\sec\theta_2)W_A,$$
$$\psi_B = \phi_{4,5} + 2k_0n_4(\sec\theta_4)W_B.$$

Before closing this chapter, we point out some of the limitations of the geometric theory. Like the classical theory of the parallel plate interferometer, it neglects the effects of diffraction. At each reflection, an incident beam (ray) is divided into a reflected and transmitted beam (ray). It is assumed that these beams maintain the shape of the incident beam over the coupling region and are thus geometric images of it in the surfaces of the film. For beams with slowly varying amplitudes, films with thickness of the order of a few wavelengths, and interaction regions of the order of the beam width, a beam transmitted across an FTR layer will maintain its shape quite well as it is multiply reflected down the guide. This can be shown by unfolding the path of such a beam in the film and testing to see if its propagation distance L satisfies the usual condition, $L \ll d^2/\lambda$, where d is the half-width of the incident beam. Obviously, diffraction effects do occur at the edges of the beam, but these are of little importance in FTR couplers. In rigorous theories of the Goos–Haenchen shift, they give rise to lateral waves (*17*, *22*) that describe trailing illumination observed beyond the edge of the reflected beam.

When the weak coupling approximation is not satisfied, a wave solution in terms of a single leaky wave is not sufficient to describe the fields, as shown in Section B. We have not yet attempted a detailed comparison of wave and ray theories in this case. It seems likely that a modified geometric theory might yet be appropriate. To test it, we would need to compare it with a rigorous solution, such as that found by Midwinter (*8*). The implications of such a comparison on studies of the Goos–Haenchen shift might be interesting.

APPENDIX

In order to evaluate the expression giving the loss rates of leaky modes or the coupling strengths between thin-film guides found in this chapter, we need expressions for the transmission and reflection coefficients of an FTR layer. We thus consider the reflectivity and transmissivity of a film of thickness s and refractive index n_2 bounded by semiinfinite media of indices n_1 and n_3. A plane wave of the form

$$E_y \quad (\text{or } H_y) = e^{i\omega t-(iux/a)-ihz}$$

is incident from the medium of index n_1 at an angle $\theta_1 > \theta_C = \sin^{-1}(n_2/n_1)$ on the lower face of the film, located at $x = 0$. We define

$$u = k_0 a n_1 \cos\theta_1; \qquad q = k_0 a(n_1{}^2 \sin^2\theta_1 - n_2{}^2)^{1/2};$$
$$u' = k_0 a n_3 \cos\theta_3; \qquad h = k_0 n_1 \sin\theta_1 = k_0 n_3 \sin\theta_3.$$

The scaling parameter a is of no consequence here but appears throughout the analysis of the book as the half-width of the waveguides. The evanescent field in the film has the form

$$E_y \quad (\text{or } H_y) = e^{i\omega t-ihz}(Ae^{-qs/a} + Be^{qs/a}).$$

The transmitted field above the film has the form

$$E_y \quad (\text{or } H_y) = t(\exp i\phi_t)\exp\{i\omega t - [iu'(x-s)/a] - ihz\}.$$

The reflected field is given by

$$E_y \quad (\text{or } H_y) = r(\exp i\phi_r)\exp[i\omega t + (iux/a) - ihz].$$

The reflection and transmission coefficients, $re^{i\phi_r}$ and $te^{i\phi_t}$, are defined in terms of E_y for TE waves and H_y for TM waves. The reflectivity R and transmissivity T, on the other hand, are defined in terms of energy flow, i.e., the Poynting vector times equivalent areas in each medium. The permeabilities of all media are assumed equal to μ_0, the permeability of free space. The indices of refraction of all media are assumed to have real values.

Accordingly,

$$t = (2\cos\phi_{12}\sin\phi_{23})/[\sinh^2(qs/a) + \sin^2(\phi_{12}+\phi_{23})]^{1/2}, \tag{3A.1}$$

$$\phi_T = \phi_t = \tan^{-1}[\coth(qs/a)\tan(\phi_{12}+\phi_{23})] - (\pi/2), \tag{3A.2}$$

$$r = \left[\frac{\sinh^2(qs/a) + \sin^2(\phi_{12}-\phi_{23})}{\sinh^2(qs/a) + \sin^2(\phi_{12}+\phi_{23})}\right]^{1/2}, \tag{3A.3}$$

$$\begin{aligned}\phi_R = \phi_r &= \tan^{-1}[\coth(qs/a)\tan(\phi_{12}+\phi_{23})] \\ &\quad + \tan^{-1}[\coth(qs/a)\tan(\phi_{12}-\phi_{23})], \end{aligned}\tag{3A.4}$$

$$R = r^2.$$

For TM waves, the energy transmittance T and the angles ϕ_{12} and ϕ_{23} are given by

$$T = (n_1^2u'/n_3^2u)t^2; \qquad \phi_{12} = \tan^{-1}(n_1^2q/n_2^2u); \qquad \phi_{23} = \tan^{-1}(n_3^2q/n_2^2u'). \tag{3A.5}$$

For TE modes, ϕ_{12} and ϕ_{23} and T are again given by Eq. (3A.5), but with n_1, n_2, and n_3 set equal to unity.

When $s \to \infty$, we have the case of total reflection at the lower face of the infinitely thick film. Equations (3A.3) and (3A.4) then reduce to

$$r = r' = 1; \qquad \phi_r = \phi_{r'} = 2\phi_{12}. \tag{3A.6}$$

REFERENCES

1. P. Leurgans and A. F. Turner, *J. Opt. Soc. Amer.* **37**, 983 (1947).
2. L. V. Iogansen, *Sov. Phys.–Tech. Phys.* **7**, 295 (1962); **8**, 985 (1964); **11**, 1529 (1967),
3. J. H. Harris and R. Shubert, *Conf. Abstracts, Int. Sc. Radio Union, Spring Meeting. Washington, D.C.* p. 71 (April 1969).
4. P. K. Tien, R. Ulrich, and R. J. Martin, *Appl. Phys. Lett.* **14**, 291 (1969).
5. J. H. Harris, R. Shubert, and J. N. Polky, *J. Opt. Soc. Amer.* **60**, 1007 (1970).
6. P. K. Tien and R. Ulrich, *J. Opt. Soc. Amer.* **60**, 1325 (1970).
7. R. Ulrich, *J. Opt. Soc. Amer.* **60**, 1337 (1970).
8. J. E. Midwinter, *IEEE J. Quant. Electron.* **QE-6**, 583 (1970).
9. A. Sommerfeld, "Optics," p. 46ff. Academic Press, New York, 1954.
10. N. S. Kapany and J. J. Burke, *J. Opt. Soc. Amer.* **51**, 1067 (1961).
11. M. Born and E. Wolf, "Principles of Optics," p. 65. Pergamon Press, New York, 1959.
12. F. Goos and H. Haenchen, *Ann. Phys.* **(5) 43**, 383 (1943); **(6) 1**, 333 (1947); **(6) 5**, 251 (1949).
13. L. Bergstein and C. Shulman, *Appl. Opt.* **5**, 9 (1966).

14. M. Kline and I. W. Kay, "Electromagnetic Theory and Geometrical Optics." Wiley (Interscience), New York, 1965.
15. P. M. Morse and H. Feshbach, "Methods of Theoretical Physics," pp. 928–929. McGraw-Hill, New York, 1953.
16. C. C. Jackson, "Field and Wave Electrodynamics," Chapter 9. McGraw-Hill, New York, 1965.
17. H. K. V. Lotsch, *Optik* **32**, 116, 189 (1970); **32**, 299, 553 (1971), *J. Opt. Soc. Amer.* **58**, 551 (1968).
18. C. Von Fragstein, *Ann. Phys.* **(6) 4**, 271 (1949).
19. K. Artmann, *Ann. Phys.* **(6) 2**, 87 (1948).
20. H. Wolter, *Z. Naturforsch.* **5a**, 143 (1950).
21. H. Maecker, *Ann. Phys.* **(6) 10**, 115 (1952).
22. B. R. Horowitz and T. Tamir, *J. Opt. Soc. Amer.* **61**, 586 (1971).

CHAPTER 4

Guided Waves along Circular Waveguides

In this chapter, we will study waves in and around a dielectric-rod guide, i.e., a circular dielectric cylinder with refractive index n_1 embedded in an infinite medium of index $n_2 < n_1$. This is an idealization of a fiber optics waveguide at optical frequencies or a dielectric antenna at microwave frequencies. We will treat, in detail, only the *true, bound, surface wave* modes. These are the cylindrical analogs of the lossless surface wave modes of the slab waveguide discussed in Chapter 2. In those cases where losses are important or interesting, we will characterize them by approximate expressions, using the exact analysis of the slab waveguide to establish their suitability. As we shall indicate at various stages of the development, the dielectric rod also supports leaky wave modes whose manifestations parallel those of the slab waveguide discussed in detail in Chapter 3. Although we will not treat such modes in detail in this chapter, an interesting special case is extensively studied in Appendix B on hollow dielectric waveguides by T. Sawatari.

In our study of the dielectric slab guide, we found that the bound modes could be viewed as a superposition of two uniform plane waves within the slab, each incident or reflected at a particular discrete angle with the normal to the guiding interfaces. Outside the slab, the modes were characterized by evanescent waves associated with total internal reflection. The elementary modes were plane polarized with their electric vectors either in the plane of incidence of the uniform plane waves (TM modes) or perpendicular to it (TE modes). In the dielectric rod, the bound modes can again be described by a superposition of uniform plane waves in the interior of the rod with evanescent waves outside. In contrast to the slab case, however, it is no longer apparent just which superpositions will provide the appropriate continuity in the tangential components of the electric and magnetic fields at the interfaces. It can be anticipated that the polarization properties of the modes will be complicated. Therefore, we will be well advised to solve the electromagnetic boundary value problem directly, and then analyze our solutions to see precisely how they do or do not fit familiar geometric and physical optics models of wave propagation. We shall be particularly interested in relating our mathematical solutions to such optical concepts as meridional and skew rays, polarization and intensity distributions. Because ray concepts are rarely encountered in the microwave literature, we will begin our study by deriving, in a conventional way, expressions for the mode fields in hollow metallic guides. These expressions will then be analyzed from an optical point of view in such a way as to preface the study of waves on fiber optics waveguides. We shall then take a quite unconventional approach to the derivation of the fields for the dielectric rod, one in which the polarization of the mode plays a major role from the outset. This derivation proves an interesting alternative to those found in most text books on electromagnetic theory.

A. WAVES IN HOLLOW, PERFECTLY CONDUCTING PIPES

1. Derivation of Expressions for the Field Vectors

Consider the waveguide formed by a circular dielectric rod of radius a and refractive index n_1, coated with a perfectly conducting film. We seek those elementary solutions of the homogeneous wave equation whose electric and magnetic fields are of the form

$$\left.\begin{matrix}\mathbf{E}(x, y, z, t)\\ \mathbf{H}(x, y, z, t)\end{matrix}\right\} = \left\{\begin{matrix}\mathbf{F}(x, y, h, \omega)\\ \mathbf{G}(x, y, h, \omega)\end{matrix}\right\} \times e^{i\omega t - ihz}. \tag{4.1}$$

We further require that the tangential components of E must vanish at the perfectly reflecting film. The polar coordinates $x = \varrho \cos \phi$ and $y = \varrho \sin \phi$ are thus well suited to the analysis. The boundary conditions accordingly become

$$E_z(\varrho = a) = E_\phi(\varrho = a) = 0. \tag{4.2}$$

Our problem is to find the vector eigenfunctions $\mathbf{F}$ and $\mathbf{G}$ and the corresponding eigenvalues h. Substituting Eq. (4.1) into the wave equation

$$\nabla^2 - \varepsilon\mu(\partial^2/\partial t^2) \left\{\begin{matrix}\mathbf{E}\\ \mathbf{H}\end{matrix}\right\} = 0 \tag{4.3}$$

we find that the eigenvectors **F** and **G** satisfy the equation

$$(\nabla_t^2 + \beta_1^2) \begin{Bmatrix} \mathbf{F}(x, y, h, \omega) \\ \mathbf{G}(x, y, h, \omega) \end{Bmatrix} = 0, \tag{4.4}$$

where

$$\beta_1^2 = k_1^2 - h^2 = \omega^2 \varepsilon_1 \mu - h^2, \tag{4.5}$$

and

$$\nabla_t^2 = \frac{\partial^2}{\partial_x^2} + \frac{\partial^2}{\partial_y^2} = \frac{\partial^2}{\partial \varrho^2} + \frac{1}{\varrho}\frac{\partial}{\partial \varrho} + \frac{1}{\varrho^2}\frac{\partial^2}{\partial \phi^2} \tag{4.6}$$

is the transverse Laplacian operator. Now, because **E** and **H** must also satisfy the first-order Maxwell equations,

$$\nabla \times \mathbf{E} = -\mu\, \partial \mathbf{H}/\partial t; \qquad \nabla \times \mathbf{H} = \varepsilon_1\, \partial \mathbf{E}/\partial t,$$

we have the additional relations between the components of F and G,

$$\begin{aligned} \nabla_t \times \mathbf{F} - ih\hat{z} \times \mathbf{F} &= -\, i\omega\mu\mathbf{G} \\ \nabla_t \times \mathbf{G} - ih\hat{z} \times \mathbf{G} &= i\omega\varepsilon_1\mathbf{F}. \end{aligned} \tag{4.7}$$

We can solve Eqs. (4.7) for the transverse components $\mathbf{G}_t$ and $\mathbf{F}_t$ in terms of the longitudinal components $\mathbf{G}_z$ and $\mathbf{F}_z$. Thus,

$$\begin{aligned} \nabla_t \times \mathbf{F}_z - ih\hat{z} \times \mathbf{F}_t &= -\, i\omega\mu\mathbf{G}_t \\ \nabla_t \times \mathbf{G}_z - ih\hat{z} \times \mathbf{G}_t &= i\omega\varepsilon_1\mathbf{F}_t. \end{aligned}$$

Multiplying the second of these equations by $-i\omega\mu$ and substituting **G** from the first, we obtain

$$\beta_1^2\mathbf{F}_t = -i\omega\mu\nabla_t \times \mathbf{G}_z - ih\, \nabla_t F_z. \tag{4.8a}$$

In a similar way, we find

$$\beta_1^2\mathbf{G}_t = i\omega\varepsilon_1\nabla_t \times \mathbf{F}_z - ih\, \nabla_t G_z. \tag{4.8b}$$

It is now apparent that if we solve the Helmholtz equation (4.5) for F_z and G_z, we can obtain the transverse components of **F** and **G** from Eqs. (4.8a) and (4.8b). (Because we have no *a priori* knowledge that such solutions will satisfy the boundary conditions, we can only regard them, at this time, as trial solutions of the boundary value problem.) Let us, therefore, solve Eq. (4.5) for F_z. Because the fields must be periodic in ϕ, we take

$F_z = f(\varrho)e^{in\phi}$ as a trial solution, where n can be any positive or negative integer or zero. We then find that the $f(\varrho)$ satisfies Bessel's differential equation

$$\frac{d^2f}{d\varrho^2} + \frac{1}{\varrho}\frac{df}{d\varrho} + \left(\beta^2 - \frac{n^2}{\varrho^2}\right)f = 0, \tag{4.9}$$

where we have dropped the subscript from β_1 for convenience. Because the field must be finite on the axis of the guide, the appropriate solution of Eq. (4.9) is the Bessel function of the first kind of order n, i.e., $J_n(\beta\varrho)$. Therefore,

$$F_z(\varrho, \phi, h, \omega) = AJ_n(\beta\varrho)e^{in\phi}, \tag{4.10}$$

where A is an arbitrary constant. From Eq. (4.8a) (with $G_z = 0$), we then find that

$$F_\phi = -(ih/\beta^2\varrho)\,\partial F_z/\partial\phi = (nh/\beta^2\varrho)AJ_n(\beta\varrho)e^{in\phi}. \tag{4.11}$$

From Eqs. (4.10) and (4.11), we see that this solution can be made to satisfy the boundary conditions, if we require that $\beta = \beta_{n,m} = u_{n,m}/a$, where $u_{n,m}$ is any one of the denumerably infinite set of roots of the equation

$$J_n(u_{n,m}) = 0; \qquad m = 1, 2, 3, \ldots . \tag{4.12}$$

This is the characteristic equation for TM modes in a hollow metallic waveguide. The transverse wave numbers $\beta_{n,m}$ defined by Eq. (4.12) determine the corresponding propagation constants $h_{n,m}$ of these modes through Eq. (4.5). The remaining components of **F** and **G** for these modes are given by Eqs. (4.8a) and (4.8b). Summarizing our results for the $\mathrm{TM}_{n,m}$ modes, we thus have

$$\left.\begin{aligned} E_z &= AJ_n(\beta\varrho) \\ E_\phi &= (nh/\beta^2\varrho)AJ_n(\beta\varrho) \\ E_\varrho &= (-ih/\beta)AJ_n{}'(\beta\varrho) \\ H_z &= 0 \\ H_\phi &= (-i\omega\varepsilon/\beta)AJ_n{}'(\beta\varrho) \\ H_\varrho &= (-\omega\varepsilon n/\beta^2\varrho)AJ_n(\beta\varrho) \end{aligned}\right\} \times e^{in\phi+i\omega t-ihz} \tag{4.13}$$

and

$$J_n(\beta a) = 0; \qquad h^2 = \omega^2\varepsilon\mu - \beta^2.$$

2. The Intensity Distribution

We now ask what these waves would look like if we were to view them with the aid of a microscope focused on the exit end of the guide. In the

Huygens–Kirchhoff approximation (the field in the aperture equals the incident field), we would expect to see an intensity distribution proportional to the time-average of the axial component of the Poynting vector. Accordingly, we calculate that

$$I(\varrho, \phi) = \tfrac{1}{2}\,\mathrm{Re}(\mathbf{E} \times \mathbf{H}^*)_z = \tfrac{1}{2}(E_\varrho H_\phi{}^* - E_\phi H_\varrho{}^*)$$
$$= (h\omega\varepsilon/2\beta^2)A^2\{[J_n{}'(\beta\varrho)]^2 + [n^2 J_n{}^2(\beta\varrho)/\beta^2\varrho^2]\}. \tag{4.14}$$

If we now use the Bessel function relations

$$J_n{}'(u) = \pm n J_n(u)/u \mp J_{n\pm 1}(u), \tag{4.15}$$

we can write Eq. (4.14) in the form

$$I(\varrho) = (h\omega\varepsilon/4\beta^2)A^2[J^2_{n+1}(\beta\varrho) + J^2_{n-1}(\beta\varrho)]. \tag{4.16}$$

From Eqs. (4.14) or (4.16), we see that the flux distribution is circularly symmetric. Because β is the mth root of $J_n(\beta_{n,m}a) = 0$, and because the functions J_{n+1} and J_{n-1} are very near their extrema when $J_n = 0$ (Fig. 4-1), we see that there are m radial maxima in the intensity distribution of the

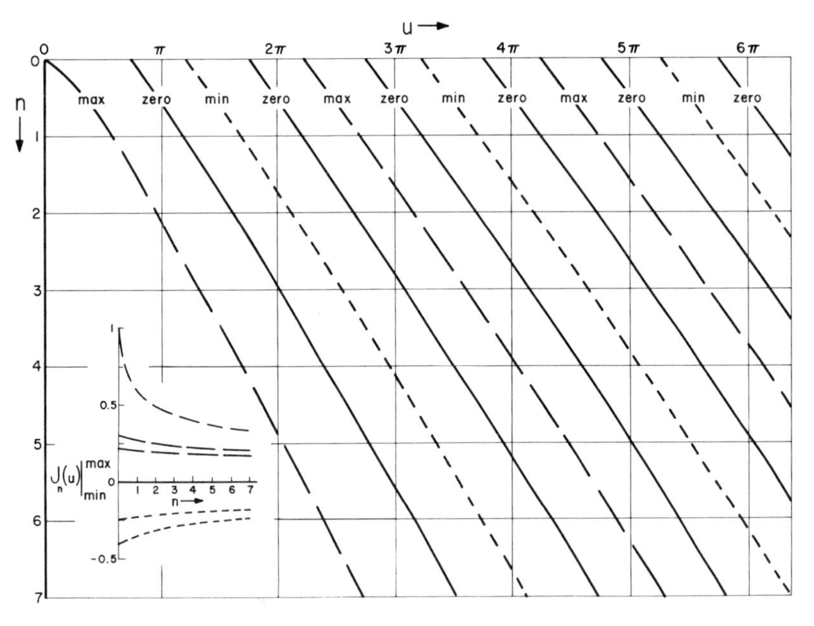

Fig. 4.1. Plots locating the maxima (— —), zeros (————) and minima (- - - - -) of the first seven Bessel functions $J_n(u)$ on the n, u plane. Inset indicates the value of $J_n(u)$ at the extrema as a function of n.

$TM_{n,m}$ mode. When $n = 0$, for example, the flux distribution is proportional to $J_1{}^2(\beta_{0,m}\varrho)$. If $m = 3$, there are three rings. The third reaches is maximum at $\varrho/a = 8.536/8.654$. The first peaks at $\varrho/a = 1.841/8.654$, and the second at $\varrho/a = 5.311/8.654$. When $n = \pm 1$, the intensity is maximum on the axis of the guide and for the $TM_{1,m}$ mode, there are m rings of lesser brightness around the central spot. For $n \neq \pm 1$, the intensity is zero on axis. For the general $TM_{n,m}$ mode, the jth radial maximum of the intensity distribution ($j \leq m$) is shifted slightly inside the jth zero of E_z and E_ϕ. More specifically, if $J_n(u_{n,m}\varrho/a) = J_n(u_{n,j}) = 0$, then the jth maximum is located at

$$\varrho/a = \beta\varrho/u_{n,m} = (u_{n,j}/u_{n,m})[1 - 1/(u_{n,j}^2 - 2n^2 - 3)]. \tag{4.17}$$

The meaning of these statements is best appreciated with the aid of Fig. 4-1. Suppose, for example, that we are studying the $TM_{2,4}$ mode. Then, $u_{n,m} = u_{2,4}$ is located at the fourth nonzero root of $J_2(u) = 0$, i.e., at $u_{2,4} \approx 5\pi - \frac{1}{4}\pi$. The longitudinal and azimuthal components of the field also vanish at three interior radii, located approximately by $u_{2,3} \approx 4\pi - \frac{1}{4}\pi$, $u_{2,2} \approx 3\pi - \frac{1}{4}\pi$, and $u_{2,1} \approx 2\pi - \frac{1}{4}\pi$. The intensity distribution reaches its successive maxima just inside each of these zeros of E_z and E_ϕ. The expression given here, which locates these maxima with respect to the zeros of E_z and E_ϕ, is accurate to about 2% for all but the first maximum, for which the approximation is an order of magnitude more coarse. The inset in Fig. 4-1 gives the values of $J_n(u)$ at its extrema as a function of n. These provide a means for obtaining estimates of the relative intensity of the m rings in the $TM_{n,m}$ mode pattern.

3. Polarization of TM Modes

We must now consider the polarization of these modes. Equation (4.16) for the intensity distribution, as well as the differential equation (4.5), suggest that we can write the transverse parts of the electric and magnetic fields in terms of orthogonal vector components, each of which is proportional to only one of the functions J_{n+1} or J_{n-1}. We note that the radial and azimuthal components of $\mathbf{E}$ are $\pi/2$ out-of-phase, a fact which suggests that circularly polarized components may be appropriate. The Bessel relations [Eq. (4.15)] indicate how this might be done. Accordingly, we form the circular components of the transverse fields and obtain

$$\begin{aligned} E_\varrho \pm iE_\phi &= \pm (ih/\beta)AJ_{n\pm1}(\beta\varrho)e^{in\phi+i\omega t-ihz} \\ H_\varrho \pm iH_\phi &= -(\omega\varepsilon/\beta)AJ_{n\pm1}(\beta\varrho)e^{in\phi+i\omega t-ihz}. \end{aligned} \tag{4.18}$$

(A detailed discussion of the meaning of these circular components of the transverse fields is given in Section B on this chapter.) It is now apparent that the polarization, like the intensity distribution, is circularly symmetric. Because the left $(E_\varrho + iE_\phi)$ and right $(E_\varrho - iE_\phi)$ circularly polarized components of the transverse fields are in-phase or out-of-phase by π, the major and minor axes of the polarization ellipse are either perpendicular or parallel to the circles of uniform intensity. The major and minor axes have lengths

$$| E_\varrho + iE_\phi | \pm | E_\varrho - iE_\phi | = (hA/\beta)(| J_{n+1} | \pm | J_{n-1} |),$$

respectively. The polarization is radial, and thus linear at $\varrho = a$ and at all interior zeros of E_z and E_ϕ. It is circular at those radii where either J_{n+1} or J_{n-1} vanish; the intensity is weak, however, on these circular zones. Near the maxima of the intensity, on the other hand, the polarization is almost linear, i.e., the length of the minor axis of the polarization ellipse, which is oriented tangent to the bright rings, is small with respect to that of the radial, major axis. If we substitute $n = 0$ in Eq. (4.18), we see that the $TM_{0,m}$ modes are linearly polarized in the radial direction everywhere inside the guide. All other $TM_{n,m}$ modes are elliptically polarized; both the eccentricity of the ellipse and its sense of rotation are functions of radius.

4. Mode Degeneracy—Circulating Waves, Standing Waves, and Angular Momentum

It is instructive, for purposes of optical interpretation, to determine the wavefronts of the waves described by the mode fields [Eqs. (4.13)]. Note that all components have the common phase factor $e^{in\phi+i\omega t-ihz}$. Evidently, the surfaces of constant phase satisfy the equation $z = (n\phi/h) + \text{const}$. Such a surface is generated by a semiinfinite line in space perpendicular to and terminating on the guide axis; the line rotates toward positive ϕ as z increases for $n > 0$ and toward negative ϕ for $n < 0$. The locus of its points of intersection with the wall of the guide is a helix. The normal to the surface $W = \omega t = hz - n\phi$ is given by $\nabla W = h\hat{\mathbf{z}} - (n/\varrho)\hat{\boldsymbol{\phi}}$. As t increases, a given surface of constant phase rotates in the negative $\hat{\phi}$ direction about the axis of the guide while advancing in the positive z direction. We may accordingly form a mental picture that suggests that the wave energy circulates about the axis of the guide as it advances along the axis. This is a characteristic of circulating waves. As we shall show later, we may accordingly associate an angular momentum with the TM modes described by Eqs. (4.13). (While this circulating property is suggestive of skew rays, we cannot, of course, localize the wave energy in such a way as to prescribe the equivalent of a geometric optics ray.)

In the engineering literature, the azimuth dependence of the fields of the TM modes is usually described by sinusoids [$\cos(n\phi)$ or $\sin(n\phi)$] rather than by exponential functions ($e^{\pm in\phi}$). Note that the $TM_{n,m}$ and $\mathrm{TM}_{-n,m}$ modes, as defined here by Eqs. (4.13) and (4.18), are degenerate, i.e., both of these vector fields have the same propagation constant, because $J_{-n}(u) = (-1)^n J_n(u)$. Hence, every different linear combination of these two modes yields a different vector field that propagates down the guide with the same phase velocity in the axial direction. In particular, the combinations $\mathrm{TM}_{n,m} \pm \mathrm{TM}_{-n,m}$ give fields proportional to $\cos(n\phi)$ and $\sin(n\phi)$; these appear to be preferred by microwave investigators. They are always linearly polarized, though never plane polarized—the direction of oscillation of the electric vector changes continuously with both ϕ and ϱ. Both of these modes are labeled $\mathrm{TM}_{n,m}$ in most of the waveguide literature, a sometimes regretable practice because it may leave the reader with the idea that the $\mathrm{TM}_{n,m}$ modes of the metallic pipe waveguide must necessarily be linearly polarized, with $2n$ azimuthal maxima and minima in their flux distributions. [For the mode combinations $\mathrm{TM}_{n,m} \pm \mathrm{TM}_{-n,m}$, for example, the axial component of the Poynting vector is proportional to $J_{n+1}^2 + J_{n-1}^2 - [2J_{n+1}J_{n-1} \cos 2n\phi]$. As we have seen, this concept is erroneous. The most elementary waves capable of satisfying the boundary conditions show no manifestations of interference in the azimuth coordinate. This is to be expected physically, because there is nothing in the waveguide to induce reflections for waves that are circulating about its axis in either direction. The fact that most fields studied at microwave frequencies exhibit azimuthal standing waves, manifested by $2n$ maxima and minima, should thus not be interpreted as a demonstration that these are the normal modes of the structure. On the contrary, it seems more appropriate to associate such fields with the bidirectional (in azimuth) character of microwave sources, which indiscriminately launch the right- and left-handed circulating waves with equal efficiency.

We can further elucidate the nature of the circulating TM waves by considering their angular momentum. In classical electromagnetic theory (*1–3*), the linear momentum density of a field is given by

$$\mathbf{P} = (\varepsilon\mu/2)\,\mathrm{Re}(\mathbf{E} \times \mathbf{H}^*). \tag{4.19}$$

The angular momentum density is

$$\mathbf{L} = (\varepsilon\mu/2)\,\mathrm{Re}(\mathbf{r} \times \mathbf{E} \times \mathbf{H}^*). \tag{4.20}$$

The energy density is

$$U = \tfrac{1}{4}\,\mathrm{Re}(\varepsilon \mathbf{E} \cdot \mathbf{E}^* + \mu \mathbf{H} \cdot \mathbf{H}^*). \tag{4.21}$$

We are interested in the z component of the angular momentum. From Eqs. (4.20) and (4.13), we find that the density of this component is given by

$$L_z = \varepsilon\mu\varrho E_z H_\varrho^*/2. \tag{4.22}$$

To obtain the total energy and z component of angular momentum in the field, we must integrate these densities over a volume that contains the entire field. The surface $\varrho = a$ completely encloses the field and thus forms one boundary of an appropriate volume. The fields extend to infinity, however, in both directions along the axis. We thus determine only the total energy per unit length and z component of angular momentum per unit length by integrating over any cylindrical volume of radius a and unit length. If we then form the ratio of these two integrals (1, p. 575), we obtain the interesting result

$$\iiint L_z\varrho\, d\varrho\, d\phi\, dz \Big/ \iiint U\varrho\, d\varrho\, d\phi\, dz = -n/\omega. \tag{4.23}$$

This purely classical calculation can be made to take on the embellishments of quantum optics simply by multiplying the numerator and denominator by $\hbar$, which is Planck's constant (6.625×10^{-34} J sec) divided by 2π. The result indicates that we may associate with each photon of energy $\hbar\omega$ precisely $-n\hbar$ units of angular momentum about the axis of the guide. The fundamental physical content of the circulating wave representation of the modes is thus apparent. The fields given by Eqs. (4.13) may accordingly be interpreted as the classical limit of a single-mode, quantized, electromagnetic field. Each mode is a member of a basis set in which the photons have a well-defined axial component of both the linear momentum and the angular momentum. (For a discussion of quantized fields, the reader is referred to the many excellent texts on quantum mechanics and electrodynamics.)

Another interesting attribute of the $\mathrm{TM}_{n,m}$ circulating modes can be noted by writing the circularly polarized transverse components of the fields [Eq. (4.18)] in terms of spatially fixed reference axes. Thus,

$$E_\pm \equiv E_x \pm iE_y = (E_\varrho \pm iE_\phi)e^{\pm i\phi} = \pm(ih/\beta)AJ_{n\pm1}(\beta\varrho)e^{i(n\pm1)\phi + i\omega t - ihz}. \tag{4.24}$$

The boundary conditions require equal contributions from each of these circular field components at $\varrho = a$. Note, however, that each of them, taken alone, still suggests a net z component of angular momentum proportional to $-n$. The E_+ component, for example, manifests through its phase factor an orbital momentum proportional to $-(n + 1)$. It also has a spin angular momentum, however, proportional to $+1$, because of its left circular polarization (*1*, p. 201). Thus, the net angular momentum in the axial direction remains proportional to $-n$. Similarly, the expressions for each of the field components, in each of the three transverse field representations we have seen [Eqs. (4.13), (4.18), and (4.24)], suggest the correct angular momentum of the field.

Note that the $\cos(n\phi)$ and $\sin(n\phi)$ representations used in engineering texts describe fields with a net axial angular momentum equal to zero. Such fields are those most commonly excited by conventional microwave sources that launch the positive and negative circulating waves with equal amplitudes. At optical frequencies, however, we can use spatial filtering techniques in concert with quarter-wave plates and polarizers to launch the more elementary, elliptically polarized, circulating waves. It appears, in fact, to be much easier to launch pure modes of this type than it is to launch the linear combinations of them that are preferred by microwave engineers. This will be discussed in Chapter 5 and in Appendix B.

5. TE Modes in Hollow Metallic Pipes

Returning to Eqs. (4.5), (4.8), and (4.9), we will now take, as trial solutions to this boundary value problem, fields for which $E_z = 0$. Solving Eq. (4.5) for G_z, we obtain

$$G_z = BJ_n(\beta\varrho)e^{in\phi}. \tag{4.25}$$

Substituting this solution into Eq. (4.8) yields

$$F_\phi = (i\omega\mu/\beta^2)\,\partial G_z/\partial\varrho = (i\omega\mu/\beta)BJ_n'(\beta\varrho)e^{in\phi}.$$

Because we have already assumed that $E_z = 0$, we see that we can satisfy the boundary conditions simply by requiring that βa be a root of

$$J_n'(\beta a) = 0. \tag{4.26}$$

This is the characteristic equation of the $\mathrm{TE}_{n,m}$ modes. From Eqs. (4.25), (4.8), and (4.9), we obtain the following expressions for the field components:

$$\left.\begin{aligned} H_z &= BJ_n(\beta\varrho) \\ H_\varrho &= (-ih/\beta)BJ_n{}'(\beta\varrho) \\ H_\phi &= (nh/\beta^2\varrho)BJ_n(\beta\varrho) \\ E_z &= 0 \\ E_\varrho &= (n\omega\mu/\beta^2\varrho)BJ_n(\beta\varrho) \\ E_\phi &= (i\omega\mu/\beta)BJ_n{}'(\beta\varrho) \end{aligned}\right\} \times e^{in\phi+i\omega t-ihz}. \tag{4.27}$$

The time average of the Poynting vector has two nonvanishing components,

$$\begin{aligned} S_z &= (h\omega\mu/2\beta^2)\mid B\mid^2\{[J_n{}'(\beta\varrho)]^2 + [n^2J_n{}^2(\beta\varrho)/\beta^2\varrho^2]\} \\ &= (h\omega\mu/4\beta^2)\mid B\mid^2[J_{n+1}^2(\beta\varrho) + J_{n-1}^2(\beta\varrho)] \\ S_\phi &= -n(\omega\mu/2\beta^2\varrho)\mid B\mid^2 J_n(\beta\varrho)J_n{}'(\beta\varrho). \end{aligned} \tag{4.28}$$

The intensity distribution (S_z) in any plane perpendicular to the axis is once more circularly symmetric. We can again refer to Fig. 4-1 to obtain the principal features of these $\mathrm{TE}_{n,m}$ mode patterns. The scale of the waveguide and the mode pattern with respect to the horizontal axis of the figure is now determined by the mth root, $u_{n,m}$, of $J_n{}'(u) = 0$. At this point on the figure, which is equivalent to $\varrho = a$ in the waveguide, E_ϕ vanishes. There are additionally $m - 1$ interior circles along which $E_\phi = 0$. Slightly outside these circles, the intensity distribution reaches its successive minima. The case $n = 0$ is an exception, because $J_0{}'(u) = -J_1(u)$, so that the minima are circles of vanishing power density at radii given by the zeros of $J_0{}'(\beta\varrho)$. More generally, the minima are located a distance

$$\Delta u_{n,j} = \frac{2/u_{n,j}}{1 + (2/n) - [(n^2 + 5)/u_{n,j}^2]}, \qquad j \leq m, \tag{4.29}$$

beyond the successive zeros of E_ϕ. The $\mathrm{TE}_{1,1}$ mode, which has its maximum power density on the axis and no subsidiary maxima or minima, is the lowest-order mode that can propagate on this waveguide. It is the last to be cut off as the diameter-to-wavelength ratio decreases.

To visualize the polarization of the TE modes, we again use circular transverse components

$$\begin{aligned} E_\varrho \pm iE_\phi &= (\omega\mu/\beta)BJ_{n\pm1}(\beta\varrho)e^{in\phi+i\omega t-ihz} \\ H_\varrho \pm iH_\phi &= \pm(ih/\beta)BJ_{n\pm1}(\beta\varrho)e^{in\phi+i\omega t-ihz}. \end{aligned} \tag{4.30}$$

Just as the TM modes, these modes have their polarization ellipses oriented with their axes radial and tangential. At $\varrho = a$, the polarization is linear (i.e., radial; $E_\phi = E_z = 0$), but the power density is at or near its minimum. Where the power density is near its maxima, on the other hand, i.e., near radii such that $J_n(\beta\varrho) = 0$, the right and left circular components are approximately π out-of-phase, and the polarization is consequently nearly linear, tangent to the circles of constant intensity.

We again have twofold degeneracy, because $J'_{-n}(u)$ vanishes for the same values of u as does $J_n'(u)$. The particular linear combinations of our circulating $TE_{n,m}$ modes given by $TE_{n,m} \pm TE_{-n,m}$ are those commonly encountered in the microwave literature. Many plots of the corresponding linearly polarized fields of low-order modes can be found in engineering texts. Our $TE_{n,m}$ modes again carry an angular momentum proportional to $-n$, as may be shown by a calculation similar to that described in connection with the TM modes. This is already suggested by the azimuthal component of the Poynting vector given in Eqs. (4.28).

6. Nonpropagating Modes

In the slab waveguide bounded by perfect conductors discussed in Chapter 2, we found that there were an infinite number of nonpropagating modes that, together with the propagating waves, formed a complete set of orthogonal vector eigenfunctions in terms of which a general field could be expressed. This is also true of the circular metallic guide. The roots of the characteristic equations [Eqs. (4.12) and (4.26)] that yield values of $\beta > \omega(\varepsilon\mu)^{1/2}$ correspond to waves that decay exponentially in the axial direction. Such modes are said to be below cutoff on the particular guide under consideration.

B. GUIDED WAVES ON CIRCULAR DIELECTRIC CYLINDERS

In analyzing the polarization of TM and TE modes in hollow, perfectly conducting waveguides, we found it advantageous to express the transverse fields in terms of circularly polarized components. A conventional analysis of the dielectric guide is also facilitated by such a representation. It is possible, in fact, to solve the boundary value problem directly in terms of these components, and we shall do so in this section. From the resulting expressions for the fields, we shall be able to characterize the polarization properties of the fields in a very simple way, one which directly suggests how the modes can be launched and/or detected optically. As a preface to the derivation of this somewhat unconventional field representation, it is appropriate to review its meaning and use. This is done in the next subsection. Subsequent subsections are devoted to the derivation and analysis of the modes themselves.

1. Circularly Polarized Components of the Transverse Fields

As is well known, the vector field **E** or **H** can be expressed as a linear superposition of any three orthogonal vector fields. The wave equation must be satisfied for each of these. The conventional decomposition of **E** is in terms of linear vibrations along axes either parallel or perpendicular to the interfaces at which boundary conditions must be imposed. Thus, for planar and circular waveguides, the decomposition is of the form

$$\mathbf{E} = \mathbf{E}_x + \mathbf{E}_y + \mathbf{E}_z = E_x\hat{\mathbf{x}} + E_y\hat{\mathbf{y}} + E_z\hat{\mathbf{z}}.$$

$$\mathbf{E} = \mathbf{E}_\varrho + \mathbf{E}_\phi + \mathbf{E}_z = E_\varrho\hat{\boldsymbol{\rho}} + E_\phi\hat{\boldsymbol{\phi}} + E_z\hat{\mathbf{z}},$$

with similar decompositions for **H**. For the dielectric rod waveguide, we have every reason to expect that we will need all three linearly polarized components of both **E** and **H** to describe a mode and that each of these components will have a different spatial variation of its amplitude. We are primarily interested in the transverse components, of course, because we can only observe the modes by viewing the end of the guide or the radiation field beyond it. Both depend only on the transverse components. Because there is nothing in the waveguide to establish a preferred reference x axis along which $\phi = 0$, it is natural to seek circulating wave solutions to the wave equation and to expect that these will be elliptically polarized in the transverse plane. To study this polarization, we will need a representation of the transverse fields that quickly yields the orientation of the polarization ellipse as a function of position, its sense of rotation, and the length of its major and minor axes. Such a representation is that in terms of circularly polarized component fields.

Let us consider, then, the two complex vector fields $\mathbf{E}_+$ and $\mathbf{E}_-$, defined as follows:

$$\mathbf{E}_\pm = E_\pm(\hat{\mathbf{x}} \mp i\hat{\mathbf{y}})e^{i\omega t - ihz}, \tag{4.31}$$

where $E_\pm$ is a complex-valued scalar quantity with magnitude $|E_\pm|$ and phase $\delta_\pm$, i.e., $E_\pm = |E_\pm| e^{i\delta_\pm}$. The associated real vector fields are

$$\operatorname{Re}\mathbf{E}_\pm = |E_\pm| \, [\hat{\mathbf{x}}\cos(\omega t - hz + \delta_\pm) \pm \hat{\mathbf{y}}\sin(\omega t - hz + \delta_\pm)]. \tag{4.32}$$

Apparently, $\mathbf{E}_+$ represents a vector of amplitude $\boldsymbol{E}_+$ rotating counterclockwise about an axis perpendicular to $\hat{\mathbf{x}}$ and $\hat{\mathbf{y}}$. This is called a left circularly polarized wave in optics, or a wave with positive helicity (and, therefore, positive angular momentum about the z axis) in physics. $\mathbf{E}_-$ gives a right circularly polarized wave, one with negative helicity. The magnetic field associated with these waves is of the form

$$\mathbf{H}_\pm = H_\pm(\hat{\mathbf{x}} \mp i\hat{\mathbf{y}})e^{i\omega t - ihz}. \tag{4.33}$$

We note that the positive and negative circularly polarized waves are orthogonal in the power sense, because

$$\mathbf{E}_\pm \times \mathbf{H}_\mp^* \propto (\hat{\mathbf{x}} \mp i\hat{\mathbf{y}}) \times (\hat{\mathbf{x}} \mp i\hat{\mathbf{y}}) = 0. \tag{4.34}$$

It should be noted, in our Eqs. (4.31) and (4.33), that we have not normalized the complex vectors $\hat{\mathbf{x}} \mp i\hat{\mathbf{y}}$, as is often done in the literature. This is because, in the nonnormalized forms of Eqs. (4.31) and (4.33), the complex amplitudes $E_\pm$ and $H_\pm$ specify both the amplitude and phase of the *real* fields, as is clear from Eq. (4.32). The amplitude of the complex fields, in which we really have no interest, is $\sqrt{2}$ times greater than that of the real fields, as can be seen by normalizing Eqs. (4.31) and (4.33). In the derivations that follow, we shall adopt the short-hand notation

$$\boldsymbol{\nu}_+ \equiv \hat{\mathbf{x}} - i\hat{\mathbf{y}} = \boldsymbol{\nu}_-^*; \qquad \boldsymbol{\nu}_- \equiv \hat{\mathbf{x}} + i\hat{\mathbf{y}} = \boldsymbol{\nu}_+^*. \tag{4.35}$$

It should be noted that $E_+ = \boldsymbol{\nu}_+^* \cdot \mathbf{E}_+/2$, rather than simply $\boldsymbol{\nu}_+^* \cdot \mathbf{E}_+$, etc., because $\boldsymbol{\nu}_+$ is not a unit vector. We observe also that our rotating vectors $\boldsymbol{\nu}_+$ and $\boldsymbol{\nu}_-$ are not the same as those employed in some physics texts (*1*, p. 205) because we are using an $e^{i\omega t}$ time dependence rather than $e^{-i\omega t}$.

With this understanding of the meaning of Eqs. (4.31) and (4.33), we can write any general electromagnetic field as a linear superposition of the form

$$\mathbf{E} = \mathbf{E}_+ + \mathbf{E}_- + \mathbf{E}_z = E_+\boldsymbol{\nu}_+ + E_-\boldsymbol{\nu}_- + E_z\hat{\mathbf{z}}, \tag{4.36}$$

with a similar superposition for $\mathbf{H}$. From the equation

$$\mathbf{E} = E_+\boldsymbol{\nu}_+ + E_-\boldsymbol{\nu}_- = E_x\hat{\mathbf{x}} + E_y\hat{\mathbf{y}},$$

we compute that

$$E_\pm = (E_x \pm iE_y)/2; \qquad E_x = E_+ + E_-; \qquad iE_y = E_+ - E_-. \tag{4.37}$$

If E_+ and E_- have the same phase, then the major axis of the polarization ellipse is the x axis. The minor axis coincides with the y axis. The ratio of the two axes is $(|E_+| - |E_-|)/(|E_+| + |E_-|)$. We can also write these circular transverse vectors in terms of their components in polar coordinates. Thus, if $\mathbf{E} = E_\varrho \hat{\boldsymbol{\rho}} + E_\phi \hat{\boldsymbol{\phi}} = E_x \hat{\mathbf{x}} + E_y \hat{\mathbf{y}}$, then

$$(E_\varrho \pm iE_\phi)/2 = e^{\mp i\phi}(E_x \pm iE_y)/2 = e^{\mp i\phi}E_\pm . \tag{4.38}$$

Accordingly, we see that if E_+ leads E_- by the phase angle of 2ϕ, then the axes of the polarization ellipse are rotated through an angle ϕ from the x axis. The ratio of the minor to the major axis remains the same. Note that the electric vector rotates in the positive (left-handed optically) sense if $|E_-| < |E_+|$ and in the negative sense if $|E_+| < |E_-|$.

The advantage of the circularly polarized component representation of elliptically polarized fields should be apparent from this discussion. They will be even more obvious when we study the modes of the dielectric guide. There, we will find that the circular components $E_\varrho \pm iE_\phi$ are precisely in phase or out of phase by π everywhere in the waveguide, so that the axes of the polarization ellipses are radial and tangential. The sense of rotation of the elliptically polarized electric vector will serve to distinguish between the two types of modes.

2. Derivation of the Mode Fields

Maxwell's equations can be easily decomposed into circular components. We first examine the operator $\boldsymbol{\nabla}$. We have

$$\begin{aligned}\boldsymbol{\nabla} &= \hat{\mathbf{x}}\nabla_x + \hat{\mathbf{y}}\nabla_y + \hat{\mathbf{z}}\nabla_z = \hat{\mathbf{x}}(\partial/\partial x) + \hat{\mathbf{y}}(\partial/\partial y) + \hat{\mathbf{z}}(\partial/\partial z)\\ &= \boldsymbol{\nu}_+\nabla_+ + \boldsymbol{\nu}_-\nabla_- + \hat{\mathbf{z}}\nabla_z\end{aligned} \tag{4.39}$$

where

$$\nabla_\pm = \tfrac{1}{2}\boldsymbol{\nu}_\pm^* \cdot \boldsymbol{\nabla} = \tfrac{1}{2}[(\partial/\partial x) \pm i(\partial/\partial y)].$$

The differential operators ∇_+ and ∇_- are well suited to the geometry of a waveguide of circular cross section. We already know that each component of the mode fields must satisfy the equation

$$(\boldsymbol{\nabla}_t^2 + \beta^2)F = 0; \qquad F = f(\varrho, \phi)e^{i\omega t - ihz}; \qquad h^2 + \beta^2 = k^2 = \omega^2\varepsilon\mu . \tag{4.40}$$

The elementary solutions of this equation in the polar coordinates appropriate to the geometry are of the form

$$F_n = Z_n(\beta\varrho)e^{in\phi + i\omega t - ihz}, \tag{4.41}$$

where Z_n is a Bessel function of order n. Next, note that, in polar coordinates, the operators $\nabla_{\pm}$ take the form

$$\nabla_{\pm} = \tfrac{1}{2}e^{\pm i\phi}[(\partial/\partial\varrho) \pm (i/\varrho)(\partial/\partial\phi)] \tag{4.42}$$

and that

$$\nabla_{\pm}F_n = \mp\beta F_{n\pm1}/2. \tag{4.43}$$

It follows from Eq. (4.43), therefore, that

$$\nabla_{-}\nabla_{+}F_n = \nabla_{+}\nabla_{-}F_n = -\beta^2 F_n/4. \tag{4.44}$$

Equation (4.44) is just the differential equation satisfied by each of the field components. It is identical with Eq. (4.40). This suggests that if we write the first-order Maxwell equations in terms of the circular components, using the operators ∇_{+} and ∇_{-}, we will be able to generate all of the components of an elementary wave from any pair of corresponding components of **E** and **H**. We will find that this is indeed the case. We accordingly write the Maxwell equation $\boldsymbol{\nabla} \times \mathbf{E} = -i\omega\mu\mathbf{H}$ in the form

$$\begin{aligned}(\boldsymbol{\nu}_{+}\nabla_{+} + \boldsymbol{\nu}_{-}\nabla_{-} + \hat{\mathbf{z}}\nabla_{z}) \times (E_{+}\boldsymbol{\nu}_{+} + E_{-}\boldsymbol{\nu}_{-} + E_{z}\hat{\mathbf{z}}) \\ = -i\omega\mu(H_{+}\boldsymbol{\nu}_{+} + H_{-}\boldsymbol{\nu}_{-} + H_{z}\hat{\mathbf{z}}).\end{aligned} \tag{4.45}$$

From the definitions of the vectors $\boldsymbol{\nu}_{\pm}$ [Eq. (4.35)], we can easily prove the following identities:

$$\boldsymbol{\nu}_{\pm} \times \boldsymbol{\nu}_{\pm} = 0; \qquad \boldsymbol{\nu}_{\pm} \times \boldsymbol{\nu}_{\mp} = \pm 2i\hat{\mathbf{z}}; \qquad \hat{\mathbf{z}} \times \boldsymbol{\nu}_{\pm} = -\boldsymbol{\nu}_{\pm} \times \hat{\mathbf{z}} = \pm i\boldsymbol{\nu}_{\pm}. \tag{4.46}$$

From the assumed form of the fields, we also have that $\nabla_z = -ih$. Using this and Eqs. (4.46), we can write Eq. (4.45) as the three scalar equations

$$\begin{aligned}&-i\nabla_{+}E_z + hE_{+} = -i\omega\mu H_{+}; \qquad i\nabla_{-}E_z - hE_{-} = -i\omega\mu H_{-}; \\ &2i\nabla_{+}E_{-} - 2i\nabla_{-}E_{+} = -i\omega\mu H_z.\end{aligned} \tag{4.47}$$

In a similar way, we can write the Maxwell equation $\boldsymbol{\nabla} \times \mathbf{H} = i\omega\varepsilon\mathbf{E}$ as the scalar equations

$$\begin{aligned}&-i\nabla_{+}H_z + hH_{+} = i\omega\varepsilon E_{+}; \qquad i\nabla_{-}H_z - hH_{-} = i\omega\varepsilon E_{-}; \\ &2i\nabla_{+}H_{-} - 2i\nabla_{-}H_{+} = i\omega\varepsilon E_z.\end{aligned} \tag{4.48}$$

The six simultaneous differential equations [Eqs. (4.47) and (4.48)], which are simply Maxwell's equations written in terms of circular components and operators, can now be solved directly by inspection. We have only to recall Eqs. (4.43) and (4.41), which show that the differential operators $\nabla_{\pm}$ are raising and lowering operators, analogous to those used to generate angular momentum states in quantum mechanics. Thus, the first of Eqs. (4.47) is clearly satisfied by setting

$$E_z = AF_n = AZ_n(\beta\varrho)e^{in\phi+i\omega t-ihz}; \qquad E_+ = CF_{n+1}; \quad H_+ = DF_{n+1}.$$

The second of Eqs. (4.47) is satisfied for the same E_z with E_- and H_- proportional to F_{n-1}. The third is satisfied for the same E_- and E_+ with H_z proportional to F_n. It is then obvious that these solutions of Eqs. (4.47) satisfy Eqs. (4.48). We have accordingly reduced Maxwell's equations to six algebraic equations in six unknown constants of proportionality which are independent of space and time. The six equations are not linearly independent, however, because we can generate two of them from the other four. Accordingly, the equations determine the values of four of the unknown coefficients in terms of the other two. Let us take

$$E_z = AF_n; \qquad H_z = BF_n. \tag{4.49}$$

Then, by combining the first of Eqs. (4.47) and (4.48), we obtain

$$E_+ = (1/2\beta)(ihA + \omega\mu B)F_{n+1}; \qquad H_+ = -(1/2\beta)(\omega\varepsilon A - ihB)F_{n+1}. \tag{4.50}$$

Similarly, the second members of Eqs. (4.47) and (4.48) yield

$$E_- = -(1/2\beta)(ihA - \omega\mu B)F_{n-1}; \qquad H_- = -(1/2\beta)(\omega\varepsilon A + ihB)F_{n-1}. \tag{4.51}$$

If we now multiply all components by $2\beta/ih$ and set $\alpha = \omega\mu B/ihA$, we obtain the following simple expressions for each of the field components:

$$\begin{aligned} &E_{\pm} = \pm(1 \pm \alpha)AF_{n\pm1}; \qquad H_{\pm} = (ih/\omega\mu)[(k^2/h^2) \pm \alpha]AF_{n\pm1} \\ &E_z = (2\beta/ih)AF_n; \qquad H_z = (ih/\omega\mu)\alpha E_z; \qquad F_n = Z_n(\beta\varrho)e^{in\phi+i\omega t-ihz}. \end{aligned} \tag{4.52}$$

Equations (4.52) are sufficiently general to describe the fields both inside the waveguide ($\varrho < a$, medium 1) and outside ($a < \varrho$, medium 2), since they are solutions to Maxwell's equations in both media. We do not yet know, however, that these solutions can be made to match each other at $\varrho = a$, so that at this stage of the derivation, they must be regarded as trial solutions to the boundary value problem. It must be understood, in particular, that we may not conclude *a priori* that the mode parameters A,

h, β, and α are equal in the two media. Because the permitivity ε is different in each medium, we know that the parameters $k^2 = \omega^2 \varepsilon \mu$ are different. We shall assume that the magnetic permeability μ is identical for the two media.

When writing equations which pertain to either the waveguide core only or to the coating only, we shall use the subscripts 1 and 2, respectively, to distinguish between the parameters in Eqs. (4.52). From the boundary conditions, we shall find that $\alpha_1 = \alpha_2 = \alpha$ and $h_1 = h_2 = h$, but that $A_1 \neq A_2$ and $\beta_1 \neq \beta_2$. Let us now apply the boundary conditions so as to determine the functions Z_n and the relations between the various parameters.

1. The fields must be periodic in ϕ. Therefore, n is an integer or zero.

2. The fields must be finite at $\varrho = 0$. Thus, $Z_{n1} = J_n(\beta_1 \varrho)$, where J_n is the Bessel function of the first kind.

3. In the medium outside the core, we can consider two cases. If we are interested in only those modes that do not radiate energy away from the guide, then the fields outside should be evanescent, their amplitudes becoming vanishingly small as ϱ increases indefinitely. The Bessel functions satisfying this condition are the modified functions of the second kind, symbolized by $K_n(q\varrho/a)$. If we are also interested in waves that radiate away from the core, then these waves must satisfy the Sommerfeld radiation condition as ϱ increases indefinitely. The appropriate Bessel functions for this condition are the Hankel functions of the second kind, labeled $H_n^{(2)}(v\varrho/a)$. Because the functions $H_n^{(2)}$ and K_n are related by the equation $H_n^{(2)}(-iq) = (-i)^{-n-1}(2/\pi)K_n(q)$, we may use these functions interchangeably as convenient. Because the functions $H_n^{(2)}$ satisfy the same recursion relations as functions J_n, it is convenient to use them in all derivations. We thus take $F_{n2} = H_n(\beta_2 \varrho)$, where we have dropped the superscript *2* on the Hankel function for convenience.

4. The tangential components of the electric and magnetic fields must be continuous at $\varrho = a$. From the longitudinal components of $\mathbf{E}$ in Eqs. (4.52), we find

$$h_1 = h_2 \equiv h; \qquad A_2 = uJ_n(u)/vH_n(v)A_1; \qquad u \equiv \beta_1 a, \quad v \equiv \beta_2 a. \tag{4.53}$$

Equating the longitudinal components of $\mathbf{H}$ at $\varrho = a$ yields $\alpha_1 = \alpha_2 = \alpha$. We must also match the azimuthal components of $\mathbf{E}$ and $\mathbf{H}$. From Eqs. (4.38) and (4.52), we have

$$\begin{aligned} E_\phi &= -i(E_+ e^{-i\phi} - E_- e^{i\phi}) = -iA[(1+\alpha)F_{n+1}e^{-i\phi} + (1-\alpha)F_{n-1}e^{i\phi}] \\ H_\phi &= -i(H_+ e^{-i\phi} - H_- e^{i\phi}) = (hA/\omega\mu)\{[(k^2/h^2) + \alpha]F_{n+1}e^{-i\phi} \\ &\qquad - [(k^2/h^2) - \alpha]F_{n-1}e^{i\phi}\}. \end{aligned} \tag{4.54}$$

Equating the expressions for these components inside and outside the core at $\varrho = a$ and using Eqs. (4.53), we find

$$(1 + \alpha)(\mathscr{J}_+ - \mathscr{H}_+) = -(1 - \alpha)(\mathscr{J}_- - \mathscr{H}_-) \tag{4.55}$$

and

$$(k_1^2/h^2)(\mathscr{J}_+ - \mathscr{J}_-) + \alpha(\mathscr{J}_+ + \mathscr{J}_-) = (k_2^2/h^2)(\mathscr{H}_+ - \mathscr{H}_-) + \alpha(\mathscr{H}_+ - \mathscr{H}_-) \tag{4.56}$$

where

$$\mathscr{J}_\pm = J_{n\pm 1}(u)/uJ_n(u) \qquad \text{and} \qquad \mathscr{H}_\pm = H_{n\pm 1}(v)/vH_n(v).$$

Equations (4.55) and (4.56) give us two different expressions for α. By equating them, we obtain the characteristic equation of the circular dielectric waveguide. From Eqs. (4.40), we also have the relation

$$u^2 - v^2 = (k_1^2 - k_2^2)a^2 = (\pi d/\lambda)^2(n_1^2 - n_2^2) \equiv R^2, \tag{4.57}$$

just as we had for the dielectric slab guide, so that the characteristic equation is a transcendental equation in u with R^2 and k_2^2/k_1^2 as parameters. Rather than proceed directly with the study of this equation, we will consider what we have already determined about the modes. This should show us what to look for in solving the characteristic equation. The fields inside the guide are as follows:

$$\begin{aligned} E_\pm &= \pm(1 \pm \alpha)AJ_{n\pm 1}(u\varrho/a)e^{i(n\pm 1)\phi + i\omega t - ihz}; \\ E_z &= (2\beta_1/ih)AJ_n(u\varrho/a)e^{in\phi + i\omega t - ihz}; \\ H_\pm &= \pm(ih/\omega\mu)[(k_1^2/h^2 \pm \alpha)/(1 \pm \alpha)]E_\pm; \\ H_z &= (ih/\omega\mu)\alpha E_z. \end{aligned} \tag{4.58}$$

The fields outside the guide are Eqs. (4.58) modified to

$$\begin{aligned} &H_p(v\varrho/a) \to J_p(u\varrho/a), \quad p = n-1,\ n,\ n+1; \qquad \beta_2 \to \beta_1; \\ &k_2 \to k_1; \qquad uJ_n(u)/vH_n(v)A \to A. \end{aligned} \tag{4.59}$$

For the bound modes in which we are primarily interested, $v = -iq$, where q is real, so that the fields are evanescent outside. The constant A is arbitrary, AA^* being proportional to the power in the mode. The intensity distribution to be observed at the output end of the guide is given by the time average of the axial component of the Poynting vector (in the Huygens–Kirchhoff approximation)

$$S_z = \mathrm{Re}(\mathbf{E} \times \mathbf{H}^*)_z/2 = \mathrm{Re}[(E_+\mathbf{v}_+ + E_-\mathbf{v}_-) \times (H_+^*\mathbf{v}_+^* + H_-^*\mathbf{v}_-^*)]/2$$
$$= \mathrm{Im}(E_+H_+^* - E_-H_-^*). \tag{4.60}$$

We note that the right and left circularly polarized fields contribute indepe nd ently to the power density. From the field equations (4.58), we then have, for $\varrho < a$,

$$S_z = (hAA^*/\omega\mu)\{(1 + \alpha)[(k_1^2/h^2) + \alpha]J_{n+1}^2(u\varrho/a)$$
$$+ (1 - \alpha)[(k_1^2/h^2) - \alpha]J_{n-1}^2(\mathrm{u}\varrho/a)\}, \tag{4.61}$$

with a similar expression for $a < \varrho$. The intensity distribution is clearly circularly symmetric. The relative contributions of the right and left circular components clearly depend on the values of α.

From Eqs. (4.58) and (4.59), we note that the circular components, $E_\varrho \pm E_\phi = e^{\mp i\phi}E_\pm$, are in phase or out of phase by π, so that the polarization of the observable transverse fields is elliptical, with the axes of the polarization ellipses radial and tangential. The polarization is clearly circularly symmetric. Because of the oscillatory character of the Bessel functions $J_{n\pm1}$, the eccentricity changes continuously with radius ϱ inside the guide. At the boundary and beyond it, however, the eccentricity is only weakly dependent on ϱ. From Eq. (4.61), we see that where the intensity is maximum the polarization ellipse is characterized almost completely by the ratio

$$\gamma = (1 - \alpha)/(1 + \alpha). \tag{4.62}$$

For obvious reasons, we refer to this as the polarization parameter. Because its values substantially determine the optically observable quantities, we shall study the characteristic equation with a view to determining them.

Before closing this subsection, we sould note that the fields [Eqs. (4.58) and (4.59)] and the intensity distribution [Eq. (4.61)] describe negative circulating waves. As we shall see later, the modes of the dielectric rod waveguide are twofold degenerate. We can describe a second degenerate vector eigenfunction of this boundary value problem simply by replacing n by $-n$ in Eqs. (4.58), (4.59), and (4.61), with the understanding that $\alpha_{-n} = -\alpha_n$. The resulting expressions describe a positive circulating wave. Because of this mode degeneracy, we can have an infinite variety of vector fields that propagate down the guide with the same velocity ω/h, each such field being represented by a particular linear superposition of our elementary circulating waves. The particular superpositions of our circulating waves that take equal contributions from each are those generally discussed in the engineering literature. Such superpositions give linearly polarized fields which are complicated functions of both ϱ and ϕ, and intensity distributions that are much more complicated than those of the elementary circulating waves. We shall not treat these superpositions in the general development, though we shall discuss them in some special cases of interest optically.

3. The Characteristic Equation

Equations (4.55) and (4.56) give us two different expressions for α. By equating them, we can generate one of the many possible (equivalent) forms of the characteristic equation. Before doing so, we will first obtain explicit expressions for γ. From Eqs. (4.55) and (4.56) we have

$$\alpha = \frac{(\mathscr{J}_- + \mathscr{J}_+) - (\mathscr{H}_- + \mathscr{H}_+)}{(\mathscr{J}_- - \mathscr{J}_+) - (\mathscr{H}_- - \mathscr{H}_+)} = \frac{(k_1^2/h^2)(\mathscr{J}_- - \mathscr{J}_+) - (k_2^2/h^2)(\mathscr{H}_- - \mathscr{H}_+)}{(\mathscr{J}_- + \mathscr{J}_+) - (\mathscr{H}_- + \mathscr{H}_+)}. \tag{4.63}$$

The explicit dependence on h can be eliminated as follows. For any Bessel function Z_n, we have the relation

$$2nZ_n(x)/x = Z_{n-1}(x) + Z_{n+1}(x). \tag{4.64}$$

If this is substituted into the denominator of the second expression for α in Eq. (4.63), we obtain the equation

$$\alpha = \frac{k_1^2(\mathscr{J}_- - \mathscr{J}_+) - k_2^2(\mathscr{H}_- - \mathscr{H}_+)}{2nh^2[(1/u^2) - (1/v^2)]}. \tag{4.65}$$

If we now substitute $(k_1^2 - \beta_1^2)/\beta_1^2a^2$ for h^2/u^2 and $(k_2^2 - \beta_2^2)/\beta_2^2a^2$ for h^2/v^2, Eq. (4.65) becomes

$$\alpha = \frac{\varepsilon_1(\mathscr{J}_- - \mathscr{J}_+) - \varepsilon_2(\mathscr{H}_- - \mathscr{H}_+)}{2n[(\varepsilon_1/u^2) - (\varepsilon_2/v^2)]} = \frac{\varepsilon_1\mathscr{J}' - \varepsilon_2\mathscr{H}'}{n[(\varepsilon_1/u^2) - (\varepsilon_2/v^2)]}, \tag{4.66}$$

where $\mathscr{J}' = J_n'(u)/uJ_n(u)$ and $\mathscr{H}' = H_n'(v)/vH_n(v)$, and we have used the Bessel relation

$$2Z_n'(x) = Z_{n-1}(x) - Z_{n+1}(x) \tag{4.67}$$

to obtain the last form of α. If we now use Eq. (4.64) again, in conjunction with Eq. (4.66), we can restate Eq. (4.63) in a form that does not depend on h. Thus,

$$\alpha = \frac{(\mathscr{J}_- + \mathscr{J}_+) - (\mathscr{H}_- + \mathscr{H}_+)}{(\mathscr{J}_- - \mathscr{J}_+) - (\mathscr{H}_- - \mathscr{H}_+)} = \frac{\varepsilon_1(\mathscr{J}_- - \mathscr{J}_+) - \varepsilon_2(\mathscr{H}_- - \mathscr{H}_+)}{\varepsilon_1(\mathscr{J}_- + \mathscr{J}_+) - \varepsilon_2(\mathscr{H}_- + \mathscr{H}_+)}. \tag{4.68}$$

From these two expressions for α, we can generate two expressions for the polarization constant γ defined by Eq. (4.62). They are

$$\begin{aligned}\gamma &= (1 - \alpha)/(1 + \alpha) = -(\mathscr{J}_+ - \mathscr{H}_+)/(\mathscr{J}_- - \mathscr{H}_-) \\ &= (\varepsilon_1 \mathscr{J}_+ - \varepsilon_2 \mathscr{H}_+)/(\varepsilon_1 \mathscr{J}_- - \varepsilon_2 \mathscr{H}_-).\end{aligned} \tag{4.69}$$

Equating these two expressions for γ yields an alternate form of the characteristic equation. We will not pursue this form directly, however. Rather, we will first derive the form most often seen in the literature and then proceed to recast that form into one which can tell us what we want to know about the parameters α and γ.

Let us first replace the sums and differences of the Bessel functions in Eq. (4.68) by expressions involving only the functions of order n or their derivatives. Thus, by employing Eqs. (4.64) and (4.67), we can rewrite Eq. (4.68) in the form

$$\alpha = n[(1/u^2) - (1/v^2)]/(\mathscr{J}' - \mathscr{H}') = (\varepsilon_1 \mathscr{J}' - \varepsilon_2 \mathscr{H}')/[(\varepsilon_1 n/u^2) - (\varepsilon_2 n/v^2)].$$

Cross-multiplying these two expressions for α yields the characteristic equation

$$(\mathscr{J}' - \mathscr{H}')(\varepsilon_1 \mathscr{J}' - \varepsilon_2 \mathscr{H}') = n^2[(1/u^2) - (1/v^2)][(\varepsilon_1/u^2) - (\varepsilon_2/v^2)]. \tag{4.70}$$

Equation (4.70), together with the relation [Eq. (4.57)] $R^2 = u^2 - v^2$, determines the eigenvalues of the boundary value problem, $u = \beta_1 a$ and $v = \beta_2 a$, in terms of the physical parameters $R^2 = (\pi d/\lambda)^2(n_1{}^2 - n_2{}^2)$ and $\varepsilon_2/\varepsilon_1$.

The characteristic equation (4.70), together with the field equations (4.58) and (4.59), apply to a rather wide range of physical problems. In deriving them, we have implied that the conductivity of both media was negligible. The equations are not so limited, of course, because we can let the parameters R^2, ε_1, and ε_2 be complex valued by setting $\varepsilon \to \varepsilon - (i\sigma/\omega)$ in each medium, as we did with the slab guide in Chapter 2. Equation (4.70) then becomes a transcendental equation in the complex variables u and v. Only a very few special cases of this equation have been treated in the literature. Among them are the following:

1. Waves on conducting wires. Here, $\sigma_1/\omega\varepsilon_1 \gg 1$ and $\sigma_2 \approx 0$. These waves are well treated by Stratton (*3*).

2. Waves in hollow, conducting pipes. We treated, in Section A of this chapter, the limiting case of $\sigma_2 = \infty$. This widely investigated case serves as the basic theory of propagation in microwave waveguides. It is also treated in Stratton (*3*) and elsewhere in the weaker approximation $1 \ll \sigma_2/\omega\varepsilon_2$.

3. Waves in hollow, dielectric tubes. In this case, the conductivities are assumed negligible in both media as a first approximation, but $\varepsilon_1 < \varepsilon_2$. The waves are not completely bound to the waveguide, power being continuously radiated from the guide into the surrounding medium as the wave propagates down the guide. This loss mechanism is represented, in the field equations, by a complex-valued propagation constant h. Accordingly, u and v are also complex. If the diameter of the tube is much greater than the wavelength, Eq. (4.70) is easily solved for the first few modes that have the real part of the propagation constant about equal to k_1. The guiding mechanism is Fresnel reflection for nearly grazing angles of incidence on the walls of the tube. These waves were first studied in detail by Marcatili and Schmeltzer (*4*) and will be treated in Appendix B by T. Sawatari.

4. Creeping waves on dielectric cylinders. In this case, σ is usually assumed negligible in both media and $\varepsilon_2 < \varepsilon_1$. These waves have been treated largely in connection with studies of the diffraction of plane waves by dielectric cylinders with diameters large with respect to the wavelength. The accent in these studies is on the evanescent waves that travel around the surface of the cylinder continuously radiating some of their power away from the surface, while exchanging power through refraction with waves inside the cylinder. For such waves, h, u, and v are complex-valued quantities. In addition, the circulating wave nature of the fields is made more explicit by transforming them, through the use of the so-called "Watson transformation," into expressions in which the parameter n becomes a complex variable ν that determines both the phase of the wave as it travels around the cylinder, and its rate of attenuation with increasing ϕ. We shall not treat these waves explicitly here. They are of considerable importance in certain microwave propagation problems and have been treated extensively in that connection (*5*, *6*).

5. Surface waves or waveguide modes on the dielectric rod. These are the principal subject of this chapter. In this case, the conductivity is assumed vanishingly small in both media as a first approximation and $\varepsilon_2 < \varepsilon_1$. The waves are completely bound to the waveguide and its immediate surround, the guiding mechanism being total internal reflection at the core–coating interface. For these waves, v is a pure imaginary-valued quantity.

Then u and h are real-valued parameters. We will now proceed to study Eq. (4.70) in detail for this case so as to be able to fully evaluate the fields described by Eqs. (4.58) and (4.59).

4. Graphical Solution of the Characteristic Equation—General Behavior of Mode Parameters

Let us begin the analysis of Eq. (4.70) by dividing both sides by ε_1 and then multiplying out the left-hand side, thus obtaining the quadratic form

$$(\mathscr{J}')^2 - [1 + (\varepsilon_2/\varepsilon_1)]\mathscr{J}'\mathscr{H}' + (\varepsilon_2/\varepsilon_1)(\mathscr{H}')^2 = n^2[(1/u^2) - (1/v^2)] \times [(1/u^2) - (\varepsilon_2/\varepsilon_1 v^2)]. \quad (4.71)$$

If we solve this equation for $\mathscr{J}'$ and make use of the short-hand notation $\bar{\varepsilon} = (\varepsilon_1 + \varepsilon_2)/2$ and $\Delta\varepsilon = (\varepsilon_1 - \varepsilon_2)/2$, we obtain the result

$$\mathscr{J}' = (\bar{\varepsilon}/\varepsilon_1)\mathscr{H}' \pm \{(\Delta\varepsilon/\varepsilon_1)^2(\mathscr{H}')^2 + n^2[(1/u^2) - (1/v^2)] \times [(1/u^2) - (\varepsilon_2/\varepsilon_1 v^2)]\}^{1/2}. \quad (4.72)$$

Before developing the implications of this equation on the possible values of α and γ, we first take a brief graphical look at it so as to gain an appreciation of its general character. We note that we can restrict our attention to values of $n \geq 0$, because the equation is invariant if n is replaced by $-n$. (This reflects the twofold degeneracy mentioned in Section B.2). In order to obtain easily interpretable plots, we will use the following identities to transform Eq. (4.72) into two equivalent equations in real valued tabulated functions. Thus substituting $-iq = v$, $\mp \mathscr{J}_{\pm} \pm n/u = \mathscr{J}'$ and $-\mathscr{K}' = -K_n'(q)/qK_n(q) = \mathscr{H}'$ into Eq. (4.72) and multiplying by u, we obtain the following results.

$EH_{n,m}$ Modes

$$\frac{J_{n+1}(u)}{J_n(u)} = \frac{\bar{\varepsilon}u\mathscr{K}'}{\varepsilon_1} + u\left\{\frac{n}{u^2} - \left[\left(\frac{-\Delta\varepsilon\mathscr{K}'}{\varepsilon_1}\right)^2 + n^2\left(\frac{1}{u^2}+\frac{1}{q^2}\right)\left(\frac{1}{u^2}+\frac{\varepsilon_2}{\varepsilon_1 q^2}\right)\right]^{1/2}\right\} \quad (4.72a)$$

for the positive sign on the radical in Eq. (4.72).

$HE_{n,m}$ Modes

$$\frac{J_{n-1}(u)}{J_n(u)} = -\frac{\bar{\varepsilon}u\mathscr{K}'}{\varepsilon_1} + u\left\{\frac{n}{u^2} - \left[\left(\frac{-\Delta\varepsilon\mathscr{K}'}{\varepsilon_1}\right)^2 + n^2\left(\frac{1}{u^2} + \frac{1}{q^2}\right)\left(\frac{1}{u^2} + \frac{\varepsilon_2}{\varepsilon_1 q^2}\right)\right]^{1/2}\right\} \quad (4.72b)$$

for the negative sign on the radical in Eq. (4.72). The reason for writing the two equations (4.72a) and (4.72b) differently is that we want to have functions on the right-hand side of the characteristic equation which are monotonic, with regular behavior near $u = 0$. If we now recall that $q = (R^2 - u^2)^{1/2}$, we see that each of these new equations is transcendental in u for fixed R, n, $\Delta\varepsilon/\varepsilon_1$, and $\bar{\varepsilon}/\varepsilon_1$. The right-hand side of Eq. (4.72a) is a monotonically decreasing function of u having value zero when $u = 0$. It resembles the function $-u/(R^2 - u^2)^{1/2}$ for $u \ll R$ and approaches $-\infty$ as u approaches $R(q \to 0)$. The left-hand side of Eq. (4.72a), on the other hand, is a multi-branched function of u resembling $\tan u$. If the two sides are plotted as functions of u, the eigenvalues u can be located by the intersections of the curves. A representative plot of Eq. (4.72a) is shown in the lower half of Fig. 4-2. The values of the parameters are $n = 2$, $R = 15.667$, $\Delta\varepsilon/\varepsilon_1 = 0.153$, $\bar{\varepsilon}/\varepsilon_1 = 0.848$. These values correspond to a fiber optical waveguide with $n_1 = 1.8$, $n_2 = 1.5$, and $d/\lambda_0 = 5$. Because the right-hand side of Eq. (4.72a) is always negative, only the negative portions of each branch of $J_{n+1}(u)/J_n(u)$ have been plotted.

Equation (4.72b) is quite similar to Eq. (4.72a). Its right-hand side is a monotonically increasing function of u, however, with zero value at $u = 0$. It resembles the function $(\varepsilon_2/\varepsilon_1)u/(R^2 - u^2)^{1/2}$ for $u \ll R$ and approaches the values $R\varepsilon_2/2(n - 1)\bar{\varepsilon}$ as u approaches $R(q \to 0)$. The left-hand side resembles the function $\cot u$. Equation (4.72b) is plotted in the upper half of Fig. 4-2. Because the right-hand side is greater than zero for all values of u, only the positive portions of each branch of $J_{n-1}(u)/J_n(u)$ are plotted. We will see presently that, unless $n = 1$, Eq. (4.72b) cannot be satisfied for values of $J_{n-1}(u)/J_n(u)$ greater than those given by the dashed line in the upper half of the figure, so that above the dashed line, the branches of J_{n-1}/J_n are shown as dashed curves.

For values of $|u| > R$, q becomes a complex variable and the fields no longer represent waves bound to the waveguide. We thus refer to the condition $u \to R(q \to 0)$ as the cutoff condition. To find explicit expressions for this condition, we examine Eqs. (4.72a) and (4.72b) for small values of q, using the approximations $K'(q) \approx -(n/q^2) - [1/2(n - 1)]$ for $n > 1$ and $K'(q) \approx -(n/q^2) - \log(2/\gamma' q)$ for $n = 1$, where γ' is Euler's constant. We then find that Eq. (4.72a) reduces to $J_{n+1}(u)/J_n(u) \approx -2n\bar{\varepsilon}/\varepsilon_1 q^2$ as q approaches zero, so that the cutoff equation for EH modes, i.e., those modes whose parameters satisfy Eq. (4.72a), is given by

$$J_n(R) = 0 \qquad \text{(EH cutoff)}. \tag{4.72a$'$}$$

In the same approximation, we find that the HE mode characteristic equation

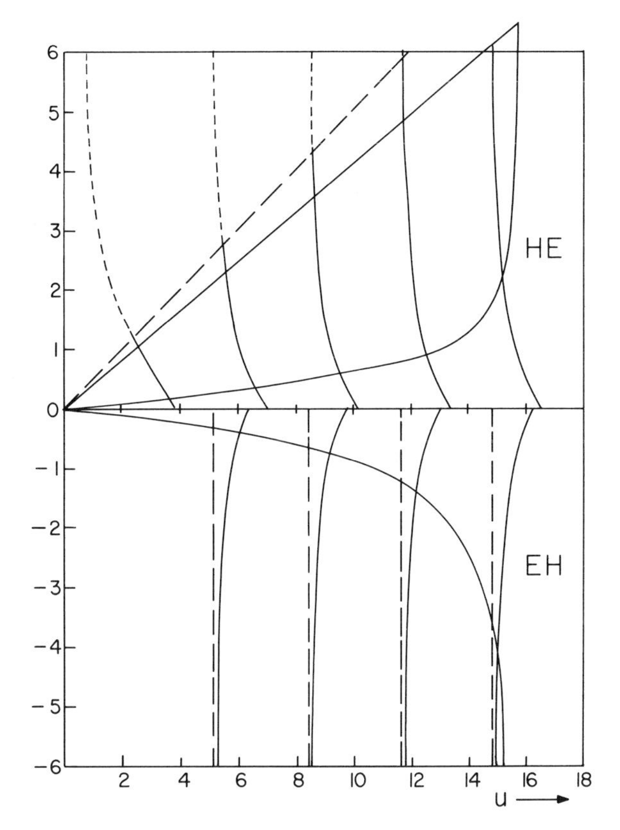

Fig. 4.2. Illustrating a graphical solution of the characteristic equations (4.72a) and (4.72b) for $n = 2$, $R = 15.667$, $\Delta\varepsilon/\varepsilon_1 = 0.153$, and $\bar{\varepsilon}/\varepsilon_1 = 0.848$. The two sides of Eq. (4.72a) for EH modes are plotted below the u axis. The vertical broken lines indicate the values of u at cutoff for EH modes. The two sides of Eq. (4.72b) are plotted above the u axis. The intersections of the negatively sloped curves with the solid straight line give the cutoff values of u for HE modes of guides with the given values of dielectric constants. Their intersections with the straight, long-dashed line give the lowest possible values of u for these modes. This obtains as $\Delta\varepsilon \to 0$.

(4.72b) reduces to $J_{n-1}(\mathrm{u})/J_n(u) \approx \varepsilon_2 u/2(n-1)\bar{\varepsilon}$ for $n > 1$ and $J_{n-1}(u)/J_n(u) \approx (\varepsilon_2 u/\bar{\varepsilon}) \log_e(2/\gamma' q)$ for $n = 1$. The cutoff equations for HE modes are, therefore,

$$J_{n-1}(R)/J_n(R) = \varepsilon_2 R/2(n-1)\bar{\varepsilon}, \quad n > 1 \qquad \text{(HE cutoff).}$$
$$\text{and} \qquad J_1(R) = 0, \quad n = 1; \tag{4.72b$'$}$$

The first of Eqs. (4.72b′) is equivalent to the equation $J_{n-2}(R)/J_n(R) =$

$-\Delta\varepsilon/\bar{\varepsilon}$, as may be seen by substituting the recursion relation

$$2(n-1)J_{n-1}(x)/x = J_{n-2}(x) + J_n(x).$$

The vertical dashed lines in Fig. 4-2 represent the solutions of Eq. (4.72a′) for $n = 2$. The solid line with slope 1/2.44 in the upper half of the figure represents the right-hand side of Eq. (4.72b′) for the chosen value of n and the physical parameters. Its intersections with the curves representing the function J_{n-1}/J_n identify the minimum values of R for which the corresponding waves can be completely bound to this waveguide. The dotted line with slope $1/2(n-1) = \frac{1}{2}$ (in this case) determines the absolute minimum values of R for each mode, regardless of the values of the dielectric constants. It corresponds to the limiting case $\Delta\varepsilon \to 0$, where, according to Eq. (4.72b′), $J_{n-2}(R) \to 0$. Thus, the statement made earlier is confirmed. The values of $J_{n-1}(u)/J_n(u)$ must be less than those given by the dashed line.

The right-hand sides of Eqs. (4.72a) and (4.72b) deviate less rapidly from the u axis as R increases and do not take on large values until u approaches R. Clearly, as $R \to \infty$, these curves coincide with the u axis. Therefore, when R is very large, the first few roots of the characteristic equations [Eq. (4.70) of (4.72)] can be coarsely approximated by the roots of $J_{n-1}(u) = 0$ and $J_{n+1}(u) = 0$. These coarse approximations can then be used to obtain good approximate solutions to the characteristic equations in the far-from-cutoff case, and thus good approximate expressions for α and γ, as we shall see subsequently.

For identification purposes, the successive roots of Eqs. (4.72) for a given value of n are assigned an integral index m. Those that satisfy Eqs. (4.72) with the positive sign on the radical [Eq. (4.72a)] are labeled $u_{n,m}^{EH}$, while those satisfying Eq. (4.72b) are labeled $u_{n,m}^{HE}$. The superscripts EH and HE distinguish between the roots corresponding to substantially different kinds of fields for the same values of n and m, as we shall see shortly. To illustrate this identification scheme, we have labeled the upper half of Fig. 4-2 HE and the lower half EH. As may be verified by studying the figure, $u_{n,m}^{HE}$ lies between the mth nonzero root of $J_{n-2}(u) = 0$ and the mth nonzero root of $J_{n-1}(u) = 0$. Similarly, $u_{n,m}^{EH}$ lies between the mth nonzero root of $J_n(u) = 0$ and the mth nonzero root of $J_{n+1}(u) = 0$.

From Eq. (4.72b′), we see that the $HE_{1,1}$ mode is never cut off, no matter how small R, and thus the diameter of the guide, becomes. We may also note that, when $n = 0$, Eqs. (4.72) simplify considerably. The EH mode, Eq. (4.72a), becomes

$$J_1(u)/J_0(u) = -uK_1(q)/qK_0(q) \qquad (TE_{0,m} \text{ modes}) \qquad (4.72a'')$$

while the HE mode, Eq. (4.72b), reduces to

$$J_1(u)/J_0(u) = -\varepsilon_2 u K_1(q)/\varepsilon_1 q K_0(q) \qquad (\mathrm{TM}_{0,m} \text{ modes}). \tag{4.72b''}$$

The $EH_{0,m}$ modes are called $TE_{0,m}$ modes because, as we shall see presently, $E_z = 0$ for these modes. Similarly, the $HE_{0,m}$ modes are labeled $TM_{0,m}$ because $H_z = 0$ for these modes. For large values of q, the modified Bessel functions $K_n(q)$ can be approximated by $(\pi/2q)^{1/2}e^{-q}$ so that the ratio of any two of these functions, as in the above equations, approaches unity. Equations (4.72) are thus graphically very similar to those encountered in treating the dielectric slab. In particular, Eqs. (4.72) should be compared with Eqs. (2.40), (2.41), (2.49), and (2.50) and Fig. 2-4 of Chapter 2.

Let us now use Eqs. (4.72) to examine the general behavior of the polarization constant γ and $\alpha = (1 - \gamma)/1 + \gamma)$. [From Eq. (4.58), we see that α is proportional to the ratio of the longitudinal component of **H** to the longitudinal component of **E**, and is thus a measure, in some sense, of the relative contributions of TM- and TE-type waves to the hybrid mode.] We first write Eqs. (4.72) in the two obviously equivalent forms

$$\mathscr{J}' - \mathscr{H}' = -(\Delta\varepsilon\mathscr{H}'/\varepsilon_1) \pm \{(\Delta\varepsilon\mathscr{H}'/\varepsilon_1)^2 + n^2[(1/u^2) - (1/v^2)] \times [(1/u^2) - (\varepsilon_2/\varepsilon_1 v^2)]\}^{1/2} \tag{4.72'}$$

$$\mathscr{J}' - (\varepsilon_2\mathscr{H}'/\varepsilon_1) = (\Delta\varepsilon\mathscr{H}'/\varepsilon_1) \pm \{(\Delta\varepsilon\mathscr{H}'/\varepsilon_1)^2 + n^2[(1/u^2) - (1/v^2)] \times [(1/u^2) - (\varepsilon_2/\varepsilon_1 v^2)]\}^{1/2}. \tag{4.72''}$$

We will now attempt to evaluate the two expressions for α,

$$\alpha = \frac{n[(1/u) - (1/v^2)]}{\mathscr{J}' - \mathscr{H}'} \tag{4.73'}$$

and

$$\alpha = \frac{\mathscr{J}' - (\varepsilon_2\mathscr{H}'/\varepsilon_1)}{n[(1/u^2) - (\varepsilon_2/\varepsilon_1 v^2)]}. \tag{4.73''}$$

by examining Eqs. (4.72′) and (4.72″). It is immediately clear that the choice of sign on the radical in Eqs. (4.72) makes a very considerable difference in the value of α. Accordingly, the nature of the fields expressed by Eqs. (4.58) and (4.59) is markedly dependent on this choice of sign. It is further to be noted that, although Eqs. (4.72′) and (4.72″) are invariant when n is replaced by $-n$, α changes sign under this substitution, and the fields are therefore quite different. [This also means, of course, that two vector eigenfunctions are associated with each solution $u_{n,m}$ of Eqs. (4.72), with $n > 0$.]

Let us now determine the range of values of α for each choice of sign in Eqs. (4.72). In accordance with the custom in the literature, we will label as EH modes all fields whose eigenvalues satisfy Eqs. (4.72) with the positive sign on the radical, and as HE modes those satisfying Eqs. (4.72) with the minus sign on the radical. Substituting Eq. (4.72″) into (4.73″), we find that, for EH modes,

$$\begin{aligned} \alpha(\text{EH}) &\geq 1 \qquad \text{for} \quad n > 0; \\ \alpha(\text{EH}) &\leq -1 \qquad \text{for} \quad n < 0; \\ \alpha(\text{EH}) &= \infty \qquad \text{for} \quad n = 0. \end{aligned} \tag{4.74}$$

These inequalities may be obtained by inspection with the realization that v is a pure imaginary quantity and that $\mathscr{H}'$ is accordingly an intrinsically positive, real quantity. If we now substitute Eq. (4.72′) into (4.73′) we find that*

$$\begin{aligned} -1 < \alpha(\text{HE}) &< 0 \qquad \text{for} \quad n > 0; \\ 0 < \alpha(\text{HE}) &< 1 \qquad \text{for} \quad n < 0; \\ \alpha(\text{HE}) &= 0 \qquad \text{for} \quad n = 0. \end{aligned} \tag{4.75}$$

Comparing these results with the field and power density expressions (4.58)–(4.60), we see that either the positive or negative circular components of **E** and **H** tend to dominate the observable intensity distribution and polarization of a mode. This is particularly true in the very common (as will be shown later) cases when $|\alpha|$ is of order unity. For a given value of n, the dominant circular component for EH modes is the weaker component for HE modes. Each mode has a degenerate counterpart, described by Eqs. (4.58) and (4.59) with n replaced by $-n$. This counterpart is characterized solely by a reversal of the relative importance of the positive and negative circularly polarized components. The physical meaning of the degeneracy is thus simply that a wave with a given phase velocity may be elliptically polarized with either sense of rotation (or by a superposition of waves with opposite senses of rotation).

We may now reasonably ask: What is the difference between EH and HE modes? We have found that either circularly polarized component of

* It is not immediately obvious that $|\alpha(HE)| < 1$. Let us assume that $|\alpha(HE)| \equiv N/D > 1$, where, from Eqs. (4.73′) and (4.72′), $N \equiv |n| [(1/u^2) - (1/v^2)]$ and

$$D \equiv (\Delta\varepsilon H'/\varepsilon_1) + \{(\Delta\varepsilon H'/\varepsilon_1)^2 + n^2[(1/u^2) - (1/v^2)][(1/u^2) - (\varepsilon_2/\varepsilon_1 v^2)]\}^{1/2};$$

then $[N - (\Delta\varepsilon H'/\varepsilon_1)]^2$ must be greater than $[D - (\Delta\varepsilon H'/\varepsilon_1)]^2$. But this implies that $-H' > |n|/v^2$ and, therefore, that $-K_{n-1}(q)/qK_n(q) > 0$, which is false.

the transverse field may be dominant for either type of mode, so that this does not provide a means of distinguishing between them. We may find one answer to the question, however, by considering the phase factors associated with the modes. The phase dependence $e^{in\phi - i\omega t}$ implies negative angular rotation about the z axis for $n > 0$ and positive angular rotation for $n < 0$. For HE waves, therefore, the dominant transverse circular components of **E** and **H** (E_- and H_- for $n > 0$; E_+ and H_+ for $n < 0$) have the same sense of rotation as the advancing wave. For EH waves, on the other hand, the stronger component rotates in a sense opposite that of the circulating wave. By a lengthy calculation that will not be reproduced here, we can show that a circulating $\mathrm{EH}_{n,m}$ or $\mathrm{HE}_{n,m}$ mode, as described by Eqs. (4.58) and (4.59), has precisely $-n\hbar$ units of field angular momentum about the z axis for each photon of energy $\omega\hbar$. [The calculation parallels that described by Eqs. (4.49)–(4.53). It is necessary to integrate both the energy density and momentum density from $\varrho = 0$ to $\varrho = \infty$ in this calculation, and to use the boundary conditions to obtain the final result.] In those special cases involving low-order modes on large guides and/or modes on guides with $\Delta\varepsilon/\varepsilon_1 \ll 1$, for which $h \approx k_1$ and the energy propagates almost axially, we will find that the dominant circular conponents of the fields are orders of magnitude larger than the longitudinal components and the weaker circular transverse components. $\mathrm{HE}_{n,m}$ modes then exhibit an orbital angular momentum (from the phase factor) proportional to $-(n-1)$ and a spin (polarization) angular momentum proportional to -1, and thus a net angular momentum proportional to $-n$. The intensity distribution is proportional to $J_{n-1}^2(u\varrho/a)$ inside the guide, where u may be coarsely approximated by a root of $J_{n-1}(u) = 0$. Under these same conditions, an $\mathrm{EH}_{n,m}$ mode exhibits an orbital angular momentum proportional to $-(n+1)$ with a spin proportional to $+1$ and an intensity distribution proportional to $J_{n+1}^2(u\varrho/a)$ inside the guide, where u may be coarsely approximated by a root of $J_{n+1}(u) = 0$. This suggests that we may think of HE modes as those for which the spin–orbit coupling is predominantly parallel, and thus additive, while, for EH modes, it is primarily nonparallel and thus subtractive. Because our language here is quite suggestive of physical processes, it is natural to want to pursue this line of thought further in the hope of uncovering a more fundamental physical understanding of the propagation. It might be conjectured, for example, that we could learn much more by quantizing the field. Unfortunately, this is not so, because our field wave functions are simply a classical description of a single mode field in which the z components of linear momentum and angular momentum have well-defined values. Because we have complete information on these two quantities, as well as the energy $\omega\hbar$ of the photons in the mode, we know nothing about the details of their motion. For engineering purposes, therefore, we may simply regard the parallel and nonparallel descriptions of the propagation given here as an aid in remembering which phase factors belong with the dominant circular components of the transverse fields.

5. Special Case I—EH and HE Waves with TE and TM Mode Characteristics

We mentioned earlier that the values of $|\alpha|$ in optical waveguides are most commonly of order unity, so that the modes of the dielectric waveguide rarely have the character of either transverse magnetic or transverse electric waves. To show this, we determine the conditions under which $|\alpha|$ is not of order unity. We are thus investigating the special cases for which the oppositely polarized components of the transverse fields [Eqs. (4.58) and (4.59)] are of the same order. From Eqs. (4.72′) and (4.73′), we see that α is of order unity whenever the second term under the radical exceeds or is of the same order as the first. To obtain values of α much greater or much less than unity, therefore, we must require that the first term be much larger than the second. This, in turn, is possible only if u and $v/i = q$ are much greater than unity, or if $n = 0$. For $1 \ll q$, we may approximate $(\mathscr{H}')^2$ by $1/q^2$. From Eqs. (4.72″) and (4.73″), we then obtain

$$\alpha(\mathrm{HE}) \approx \frac{-n^2[(1/u^2) + (1/q^2)][(1/u^2) + (\varepsilon_2/\varepsilon_1 q^2)]}{2n(\Delta\varepsilon/\varepsilon_1 q)[(1/u^2) + (\varepsilon_2/\varepsilon_1 q^2)]} = -\frac{n\varepsilon_1 R^2}{2\Delta\varepsilon u^2 q}. \tag{4.76}$$

From Eqs. (4.72′) and (4.73′), we obtain, for the same conditions,

$$\alpha(\mathrm{EH}) \approx \frac{2n[(1/u^2) + (1/q^2)](\Delta\varepsilon/\varepsilon_1 q)}{n^2[(1/u^2) + (1/q^2)][(1/u^2) + (\varepsilon_2/\varepsilon_1 q^2)]} = \frac{2\Delta\varepsilon q}{n\varepsilon_1[(q^2/u^2) + (\varepsilon_2/\varepsilon_1)]}. \tag{4.77}$$

From Eqs. (4.76) and (4.77), we then have

$$\gamma(\mathrm{HE}) \approx 1 + (n\varepsilon_1/\Delta\varepsilon q)[(q^2/u^2) + 1] \tag{4.76a}$$

and

$$\gamma(\mathrm{EH}) \approx -1 + (n\varepsilon_1/\Delta\varepsilon q)[(q^2/u^2) + (\varepsilon_2/\varepsilon_1)]. \tag{4.77a}$$

By differentiating these expressions with respect to u, we can determine their extreme values and the values of u and q (in terms of R) at the extrema. For HE modes, we find

$$\gamma(\mathrm{HE})_{\min} \approx 1 + 3^{3/2}(n\varepsilon_1/2\Delta\varepsilon R); \qquad u \approx (2/3)^{1/2}R; \qquad q \approx R/3^{1/2}. \tag{4.76b}$$

For EH modes, the expressions are more complex. In the case of fiber optical waveguides, however, the approximation $\Delta\varepsilon/\varepsilon_1 \ll 1$ can often be made. When this is so, we find

$$\gamma(\mathrm{EH})_{\max} \approx -1 + 3^{3/2}(n\varepsilon_2/2\Delta\varepsilon R); \quad u \approx (2/3)^{1/2}R; \quad q \approx R/3^{1/2}. \quad (4.77\mathrm{b})$$

For $\Delta\varepsilon \approx \varepsilon_2$, on the other hand, we find

$$\gamma(\mathrm{EH})_{\max} \approx -1 + (10n/3R); \qquad u \approx 3^{1/2}R/2; \qquad q \approx R/2. \quad (4.77\mathrm{b}')$$

Finally, for $\varepsilon_2 \ll \Delta\varepsilon$, the result is

$$\gamma(\mathrm{EH})_{\max} \approx -1 + n(\varepsilon_2/\Delta\varepsilon)^{1/2}/R; \qquad u \approx R[1 - (\varepsilon_2/\Delta\varepsilon)]^{1/2}; \\ q \approx R(\varepsilon_2/\Delta\varepsilon)^{1/2}. \quad (4.77\mathrm{b}'')$$

Referring back to the fields equations (4.58) and (4.59), we see that, as the second terms in these expressions for γ decrease, the two transverse, circularly polarized components of the fields have nearly equal amplitude. Also, as $\gamma(\mathrm{HE}) \to 1$, $\alpha(\mathrm{HE}) \to 0$ and $H_z \to 0$. Thus, HE waves behave like TM waves in this approximation. Conversely, as $\gamma(\mathrm{EH}) \to -1$, $\alpha(\mathrm{EH}) \to \infty$ and $E_z \to 0$. Thus, EH waves take on the characteristics of TE waves. It is also clear that, for $n = 0$, the relations developed here are exact; the corresponding waves have accordingly been labeled $\mathrm{TM}_{0,m}$ and $\mathrm{TE}_{0,m}$ modes in the literature.

For general n, however, the approximations made here are quite restrictive. They are valid only if the following two conditions are met:

$$n\varepsilon_1/\Delta\varepsilon \ll q; \qquad u,\ q \quad \text{of order} \quad R \gg 1. \quad (4.78)$$

Thus, in general, HE and EH waves do not have the character of TM and TE modes, respectively, but exhibit rather the dominance of one circular polarization over the other. This is particularly true in the important cases of propagation near cutoff ($q \ll u \approx R$; $h \approx k_2$) and far from cutoff ($u \ll q \approx R$; $h \approx k_1$). In the former case, most of the mode energy is in the surface wave outside the guide wall, while in the latter, it is essentially confined to the interior of the guide. For later use, we shall want to know the behavior of the fields in both these cases. We therefore examine them now in some detail.

6. Special Case II—The Fields Far from Cutoff

We define this case, involving nearly axial propagation of the energy inside the guide, by the equations $u \ll q \approx R \gg 1$; $h \approx k_1$. Because the eigenvalues u and q are solutions of the complicated transcendental equations (4.72), the inequalities above are not an explicit definition of the far-from-cutoff case. To obtain a useful approximation for u when $u \ll R$, we expand the left-hand sides of Eqs. (4.72a) and (4.72b) about the points U_+ and U_-, respectively. These points are defined as the roots of these equations when $R = \infty$. Thus, U_+ satisfies $J_{n+1}(U_+) = 0$, while U_- satisfies $J_{n-1}(U_-) = 0$. We approximate the right-hand sides of Eqs. (4.72a) and (4.72b) by neglecting terms of order higher than u/R. Thus, $q = R(1 - u^2/R^2)^{1/2}$ is set equal to R, and only the first terms in the asymptotic expansions of the Hankel functions are retained. The resulting approximate characteristic equations are

$$\mathscr{J}_\pm'(U_\pm)(u_\pm - U_\pm) + \tfrac{1}{2}\mathscr{J}''(U_\pm)(u_\pm - U_\pm)^2 = \mp\, \bar{\varepsilon}/\varepsilon_1 R,$$

where $\mathscr{J}_\pm'(U_\pm)$ and $\mathscr{J}_\pm''(U_\pm)$ are the first and second derivatives of $\mathscr{J}_\pm(u)$ evaluated at $U_\pm$, and $u_\pm$ is the approximate root of Eqs. (4.72a) and (4.72b), respectively, which we seek. By evaluating these derivatives and using the Bessel recursion relations, and the fact that $\mathscr{J}_\pm(U_\pm) = 0$ (by definition of $U_\pm$), we find that

$$\mathscr{J}_\pm'(U_\pm) = \pm\, 1/U_\pm; \qquad \mathscr{J}_\pm''(U_\pm) = -(2n \pm 3)/U_\pm^{\,2}.$$

Substituting these expressions into the above approximate characteristic equations and solving for the eigenvalues $u_\pm$, we obtain the following expressions, accurate to second order in R^{-1}:

$$u_\pm \approx U_\pm\{1 - (\bar{\varepsilon}/\varepsilon_1 R)(1 \mp [(2n \pm 3)\bar{\varepsilon}/8\varepsilon_1 R)]\}; \qquad q \approx R. \quad (4.79)$$

With these explicit expressions for u and q, we can now find expressions for $\beta_1 = u/a$, h, and α, accurate to first order in u/R. From the definitions of u and R, we have

$$u/R = \beta_1 a/(k_1^{\,2} - k_2^{\,2})^{1/2} a = (\beta_1/k_1)(\varepsilon_1/\Delta\varepsilon)^{1/2}.$$

Therefore,

$$\beta_1 = (2\Delta\varepsilon/\varepsilon_1)^{1/2} u k_1/R; \qquad h = (k_1^{\,2} - \beta_1^{\,2})^{1/2} \approx k_1[1 - (\Delta\varepsilon u^2/\varepsilon_1 R^2)]. \quad (4.80)$$

(We retain the second order term in u/R in the expression for h so as to be able to retain an important distinction between EH and HE modes in this approximation, as will be discussed later.) If we now substitute these approximate expressions into the general expressions for α given in Eqs.

(4.73′) and (4.73″), we obtain

$$\alpha_\pm = \pm 1 + (\Delta\varepsilon u_\pm{}^2/\varepsilon_1 nR). \tag{4.81}$$

If we now substitute these expressions for α into the field equations (4.58), we obtain, for EH modes with $n > 0$,

$$\begin{aligned} E_+ &\approx [2 + (\Delta\varepsilon u_+{}^2/n\varepsilon_1 R)]AJ_{n+1}(u_+\varrho/a)e^{i(n+1)\phi+i\omega t-ih_+z}, \\ E_- &\approx (\Delta\varepsilon u_+{}^2/n\varepsilon_1 R)AJ_{n-1}(u_+\varrho/a)e^{i(n-1)\phi+i\omega t-ih_+z}, \\ E_z &\approx -(2iu_+/R)(2\Delta\varepsilon/\varepsilon_1)^{1/2}AJ_n(u_+\varrho/a)e^{in\phi+i\omega t-ih_+z}, \end{aligned} \tag{4.82}$$

and

$$H_\pm \approx \pm\, i(\varepsilon_1/\mu)^{1/2}E_\pm; \qquad H_z \approx i(\varepsilon_1/\mu)^{1/2}E_z.$$

Similarly, for $\mathrm{EH}_{n,m}$ modes with $n > 0$, we obtain

$$\begin{aligned} E_+ &\approx (\Delta\varepsilon u_-{}^2/n\varepsilon_1 R)AJ_{n+1}(u_-\varrho/a)e^{i(n+1)\phi+i\omega t-ih_-z}, \\ E_- &\approx [-\,2 + (\Delta\varepsilon u_-{}^2/n\varepsilon_1 R)]AJ_{n-1}(u_-\varrho/a)e^{i(n-1)\phi+i\omega t-ih_-z}, \\ E_z &\approx -(2iu_-/R)(2\Delta\varepsilon/\varepsilon_1)^{1/2}AJ_n(u_-\varrho/a)e^{in\phi+i\omega t-ih_-z}, \end{aligned} \tag{4.83}$$

and

$$H_\pm \approx +i(\varepsilon_1/\mu)E_\pm; \qquad H_z \approx i(\varepsilon_1/\mu)E_z.$$

The fields for the degenerate EH and HE modes are obtained from these by replacing ϕ by $-\phi$ and interchanging E_+ and E_-. By substituting these expressions into Eq. (4.61), we find the following simple expressions for the observable mode patterns, accurate to order u/R:

$$S_z \approx 4(\varepsilon_1/\mu)^{1/2}AA^*[1 + (\Delta\varepsilon u_+{}^2/n\varepsilon_1 R)]J_{n+2}^2(u_+\varrho/a) \qquad \text{(EH modes)}, \tag{4.84}$$

and

$$S_z \approx 4(\varepsilon_1/\mu)^{1/2}AA^*[1 - (\Delta\varepsilon u_-{}^2/n\varepsilon_1 R)]J_{n-1}^2(u_-\varrho/a) \qquad \text{(HE modes)}. \tag{4.85}$$

These expressions manifest an important near-degeneracy between the $\mathrm{EH}_{n-1,m}$ and the $\mathrm{HE}_{n+1,m}$ modes. From Eq. (4.79), we see that $u_{+,n-1,m}$ and $u_{-,n+1,m}$ are equal to order $1/R$. Thus, it would appear at first that these two modes would be virtually indistinguishable. We notice, however, that if both modes were propagating at the same time, their phase difference would lead to a modulation of the power density in the direction of the axis of the guide. We could observe this modulation on a guide of fixed length by varying the frequency of the source that launches the waves. We shall examine these frequently observed effects when we discuss mode combinations. For the present, we observe, from Eq. (4.79), that the difference between the propagation constants of the $EH_{n-1,m}$ mode and the $HE_{n+1,m}$ mode is given by

$$h_+ - h_- \approx (k_1 \Delta\varepsilon/\varepsilon_1 R^2)(u_+^2 - u_-^2)$$
$$\approx k_1 n\bar{\varepsilon}^2(\Delta\varepsilon U^2/\varepsilon_1^3 R^4) + O(1/R^5), \tag{4.86}$$

where U satisfies $J_n(U) = 0$.

7. Special Case III—The Fields Near Cutoff

Thus far, we have not paid much attention to the fields in the surface wave outside the waveguide core. In the far-from-cutoff case just considered, substantially all of the power is carried inside the guide, and the surface wave is thus of little consequence. The opposite is true, however, as the mode approaches cutoff. We can observe this approach to cutoff by using a white-light source in combination with a monochromator and spatial filter to launch a particular mode on a fiber optical waveguide. As the wavelength of excitation is increased, the mode pattern, as observed with the aid of a microscope focused on the end of the guide, will appear to spread out until a wavelength is reached at which the pattern vanishes, more or less abruptly. (An accurate description of how the pattern vanishes would have to take into account the nature and radial extent of the coating glass of index n_2, and the way in which the fiber, which is made up of core and coating glasses, is bounded by other materials that act to support it physically.) When the guide is near cutoff, most of the power is in the coating. If the coating glass is less absorbing than the core, the power in a mode near cutoff would be less rapidly attenuated. Conversely, if the coating glass had an inverted atomic population, a mode near cutoff would be amplified. Both of these phenomena have been observed and will be discussed in Chapter 7. It is thus of some importance that we consider approximate expressions for the fields near cutoff for the EH and HE modes.

Let us return to the discussion immediately preceding Eq. (4.72a′) so as to recall the appropriate approximations for the near-cutoff case. The characteristic equation for EH modes is, accordingly,

$$\mathscr{J}_+ \approx -2n\bar{\varepsilon}/\varepsilon_1 q^2, \tag{4.87}$$

while, for HE modes, it is

$$\begin{aligned} \mathscr{J}_- &\approx \varepsilon_2/2(n-1)\bar{\varepsilon} && \text{for} \quad n > 1 \\ \mathscr{J}_- &\approx (\varepsilon_2/\bar{\varepsilon}) \log_e(2/\gamma' q) && \text{for} \quad n = 1. \end{aligned} \tag{4.88}$$

If we substitute these approximations into Eqs. (4.73), we find that, as q approaches zero,

$$\begin{aligned} \alpha(\mathrm{EH}) &\approx \varepsilon_1/\varepsilon_2 \\ \alpha(\mathrm{HE}) &\approx -1 + [\Delta\varepsilon q^2/2\bar{\varepsilon}n(n-1)] && \text{for} \quad n > 1 \\ \alpha(\mathrm{HE}) &\approx -1 && \text{for} \quad n = 1. \end{aligned} \tag{4.89}$$

To evaluate the fields outside the core, it is necessary to replace all factors that depend on q (including the factors $1 \pm \alpha$) by the first term of their small-argument approximations. When the resulting expressions are reduced to their simplest terms, we obtain the following expressions for the fields outside the guide at cutoff ($q \approx 0$).

$HE_{1,m}$ Modes

$$\begin{aligned} E_+ &\approx [1 - (\varepsilon_1/\varepsilon_2)] J_0(R)(a/\varrho)^2 e^{2i\phi + i\omega t - ik_2 z}, \\ E_- &\approx -[1 + (\varepsilon_1/\varepsilon_2)] J_0(R) e^{i\omega t - ik_2 z}, \\ E_z &\approx H_z \approx 0; \qquad H_\pm \approx \pm(ik_2/\omega\mu)E_\pm, \end{aligned} \tag{4.90}$$

where R satisfies $J_1(R) = 0$. It is apparent that the fields do not, in fact, suddenly vanish at the cutoff wavelength. What actually happens in a given experiment depends very little on the mode, in fact, but depends almost entirely on how the radiation is being coupled into the guide and on how the guide is supported and bounded. In fiber optics guides, it is common practice to use a coating that is 50–100 wavelengths thick, and to coat this coating with a highly absorbing material. In this case, as is clear from Eqs. (4.90), the intensity vanishes because the surface wave is being attenuated in the absorbing coating. If the fiber core was indeed embedded in an infinite dielectric medium, the $\mathrm{HE}_{1,m}$ modes near cutoff could clearly be launched very efficiently by an infinite plane wave that is circularly polarized, because the E_- component, above, extends indefinitely into the coating with no radial decay in its amplitude. This behavior is peculiar to the $\mathrm{HE}_{1,m}$ modes. All other modes have fields whose amplitudes decrease with increasing radius, as indicated below.

$HE_{n,m}$ Modes. For the general $\mathrm{HE}_{n,m}$ mode with $n \neq 1$, the fields outside the core may be approximated, when $q \approx 0$ and $u \approx R$, by

$$\begin{aligned}
E_+ &\approx [1 - (\varepsilon_1/\varepsilon_2)]J_{n-1}(R)(a/\varrho)^{n+1}e^{i(n+1)\phi+i\omega t-ik_2z}, \\
E_- &\approx -[1 + (\varepsilon_1/\varepsilon_2)]J_{n-1}(R)(a/\varrho)^{n-1}e^{i(n-1)\phi+i\omega t-ik_2z}, \\
E_z &\approx -i(\cot\theta_c)J_n(R)(a/\varrho)^n e^{in\phi+i\omega t-ik_2z}, \\
H_+ &\approx (ik_2/\omega\mu)\{[1 - (\varepsilon_1/\varepsilon_2)]J_{n-1}(R) + [2nJ_n(R)/k_2{}^2a^2]\} \\
&\quad \times (a/\varrho)^{n+1}e^{i(n+1)\phi+i\omega t-ik_2z}, \\
H_- &\approx (ik_2/\omega\mu)(1 + \varepsilon_1/\varepsilon_2)J_{n-1}(R)(a/\varrho)^{n-1}e^{i(n-1)\phi+i\omega t-ik_2z}, \\
H_z &\approx -(k_2/\omega\mu)(\tan\theta_c)J_n(R)(a/\varrho)^n e^{in\phi+i\omega t-ik_2z},
\end{aligned} \tag{4.91}$$

where θ_c is the critical angle.

In the same approximation, the $\mathrm{EH}_{n,m}$ mode fields outside the guide have the form

$$\begin{aligned}
E_+ &\approx (2\varepsilon_1/\varepsilon_2)J_{n+1}(R)(a/\varrho)^{n+1}e^{i(n+1)\phi+i\omega t-ik_2z}, \\
E_- &= H_- = E_z = H_z \approx 0, \\
H_+ &\approx (ik_2/\omega\mu)E_+.
\end{aligned}$$

Thus, in summary, the $\mathrm{EH}_{n,m}$ modes are circularly polarized outside the guide at cutoff. Inside, only H_- vanishes, except at $\varrho = a$. The $\mathrm{HE}_{n,m}$ modes with $n > 1$ are elliptically polarized outside; none of the six field components vanish. For the $\mathrm{HE}_{1,m}$ modes, the longitudinal components vanish, while the stronger circularly polarized transverse component does not decrease with radius. Inside the guide, only E_+ vanishes, all other components are nonzero except at $\varrho = a$. The HE waves become circularly polarized in the transverse plane inside the guide at cutoff, while the EH waves become circularly polarized outside.

C. PLANE-WAVE DECOMPOSITION OF THE DIELECTRIC WAVEGUIDE MODES

It is sometimes helpful to think of the modes of the dielectric waveguide in terms of their component plane waves. When the fields are expressed in this form, we may associate rays with the normals to the plane waves and view the propagation of the wave energy down the guide in terms of the familiar optical model of multiple total internal reflection. With each of the component plane waves, we will be able to associate a particular phase, relative to all the other plane waves and a particular polarization. The intensity distributions and polarizations of the mode patterns discussed earlier can then be viewed simply as the interference patterns arising from the superposition of these plane waves. As we shall see in the next chapter, this model of the propagation directly suggests, in familiar optical terms, the characteristics of the far-field radiation patterns of the modes, as well as optical techniques by which modes can be launched or detected. It also indicates, in an obvious way, how a conical lens can be used with a coated circular waveguide to effect the same kind of coupling into the waveguide modes as has been obtained with prisms for slab waveguides. This was described in some detail in Chapter 3.

Let us consider the well-known integral representation of the Bessel function (7)

$$2\pi J_n(Z) = \int_{-\pi}^{\pi} e^{iZ\cos w} e^{in(w-\pi/2)}\, dw. \tag{4.92}$$

Let $w = \phi' + \pi - \phi$, with ϕ' as the new variable of integration. Because the integrand in w or ϕ' is periodic with period 2π, all integrations over intervals of length 2π are equivalent, so that the end points need not be altered under this substitution. We then have

$$2\pi J_n(Z) \exp in\phi = \int_{-\pi}^{\pi} \exp[-iZ\cos(\phi' - \phi)] \exp[in(\phi' + \tfrac{1}{2}\pi)]\, d\phi'. \tag{4.93}$$

We now let $Z = \beta\varrho$. Then, because $x = \varrho \cos \phi$ and $y = \varrho \sin \phi$, we may write

$$\beta\varrho \cos(\phi' - \phi) = \beta x \cos(\phi') + \beta y \sin(\phi'). \tag{4.94}$$

Let us now substitute $\beta = k \sin \theta$, $h = k \cos \theta$. After multiplying both sides of Eq. (4.93) by $e^{i\omega t - ihz}$, we then obtain

$$\begin{aligned} &2\pi J_n(\beta\varrho) \exp(in\phi + i\omega t - ihz) \\ &\quad = \int_{-\pi}^{\pi} \exp[-ik(\sin\theta)(\cos\phi')x - ik(\sin\theta)(\sin\phi')y - ik(\cos\theta)z + i\omega t] \\ &\qquad \times \exp[in(\phi' + \tfrac{1}{2}\pi)]\, d\phi'. \end{aligned} \tag{4.95}$$

It is apparent that the integral on the right of Eq. (4.95) describes the sum of uniform plane waves whose wave normals form a cone of apex angle θ about the z axis. The magnitude of all these wave vectors is the same and equal to k. The wave whose wave normal makes an angle ϕ' with the x axis has a phase, relative to the wave in azimuth direction $\phi' = 0$, of $n\phi'$.

The mode fields inside the waveguide core [Eqs. (4.58)] describe a vector wave, rather than the simple scalar wave described by Eq. (4.95). Therefore, if we translate Eqs. (4.58) according to Eq. (4.95), we should be able to characterize the polarization of the component plane waves. In writing the integral representations of the various components of **E** and **H**, we will eliminate all reference to the parameters u/a and h by substituting $u/a = k_1 \sin\theta$ and $h = k_1 \cos\theta$, so that $h/\omega\mu$ becomes $(\varepsilon_1/\mu)^{1/2}\cos\theta$. Then, using Eq. (4.95), we obtain from Eqs. (4.58) the following integral representation of the mode fields inside the core of the guide

$$\begin{aligned} E_\pm &= \pm(1 \pm \alpha) \int_{-\pi}^{\pi} \exp[-i\mathbf{k}_1 \cdot \mathbf{r} + i\omega t + i(n \pm 1)(\phi' + \tfrac{1}{2}\pi)]\, d\phi', \\ E_z &= -\, 2i(\tan\theta) \int_{-\pi}^{\pi} \exp[-i\mathbf{k}_1 \cdot \mathbf{r} + i\omega t + in(\phi' + \tfrac{1}{2}\pi)]\, d\phi', \\ H_\pm &= i(\varepsilon_1/\mu)^{1/2}(\cos\theta)[(\sec^2\theta) \pm \alpha] \\ &\quad \times \int_{-\pi}^{\pi} \exp[-i\mathbf{k}_1 \cdot \mathbf{r} + i\omega t + i(n \pm 1)(\phi' + \tfrac{1}{2}\pi)]\, d\phi', \\ H_z &= 2(\varepsilon_1/\mu)^{1/2}(\sin\theta) \int_{-\pi}^{\pi} \exp[-i\mathbf{k}_1 \cdot \mathbf{r} + i\omega t + in(\phi' + \tfrac{1}{2}\pi)]\, d\phi'. \end{aligned} \tag{4.96}$$

We have neglected the common factor $1/2\pi$ in these expressions. Let us specialize now to the particular uniform plane wave that is propagating at the angles (θ, ϕ') as shown in Fig. 4-3. Let the projection of the wave vector k_1 on the x, y plane define the x' axis of a new rectangular coordinate system (x', y') generated by rotating the x axis through an angle ϕ'. We wish to find the rectangular (x', y', z) components of **E** and **H** for this plane wave, which has azimuth direction ϕ'. From Eq. (4.38), we see that we can write

$$\tfrac{1}{2}(E_{x'} \pm E_{y'}) = \exp(\mp\phi')E_\pm; \qquad \tfrac{1}{2}(H_{x'} \pm H_{y'}) = \exp(\mp\phi')H_\pm. \tag{4.97}$$

Thus, the x', y', and z components of the plane wave at azimuth direction ϕ' have the common phase factor $\exp[-i\mathbf{k}_1 \cdot \mathbf{r} + i\omega t + in(\phi' + \frac{1}{2}\pi] \equiv P$. From Eqs. (4.96) and (4.97), we thus obtain

$$\begin{aligned} &E_{y'} = 2\alpha P, \quad H_{x'} = -2(\varepsilon_1/\mu)\alpha(\cos\theta)P, \quad H_z = 2(\varepsilon_1/\mu)\alpha(\sin\theta)P \quad \text{(TE)} \\ &H_{y'} = 2i(\varepsilon_1/\mu)(\sec\theta)P, \quad E_{x'} = 2iP, \quad E_z = -2i(\tan\theta)P \quad \text{(TM)} \end{aligned} \tag{4.98}$$

As indicated, the upper three components of Eqs. (4.98) describe a uniform plane TE wave, e.g., one that is polarized perpendicular to the plane of incidence, which is the plane containing the wave normal to k_1 and the normal to the cylindrical interface separating the waveguide core from its surround. The lower three components describe a TM wave, e.g., one that is polarized in the plane of incidence. These fields are shown schematically in Fig. 4-3. The electric field of the TM wave has amplitude 2 sec θ, while that of the TE wave is 2α. It is thus the product $\alpha \cos \theta$, rather than α alone, which specifies the ratio of the TE to TM contributions to the modes of the circular dielectric guide. We note also that the TE and TM contributions

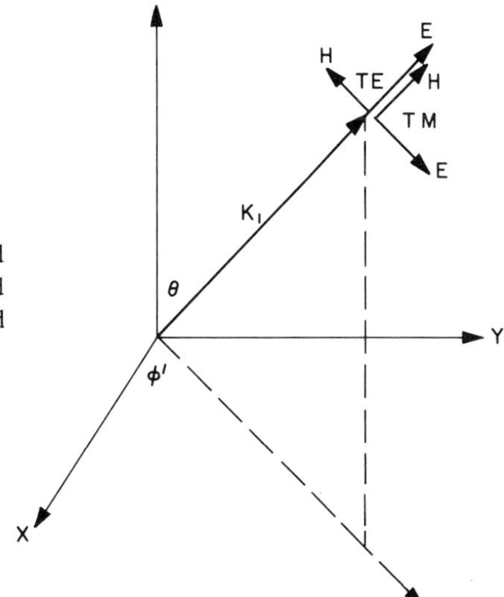

Fig. 4.3. Illustrating the electric and magnetic vectors of the component TE and TM plane waves that comprise a hybrid mode of a circular waveguide.

to the uniform plane wave are $\pi/2$ out of phase. The polarization is thus elliptical with major and minor axes in and perpendicular to the plane of incidence. When θ is small, so that the propagation is almost axial, we know that $\alpha \approx +1$ for EH modes with $0 < n$ and $\alpha \approx -1$ for HE modes with $0 < n$. The component uniform plane waves then become circularly polarized, as described in the last section. It is interesting that, in this representation, the angular momentum of the fields appears explicitly only in the rotation of the electric vectors of the uniform plane waves. The representation does not directly suggest anything analogous to the skew rays of geometric optics.

Equations (4.96) are quite suggestive of techniques by which the mode fields might be synthesized. Thus, from a two-path optical system whose input is a plane polarized plane wave, we can generate two circularly polarized conical waves of appropriate relative amplitude and phase by proper combinations of quarter-wave plate, amplitude and phase filters, and conical or spherical lenses. We will discuss such techniques at length in Chapter 5.

D. MODE COMBINATIONS AND THE CONTINUOUS MODE SPECTRUM

In the previous sections of this chapter, we have discussed at some length the properties of those vector eigenfunctions of the homogeneous Maxwell equations that represent circulating waves bound to the circular dielectric waveguide. At large distances along its axis from any sources or scatterers of energy, all possible observed disturbances on a guide must be representable as linear combinations of these waves. (This is, in fact, an accurate definition of large distances along a waveguide.) One of our interests in this section is in the more important superpositions of the bound modes, specifically those that arise from the degeneracy or near degeneracy of the $\mathrm{HE}_{n+1,m}$ and $\mathrm{EH}_{n-1,m}$ modes propagating nearly axially $(h \approx k_1)$ along a guide. These are important because they are launched relatively efficiently by plane polarized or circularly polarized monochromatic waves incident on the entrance face of a guide. They are also similar, in many respects, to the modes of the laser interferometer, as we shall indicate. These are not, of course, the only observable mode combinations. On a guide that can support many bound modes, the variety of possible observable intensity distributions is substantially unlimited.

In principle, each bound mode on a waveguide can serve as a carrier wave for communication, so that a multimode waveguide represents a multichannel communications system. Providing that the propagation distance z and the bandwidth $\Delta\omega$ of a signal modulating a carrier at frequency ω_0 are such that $(\Delta\omega)^2(d^2h/d\omega^2)_{\omega=\omega_0}z \ll 1$, the modulated wave will travel undistorted down the guide with the well-defined group velocity of the mode at the carrier frequency, i.e., $v_g = [(dh/d\omega)_{\omega=\omega_0}]^{-1}$. The signals imposed on different carriers (modes) can then be separated at the receiver by mode filtering techniques, as described in Chapter 6. In practice, however, intermode coupling caused by bends or scattering discontinuities in the guide frustrates this apparent multichannel potential, so that a waveguide designed for communications applications is generally capable of supporting only one mode. (The interested reader is referred to C. C. Johnson, "Field and Wave Electrodynamics," Chapter 4, McGraw-Hill, New York, 1965, for an extensive discussion on the transmission line characteristics of the modes of perfectly conducting waveguides.)

Our description of a general disturbance on a circular dielectric waveguide cannot be completed without the inclusion of the radiation field $\mathbf{E}_R$, $\mathbf{H}_R$) which characterizes those components of a general field that describe radiation away from the guide axis. Such components will usually be present at moderate distances along the guide from sources or scatterers of energy. The spatial spectrum (in h or β) of the radiation field is a continuous distribution of modes, unlike the discrete spectrum of the bound modes. (It is somewhat analogous to the plane-wave spectrum representing the modes of free space.) To obtain some understanding of the continuously distributed modes of the radiation field, we shall examine in this section the response of the circular waveguide to an obliquely incident plane wave. By studying all possible angles of incidence, we shall be able to generate the complete spatial spectrum of the radiation field. This spectrum, together with the discrete spectrum of the bound modes, provides the complete, orthogonal set of eigenfunctions needed to represent a general monochromatic disturbance in any plane transverse to the axis of the guide. This spectral representation of a general field, in turn, serves as a starting point for obtaining other field representations. Among these are the saddle-point representations exhibiting leaky wave fields and asymptotic expansions in terms of creeping waves (*5*), yielding the parameters of the diffracted rays of the geometric theory of diffraction (*8*). We shall indicate how these arise after we have found the spectral representations.

1. Linear Combinations of Degenerate and Nearly Degenerate Modes

When the anisotropy of a dielectric rod is small and the ratio of its diameter to the free-space wavelength is large, the waveguide is similar to the (unfolded) optical interferometer considered by Fox and Li (*9*) and the hollow waveguides studied by Marcatili and Schmeltzer (*4*). The latter authors recognized the similarity between the modes of the hollow dielectric and metallic waveguides that they analyzed, and those of optical interferometers. However, they did not relate their work to the extensive literature of the isotropic rod waveguide. Snitzer (*10*) has noted the similarities between the dielectric rod resonator and the unbounded laser interferometer.

There are, of course, significant differences. In one case, the guide walls determine the transverse dependence of the modes, while in the other, the end mirrors serve this purpose. It is instructive to determine how the polarization degeneracies of the unbounded optical interferometer would be removed by internal reflections at its side walls. In the approximation used by Marcatili and Schmeltzer in their study of optical waveguides and resonators, the distinction between the $HE_{n+1,m}$ and $EH_{n-1,m}$ modes far from cutoff was not retained in considering mode combinations. Consequently, these modes (labeled $EH_{|n|+1,m}$ and $EH_{-(|n|-1),m}$ by them) appeared to be degenerate. The fact that they are not, however, can lead to some interesting rotational effects, both in the intensity distribution and the polarization, as these modes propagate at optical frequencies down a long waveguide. We here examine the intensity distributions and polarizations of several mode combinations that commonly occur in isotropic or near-isotropic waveguides and resonators far from cutoff, and compare them with the Fox–Li and Marcatili–Schmeltzer modes.

As noted in Section B.6, $\alpha \to \pm 1$ in the isotropic guide when the mode is far from cutoff. The longitudinal components are then insignificant and the weaker of the circular transverse components ($+$ or $-$) can also be neglected. Under these conditions, there are four modes whose radial and azimuthal dependences are almost identical and whose propagation constants are nearly equal. From Eqs. (4.83) and (4.82), we see that these modes can be distinguished by their circularly polarized transverse electric vectors, as in the following equations, where each is identified by its historical label (to simplify the notation, let $p = |n| > 1$, to avoid the special cases $|n| = 0, 1$):

$$\begin{aligned}
&EH_{p-1,m}, \qquad && E_+ = J_p(U\varrho/a)e^{i(p\phi+\omega t-h_+z)}, \qquad && E_- \approx 0 \\
&EH_{-p+1,m}, \qquad && E_- = J_p(U\varrho/a)e^{i(-p\phi+\omega t-h_+z)}, \qquad && E_+ \approx 0 \\
&HE_{p+1,m}, \qquad && E_- = J_p(U\varrho/a)e^{i(p\phi+\omega t-h_-z)}, \qquad && E_+ \approx 0 \\
&HE_{-p-1,m}, \qquad && E_+ = J_p(U\varrho/a)e^{i(-p\phi+\omega t-h_-z)}, \qquad && E_- \approx 0
\end{aligned} \tag{4.99}$$

where U can be approximated by the mth root of $J_p(U) = 0$. Note that the two EH modes are degenerate and thus have the same propagation constant h_+. The propagation constant of the two degenerate HE modes is denoted by h_-, to indicate that it is not equal, though it is nearly so, to that of the EH modes. The difference between h_+ and h_- is given by Eq. (4.86).

In this approximation, each of these modes, taken alone, will have an intensity distribution of m concentric rings in any cross section of the guide and will be circularly polarized. In the Fox and Li interferometer with plane mirrors, the analogous modes are the so-called $TEM_{p,m-1}$ transverse modes. There are four such modes in the interferometer also, but they are all degenerate. The fact that they are not all degenerate in the dielectric rod and hollow dielectric waveguides leads to the major optically observable differences between the bounded and unbounded systems.

Consider how these four modes appear when simultaneously propagating in a waveguide at optical frequencies, where only the time-average power distributions and polarization are observable. Five physically significant combinations of the modes of Eqs. (4.99) are analyzed here. In each case, the x and y electric field components and the z component of the Poynting vector are found. These quantities are sufficient to specify the polarization and power distribution. Multiplying constants and the common factor $e^{i\omega t}$ have been omitted.

a. $\mathrm{EH}_{p-1,m} + \mathrm{EH}_{-p+1,m}$

$$\begin{aligned} E_x &= \tfrac{1}{2}(E_+ + E_-) \approx J_p(U\varrho/a)(\cos p\phi)e^{-ih_+z}, \\ E_y &= -\tfrac{1}{2}\, i(E_+ - E_-) \approx J_p(U\varrho/a)(\sin p\phi)e^{-ih_+z}, \\ S_z &= \tfrac{1}{4}\,\mathrm{Re}[i(E_+H_+{}^* - E_-H_-{}^*)] = J_p{}^2(U\varrho/a). \end{aligned} \tag{4.100}$$

The polarization of this combination of the degenerate EH modes is linear, but its direction varies with azimuth, rotating toward positive ϕ as ϕ is increased. This combination describes what is generally called the $\mathrm{EH}_{p-1,m}$ mode in dielectric-rod literature. Marcatili and Schmeltzer have presented field plots of several analogous modes on hollow dielectric guides and resonators. In their notation, these are the $\mathrm{EH}_{-p+1,m}$ modes. The intensity distributions are composed of m concentric rings.

b. $\mathrm{HE}_{p+1,m} + \mathrm{HE}_{-p-1,m}$

$$\begin{aligned} E_x &= J_p(U\varrho/a)(\cos p\phi)e^{-ih_-z}, \\ E_y &= -J_p(U\varrho/a)(\sin p\phi)e^{-ih_-z}, \\ S_z &= J_p{}^2(U\varrho/a). \end{aligned} \tag{4.101}$$

The polarization of this combination of degenerate HE modes is also linear, but its direction rotates toward negative ϕ as ϕ increases. This is the customary $\mathrm{HE}_{p+1,m}$ mode of the literature, called the $\mathrm{EH}_{p+1,m}$ mode by Marcatili and Schmeltzer, who present several appropriate field plots. Note that replacing p by $-p$ in either of the combinations (a) or (b) turns it into the other combination. There is, therefore, no need, as recognized by Marcatili and Schmeltzer, to use different labels (i.e., EH and HE) to describe the modes. The $\mathrm{HE}_{1,1}$ and $\mathrm{HE}_{2,1}$ modes of the literature are given by this combination with $p = 0$ and 1, respectively. In this approximation, the $HE_{1,1}$ mode is linearly polarized in a constant direction (i.e., plane polarized).

c. $EH_{p-1,m} + HE_{p+1,m}$

$$
\begin{aligned}
E_x &= J_p(U\varrho/a)\cos[\tfrac{1}{2}(h_- - h_+)z]\exp i[p\phi - \tfrac{1}{2}(h_- + h_+)z],\\
E_y &= J_p(U\varrho/a)\sin[\tfrac{1}{2}(h_- - h_+)z]\exp i[p\phi - \tfrac{1}{2}(h_- + h_+)z],\\
S_z &= J_p^2(U\varrho/a).
\end{aligned}
\tag{4.102}
$$

Equation (4.86) gives an expression for $h_- - h_+$ in terms of the physical parameters of the guide. The polarization of this combination is linear and in the same direction everywhere in any transverse plane, but this direction rotates with distance down the guide. The intensity distribution again has circular symmetry. The combination of the modes $EH_{-p+1,m}$ and $HE_{-p-1,m}$ yields the same polarization and intensity distribution. Such combinations are clearly possible in the Fox and Li interferometer, where, however, no rotation of the direction of polarization would occur. In the Marcatili and Schmeltzer guide, this combination should also be possible, with the rotational effect. The intensity distributions are again circularly symmetric.

d. $EH_{p-1,m} + HE_{-p-1,m}$

$$
\begin{aligned}
E_x &= J_p(U\varrho/a)\cos[p\phi - \tfrac{1}{2}(h_- - h_+)z]\,e^{-i(h_-+h_+)z/2},\\
E_y &= E_x e^{-i\pi/2},\\
S_z &= J_p^2(U\varrho/a)\cos^2[p\phi - \tfrac{1}{2}(h_- - h_+)z].
\end{aligned}
\tag{4.103}
$$

For this combination (also $EH_{-p+1,m} + HE_{p+1,m}$), there are again m radial maxima, but the circular symmetry of the elementary modes has been lost. On each ring of maximum intensity, there are now $2p$ azimuthal nodes. The position of these nodes rotates with distance down the guide. The wave is circularly polarized. This same distribution and polarization can occur in hollow dielectric guides. A similar distribution is also possible in unbounded interferometers. In that case, however, there is no rotation of the pattern, since the modes are fourfold degenerate.

e. $EH_{p-1,m} + HE_{-p-1,m} + EH_{-p+1,m} + HE_{p+1,m}$

$$
\begin{aligned}
E_x &\approx J_p(U\varrho/a)(\cos p\phi)\cos[\tfrac{1}{2}(h_- - h_+)z]e^{-i(h_-+h_+)z/2},\\
E_y &\approx J_p(U\varrho/a)(\sin p\phi)\sin[\tfrac{1}{2}(h_- - h_+)z]e^{-i(h_-+h_+)z/2},\\
S_z &\approx J_p^2(U\varrho/a)\{(\cos^2 p\phi)\cos^2[\tfrac{1}{2}(h_- - h_+)z]\\
&\quad + (\sin^2 p\phi)\sin^2[\tfrac{1}{2}(h_- - h_+)z]\}.
\end{aligned}
\tag{4.104}
$$

The intensity distribution of this combination changes continuously with distance down the guide. In the plane $z = 2p\pi/(h_- - h_+)$, there are $2p$ azimuthal nodes and maxima, and the polarization is linear and constant in direction ($E_y = 0$). In the planes $z = (2p + 1)\pi/2(h_- - h_+)$, the intensity has no azimuthal dependence, and the polarization is as in combinations (a) and (b). In the planes $z = (2p + 1)\pi/(h_- - h_+)$, $2p$ nodes again occur, but at the azimuth positions where maxima occurred in the planes $z = 2p\pi/(h_- - h_+)$. The polarization is again linear and constant in direction, but now $E_x = 0$. Generally, the intensity has $2p$ azimuthal maxima and minima, the angular positions of the minima and the direction of the linearly polarized transverse field appearing to rotate with distance down the guide. Thus, a phenomenon like rotation of the plane of polarization occurs.

Such mode combinations have been described qualitatively by Snitzer and Osterberg (*11*, p. 575) who observed them experimentally in fiber optical waveguides excited by plane polarized waves. Marcatili and Schmeltzer have also recognized the possibility of obtaining such mode combinations in hollow guides and have used sketches of the fields to describe them. They did not, however, note any of the rather complex rotatory effects arising from the slight difference in the propagation constants of the modes. In the laser interferometer, on the other hand, these four modes are degenerate, and hence no rotatory effects occur. The simple distribution in the plane $z = 0$ is preserved down the guide, and the polarization is constant in direction.

All of the five mode combinations described here can exist in the laser interferometer with plane circular mirrors, since they are all forms of the degenerate $\text{TEM}_{p,m-1}$ transverse modes. The description of this mode in terms of the principal circular component of the field and the attendant exponential field dependence facilitates explanation of the various intensity distributions and polarizations which can be observed. Examples of these mode combinations are illustrated schematically in Figs. 6-1 and 6-2 of Chapter 6.

2. The Continuous Mode Spectrum

Let us consider now a monochromatic plane wave incident obliquely on an infinitely long dielectric cylinder of radius a and index of refraction n_1 from the embedding medium of index n_2. We wish to determine the fields inside and outside the cylinder that are induced by this incident field. In analogy with the familiar analyses of the reflection and transmission of plane waves at plane interfaces, we shall refer to the induced interior and exterior fields as the transmitted (T) and reflected (R) fields, respectively, and use

(I) for the incident fields. As in Fig. 4-3, let the $\mathbf{k}_2$ for the incident wave lie in the z, x' plane at an angle θ with the z axis. Its projection on the x, y plane makes an angle ϕ' with the x axis. The axial (z) component of $\mathbf{k}_2$ is $h = k_2 \cos\theta$. Its transverse component is $\beta_2 = k_2 \sin\theta$. We consider first an incident TM plane wave whose electric vector lies in the x', z plane; the magnetic field is perpendicular to this plane, as illustrated in Fig. 4-3. The incident field is thus described by

$$\begin{aligned} E_z\,(\mathrm{I}) &= -E_0(\sin\theta)\exp(i\omega t - i\mathbf{k}_2\cdot\mathbf{r}), \\ E_{x'}\,(\mathrm{I}) &= E_0(\cos\theta)\exp(i\omega t - i\mathbf{k}_2\cdot\mathbf{r}), \\ H_{y'}\,(\mathrm{I}) &= (\varepsilon_2/\mu)^{1/2}E_0\exp(i\omega t - i\mathbf{k}_2\cdot\mathbf{r}). \end{aligned} \tag{4.105}$$

where E_0 is an arbitrary amplitude coefficient. We now use the definitions of h and β_2 to write

$$\begin{aligned} \mathbf{k}_2\cdot\mathbf{r} &= k_2(\sin\theta)(\cos\phi')x + k_2(\sin\theta)(\sin\phi')y + k_2(\cos\theta)z \\ &= \beta_2\varrho[\cos(\phi' - \phi)] + hz. \end{aligned} \tag{4.106}$$

Referring now to Eq. (4.93), we recognize that the function $i^{-n}J_n(Z)e^{in\phi}$ is just the amplitude of the nth component of the Fourier series representation of the function $\exp[-iZ\cos(\phi' - \phi)]$. We may thus write

$$\exp[-iZ\cos(\phi' - \phi)] = \sum_{-\infty}^{\infty} i^{-n}J_n(Z)\exp in(\phi - \phi'). \tag{4.107}$$

Equation (4.107) permits us to describe the incident plane wave [Eq. (4.105)] as the superposition of an infinite number of conical waves, one for each integer value of n, all incident at the same angle θ with the cylinder axis. Accordingly, we obtain

$$E_z(\mathrm{I}) = -E_0(\sin\theta)\sum_{-\infty}^{\infty} i^{-n}J_n(\beta_2\varrho)e_n, \tag{4.108}$$

where $e_n = \exp[in(\phi - \phi') + i\omega t - ihz]$. Then, from Eqs. (4.8) or (4.13) we have

$$E_\phi(\mathrm{I}) = (-ih/\beta_2{}^2\varrho)\,\partial E_z(\mathrm{I})/\partial\phi = iE_0(\cos\theta)\sum_{-\infty}^{\infty} i^{-n}e_n nJ_n(\beta_2\varrho)/\beta_2\varrho \tag{4.109}$$

$$H_\phi(\mathrm{I}) = (-i\omega\varepsilon_2/\beta_2{}^2)\,\partial E_z(\mathrm{I})/\partial\varrho = i(\varepsilon_2/\mu)^{1/2}E_0\sum_{-\infty}^{\infty} i^{-n}e_n J_n{}'(\beta_2\varrho). \tag{4.110}$$

The radial components of $\mathbf{E}$ and $\mathbf{H}$ for the incident plane wave can be similarly derived from $E_z(\mathrm{I})$ with the aid of Eqs. (4.8) and (4.13), if desired. We will not need them here. Now, because the conical waves are elements of a Fourier series, they are obviously linearly independent. Similarly, the response of the cylinder to the incident conical wave of order n is independent of its response to the incident conical waves of order $m \neq n$. We may thus treat each incident conical wave of order n separately and then form the final result by superposition. Let us, therefore, simplify our problem by considering only the incident conical wave of order n. Its tangential field components are given by

$$\begin{aligned} E_z(\mathrm{I}) &= -E_0(\sin\theta)i^{-n}e_n J_n(\beta_2\varrho); \\ E_\phi(\mathrm{I}) &= iE_0(\cos\theta)i^{-n}e_n nJ_n(\beta_2\varrho)/\beta_2\varrho; \\ H_z = 0; \qquad H_\phi(\mathrm{I}) &= i(\varepsilon_2/\mu)^{1/2}E_0 i^{-n}e_n J_n'(\beta_2\varrho). \end{aligned} \tag{4.111}$$

Now, because conical waves are the natural wave functions of the circular cylinder, our general equations (4.52) are ideally suited to the description of the transmitted (interior) and reflected (exterior) fields. We have only to determine the appropriate amplitude coefficients A_{n1}, A_{n2}, α_{n1}, and α_{n2} in terms of the amplitude of the incident conical wave. Inside the guide, the appropriate Bessel functions are again those of the first kind, $J_n(\beta_1\varrho)$, where $k_1^2 = \beta_1^2 + h^2$. Outside the guide, we require waves propagating away from the cylinder. At large radial distances, these should be of the form $\exp[-i\beta_2\varrho]/(\beta_2\varrho)^{1/2}$, because of our assumed $\exp(+i\omega t)$ time dependence. The appropriate Bessel functions are thus the Hankel functions of the second kind, $H_n^{(2)}(\beta_2\varrho)$. We will drop the superscript (2) in the equations for convenience. Then, from Eqs. (4.53) and (4.54), we obtain the required tangential components of the transmitted and reflected fields

$$\begin{aligned} E_z(\mathrm{T}) &= (2\beta_1/ih)e_n A_{n1}J_n(\beta_1\varrho), \\ H_z(\mathrm{T}) &= (2\beta_1/\omega\mu)e_n A_{n1}\alpha_{n1}J_n(\beta_1\varrho), \\ E_\phi(\mathrm{T}) &= 2ie_n[A_{n1}\alpha_{n1}J_n'(\beta_1\varrho) - A_{n1}nJ_n(\beta_1\varrho)/\beta_1\varrho], \\ H_\phi(\mathrm{T}) &= (2he_n/\omega\mu)\{[A_{n1}\alpha_{n1}nJ_n(\beta_1\varrho)/\beta_1\varrho] - (k_1^2h^2)A_{n1}J_n'(\beta_1\varrho)]\}. \end{aligned} \tag{4.112}$$

where we have used the Bessel relations (4.64) and (4.67) to simplify Eqs. (4.54). Similarly, for the reflected field outside the guide, we obtain from Eqs. (4.52) and (4.54)

$$\begin{aligned} E_z(\mathrm{R}) &= (2\beta_2/ih)e_n A_{n2}H_n(\beta_2\varrho), \\ H_z(\mathrm{R}) &= (2\beta_2/\omega\mu)e_n A_{n2}\alpha_{n2}H_n(\beta_2\varrho), \\ E_\phi(\mathrm{R}) &= 2ie_n\{A_{n2}\alpha_{n2}H_n'(\beta_2\varrho) - [A_{n2}nH_n(\beta_2\varrho)/\beta_2\varrho]\}, \\ H_\phi(\mathrm{R}) &= (2he_n/\omega\mu)\{[A_{n2}\alpha_{n2}nH_n(\beta_2\varrho)/\beta_2\varrho] - (k_2^2/h^2)A_{n2}H_n'(\beta_2\varrho)]\}. \end{aligned} \tag{4.113}$$

At the boundary, we must have continuity of the tangential components, so that, for example, $E_z(\mathrm{I}) + E_z(\mathrm{R}) = E_z(\mathrm{T})$ at $\varrho = a$. We thus obtain from Eqs. (4.111)–(4.113) four inhomogeneous linear equations in the four unknown coefficients A_{n1}, A_{n2}, α_{n1}, and α_{n2}. Solving these, we obtain, with $u = \beta_1 a$ and $v = \beta_2 a$,

$$\begin{aligned} A_{n1} &= \frac{i^{-n}E_0 \cos\theta}{2i(\alpha_\mathrm{H} - \alpha_\mathrm{E})} \frac{J_n'(v) - H_n'(v)}{uJ_n(u)[(\varepsilon_1/\varepsilon_2 u^2) - (1/v^2)]}, \\ \alpha_{n1} &= \alpha_\mathrm{E}, \\ 1 - \frac{\alpha_{n1}}{\alpha_{n2}} &= \frac{n(\alpha_\mathrm{H} - \alpha_\mathrm{E})[(\varepsilon_1/\varepsilon_2 u^2) - (1/v^2)]}{[J_n'(v)/vJ_n(v)] - [H_n'(v)/vH_n(v)]}, \\ A_{n2}\alpha_{n2} &= \frac{A_{n1}\alpha_{n1}uJ_n(u)}{vH_n(v)}, \end{aligned} \tag{4.114}$$

where α_H is given by Eqs. (4.65) or (4.66) and α_E is given by the first expression in Eqs. (4.68) or (4.63). We now note the presence of the resonance denominator, $\alpha_\mathrm{H} - \alpha_\mathrm{E}$, in the expression for the amplitude, A_{n1}, of the field inside the cylinder. (Recall that the characteristic equation of the waveguide modes may be written $\alpha_\mathrm{H} - \alpha_\mathrm{E} = 0$.) We need not be concerned about an infinite field, of course, because the characteristic equation has no real roots for $h < k_2$, and we cannot have uniform plane waves with $h = k_2 \cos\theta > k_2$. The characteristic equation does have an infinite number of complex roots, however, yielding leaky waves, as discussed in Chapter 2. One or more of these, for certain values of n, may have propagation constants whose real parts are equal to $h = k_2 \cos\theta$. If, in addition, their imaginary parts are relatively small, then the corresponding amplitude coefficients A_{n1} will be large. This can occur, for example, with leaky modes near the critical angle inside a guide driven by a plane or conical wave near grazing incidence.

A similar analysis can be carried out for a TE plane wave incident at the spherical angles (θ, ϕ) as illustrated also in Fig. 4-3. For this case, we have

$$\begin{aligned} H_z(I) &= H_0(\sin\theta)\exp\{i\omega t - i\beta_2\varrho[\cos(\phi - \phi')] - ihz\} \\ &= H_0(\sin\theta)\sum_{-\infty}^{\infty} i^{-n}J_n(\beta_2\varrho)e_n, \\ E_\phi(I) &= (i\omega\mu/\beta_2^2)\,\partial H_z(I)/\partial\varrho = i(\mu/\varepsilon_2)^{1/2}H_0\sum_{-\infty}^{\infty} i^{-n}J_n'(\beta_2\varrho)e_n, \\ H_\phi(I) &= (-ih/\beta_2^2\varrho)\,\partial H_z(I)/\partial\phi = H_0(\cos\theta)\sum_{-\infty}^{\infty} i^{-n}e_n nJ_n(\beta_2\varrho)/\beta_2\varrho. \end{aligned} \tag{4.115}$$

We may again consider each conical wave of order n separately and form the final result by superposition. The response of the cylinder to a particular conical TE wave is again described by the general equations (4.52), so that Eqs. (4.112) and (4.113) once more yield the appropriate expressions for the tangential components of the interior (transmitted) and exterior (reflected) fields. By equating these components inside and outside at $\varrho = a$, we find the unknown coefficients

$$\begin{aligned} A_{n1} &= \frac{i^{-n}(\mu/\varepsilon_2)^{1/2}H_0}{2(\alpha_{\mathrm{H}} - \alpha_{\mathrm{E}})} \; \frac{J_n'(v) - H_n'(v)}{uJ_n(u)\{[J_n'(u)/uJ_n(u)] - [H_n'(v)/vH_n(v)]\}}, \\ \alpha_{n1} &= \alpha_{\mathrm{H}}, \\ \alpha_{n1} - \alpha_{n2} &= (\alpha_{\mathrm{H}} - \alpha_{\mathrm{E}}) \frac{[J_n'(u)/uJ_n(u)] - [H_n'(v)/vH_n(v)]}{[J_n'(v)/vJ_n(v)] - [H_n'(v)/vH_n(v)]}, \\ A_{n2} &= \frac{A_{n1}uJ_n(u)}{vH_n(v)}. \end{aligned} \tag{4.116}$$

Equations (4.116) give the amplitude coefficients to be inserted in Eqs. (4.52) to describe the response of the system to the incident conical TE wave of order n. By summing Eqs. (4.52) over n, we then have the response of the cylinder to the incident TE plane wave described by Eqs. (4.115). We again note the resonance denominator, $\alpha_{\mathrm{H}} - \alpha_{\mathrm{E}}$. We may, of course, combine Eqs. (4.114) and (4.116) to describe the cylinder's response to conical waves of any desired elliptical polarization, or, by summation over n, its response to a plane wave of the same elliptical polarization.

Let us now look at the range of possible values of $h = k_2 \cos\theta$ for the incident plane or conical waves. Clearly, we may have $0 < h < k_2$, corresponding to all uniform plane waves with real angles of incidence. We may also have purely imaginary values of h in the range $-i\infty < h < i0$. These would correspond to nonuniform evanescent plane waves propagating in directions ϕ', $\theta = 0$, perpendicular to the cylinder axis. Such waves would arise, for example, in a case where the dielectric rod of index n_1 and its surround of index n_2 occupied the half-space $z > 0$, while the space $z < 0$ was filled with a medium with index $n_3 > n_2$. A plane wave incident from medium 3 on the plane $z = 0$ at an angle greater than the critical angle, $\sin^{-1}(n_2/n_3)$, would launch a wave which is evanescent in the z direction in medium 2. We would then need fields of the form in Eqs. (4.52), with imaginary values of h to describe the total fields in the space $z > 0$.

We may also consider nonuniform plane waves incident on the cylinder that are evanescent in radial directions toward the cylinder, with real values of $h > k_2$. These would be described by Eqs. (4.110) or (4.115) with imaginary values of β_2, so that the appropriate radial functions would be the modified Bessel functions of the first kind, $I_n[(h^2 - k_2{}^2)^{1/2}\varrho]$. These increase approximately exponentially with increasing radius, however, and thus represent incident waves that cannot be realized if the medium in which the cylinder is embedded is infinite. Therefore, if medium 2 fills all space not occupied by the infinitely long rod, there can be no continuous spectrum of modes with $h > k_2$. Only the discrete bound mode spectrum applies in this range. We have thus shown that a spectral representation of a general monochromatic field in any transverse plane will be in the form of a finite sum over a continuous distribution of modes of the form of Eqs. (4.52) with all real values of $h < k_2$, and all imaginary values of h. A further summation over all integral values of n for every value of h in the continuous spectrum is required. As shown by Collin (*12*), the modes of the continuous spectrum, as well as those of the discrete spectrum are orthogonal to one another in the power sense.

Let us now suppose that we have a dielectric waveguide of index n_1 coated with a film of index $n_2 < n_1$, and that this assembly is embedded in an infinite medium whose index of refraction matches that of the rod. For this three-medium system, it is clearly possible to have waves of the form $I_n[(h^2 - k_2{}^2)^{1/2}\varrho)e^{i\omega t - ihz}$ incident on the rod from the coating. Any value of h in the range $k_2 < h < k_1$ is allowed for every value of n. The complete spectrum of eigenfunctions is thus an entirely continuous one in h, with $0 < h < k_1$ and $-i\infty < h < -i0$; there are no true, bound modes. An infinite number of leaky modes exist, however. If the coating thickness is of the order of a wavelength or greater, some of these will satisfy the resonance condition $\alpha_{\mathrm{H}} - \alpha_{\mathrm{E}} \ll 1$, for appropriate values of h and n. (Note that the equation $\alpha_{\mathrm{H}} - \alpha_{\mathrm{E}} = 0$ is not the characteristic equation of the three-medium system.) We can efficiently couple radiation into these modes from conical waves of order n in the embedding medium 3. These conical waves would have apex angles $\theta_{n,m}$ satisfying the equation $\alpha_{\mathrm{H}} = \alpha_{\mathrm{E}}$. This is entirely analogous to the FTR prism coupling discussed in Chapter 3. Instead of using a prism, however, we use a cone of the proper apex angle around the coated rod. A uniform plane wave incident on the cone parallel to its axis becomes a conical wave inside the cone. If we can impose a phase modulation $\exp(in\phi')$ on the plane wave, then a conical wave of order n will be generated on refraction of the plane wave into the cone. Up to the time of this writing, however, no experimental technique yielding a good approximation to such a phase modulation has been demonstrated. For the important $\mathrm{HE}_{1,m}$ modes far from cutoff, on the other hand, no angular modulation is required. Accordingly, a left circularly polarized plane wave would, after refraction into the cone, efficiently drive the $\mathrm{HE}_{1,m}$

circulating leaky wave. A right circularly polarized plane wave would drive the $HE_{-1,m}$ mode, while a linearly polarized wave would drive both of these degenerate modes equally. A more versatile, though less efficient method of launching conical waves is discussed in Chapter 5.

The subject of the diffraction of plane waves by cylinders, with which we started this section, has an extensive literature. Some of its most interesting results, from an optical point of view, are those found from asymptotic evaluation of the infinite summations over n that we have derived. These lead to the diffraction coefficients needed to apply the geometric theory of diffraction (*8*). To obtain these asymptotic approximations, it is necessary to transform the infinite summations over n into integrals over a complex-valued variable ν that replaces n as the order of the Bessel functions. These integrals are evaluated, in part, in terms of the residues at the poles ν_m of the integrands. These occur, in turn, at the zeros of $\alpha_{\mathrm{H}} - \alpha_{\mathrm{E}}$ in our Eqs. (4.114) and (4.116), with n replaced by ν. A portion of the resulting asymptotic fields describes circulating leaky surface waves outside the rod and associated rays at the critical angle in the interior. These provide abundant correlations between geometric and wave concepts of the propagation of light in and around cylinders and other smooth surfaces. More extensive asymptotic representations of the fields that result when a plane wave is diffracted by a transparent sphere (13) and a transparent cylinder (*5*) provide further interesting correlations between geometric and wave concepts.

E. ANISOTROPIC CIRCULAR DIELECTRIC WAVEGUIDE

In applications of dielectric waveguides and resonators, it is sometimes necessary or desirable to use dielectric media that are anisotropic (*14*), either naturally so or as a consequence of imposed external fields. Examples include such systems as the ruby laser and KDP light modulators. As an initial study of the consequences of anisotropy, it is instructive to examine the modes of the anisotropic dielectric-rod waveguide. This system also offers an additional interesting opportunity to relate familiar physical optical concepts to waveguide mode analysis. It further provides us with a broader framework from which to examine the modes of the isotropic guide. However, we will find that the modes of the more general guide cannot be logically separated into two general classes analogous to the HE and EH modes of the isotropic guide. We will nevertheless retain the same nomenclature, labeling the general modes according to the wave into which they degenerate when the guide is isotropic.

1. Fields of the General Mode

Let us consider, then, a uniaxial dielectric crystal in the form of a circularly cylindrical rod with its optical axis oriented along the axis of the rod. It is embedded in an isotropic medium with index of refraction n_2. Let n_o and n_e be the ordinary and extraordinary indices of refraction of the rod, respectively. The constituitive relation between the electric displacement vector **D** and the electric field vector **E** inside the rod is then

$$\mathbf{D} = \varepsilon_0 n_o^2 \mathbf{E}_t + \varepsilon_0 n_e^2 \mathbf{E}_z, \qquad (4.117)$$

where $\mathbf{E}_t$ represents the vector component of **E** that is perpendicular to the axis of the guide and $\mathbf{E}_z$ is its vector component parallel to the axis. $\mathbf{E}_t$ may, of course, be further decomposed into any orthogonal pair of transverse components, in particular, the circularly polarized components used previously.

Just as in Section B, we must now solve Maxwell's equations in both media and match the solutions at the boundary $\varrho = a$. We will again find it useful to consider the circular component form of Maxwell's equations given by Eqs. (4.47) and (4.48). Appropriate solutions for the isotropic medium may again be found by inspection. They are identical with the general equations (4.52). Inside the rod, on the other hand, the first two of Eqs. (4.48) require the substitution $\varepsilon = \varepsilon_0 n_o^2$, while the third is appropriate only if $\varepsilon = \varepsilon_0 n_e^2$. Equations (4.47) and (4.48) may then be solved by inspection, with the aid of Eq. (4.43), if we take

$$E_z = AF_n(\beta_e \varrho); \qquad H_z = BF_n(\beta_o \varrho); \qquad F_n(\beta\varrho) = Z_n(\beta\varrho)e^{in\phi + i\omega t - ihz}, \qquad (4.118)$$

where Z_n is an appropriate Bessel function of order n. The choice of different transverse wave numbers β_e and β_o for the TM and TE contributions to the hybrid fields has the following physical basis. We showed, in the last section, that the mode fields inside the isotropic guide could be viewed as resulting from the superposition of uniform plane waves, all of which propagate at the same angle θ with the guide axis, where θ satisfies $h = k_0 n_1 \cos\theta$ and h is the longitudinal wave number. From familiar optical analyses of anisotropic media, on the other hand, we know that when a plane wave from an isotropic medium is refracted into an anisotropic medium, two waves are generated in the latter. Although all three waves propagate at different angles from the normal to the interface separating the media, their wave vectors all lie in the plane of incidence and have equal components (h, say) parallel to the interface. If we consider the dielectric rod with this in mind, we see that the evanescent surface wave in the isotropic medium outside the core may be viewed as an incident (nonuniform) plane wave with wave number h propagating parallel to the guide axis. Two waves (rays) are thus required in the anisotropic core to match the surface wave at the interface. Both must have the same longitudinal wave number h. The ordinary wave (ray), for which $\mathbf{E}$ is perpendicular to the plane of incidence (and thus a TE wave), will then have the transverse wave number β_o satisfying

$$h^2 + \beta_o^2 = \omega^2 \varepsilon_0 n_o^2 \mu_0 = k_0^2 n_o^2. \tag{4.119}$$

The extraordinary (TM) wave, one the other hand, must have transverse wave number β_e satisfying

$$(\beta_e^2/n_e^2) + (h^2/n_o^2) = \omega^2 \varepsilon_0 \mu_0 = k_0^2. \tag{4.120}$$

Equations (4.119) and (4.120) follow from conventional optical analyses of double refraction in optics (see, for example "Radiation and Optics," p. 429, by J. M. Stone, McGraw-Hill, New York, 1963) and are in no way peculiar to waveguide studies. We thus see that we must anticipate that the TE and TM contributions to the hybrid modes will travel at different angles inside the rod. The characteristic equation will determine the proper values of these angles, and thus also the magnitude of the wave vector $\mathbf{k}_e$ for the extraordinary (TM) rays. From Eqs. (4.119) and (4.120), we see also that β_e must satisfy

$$\beta_e^2 = (n_e^2/n_o^2)\beta_o^2. \tag{4.121}$$

If we now substitute Eq. (4.118) into Eqs. (4.47) and (4.48), we generate six algebraic equations that may be solved in the same manner as was used for the isotropic guide. Thus, analogous to Eqs. (4.50) and (4.51), we obtain

$$\begin{aligned} E_\pm &= \pm(1/2\beta_o^2)[ihA\beta_e F_{n\pm1}(\beta_e\varrho) \pm \omega\mu B\beta_o F_{n\pm1}(\beta_o\varrho)] \\ H_\pm &= -(1/2\beta_o^2)[\omega\varepsilon_0 n_o^2 \beta_e A F_{n\pm1}(\beta_e\varrho) \mp ih\beta_o B F_{n\pm1}(\beta_o\varrho)]. \end{aligned} \tag{4.122}$$

We note that Eqs. (4.118) and (4.122) are sufficiently general to satisfy Maxwell's equations both inside and outside the core of the guide, even if both media are anisotropic, provided only that their optical axes are parallel to each other and to the guide axis. We need only append the proper subscript 1 or 2 to the unspecified constants and functions to discriminate between the media. The appropriate Bessel functions inside the guide are clearly those of the first kind, so that $Z_n = J_n$ for $\varrho < a$. As discussed above, the appropriate functions outside the guide core are the Hankel functions $H_n^{(2)}(v)$, where v is imaginary. We now apply Eqs. (4.118) and (4.122) specifically to the anisotropic core of the guide. After a little straightforward algebra entirely analogous to that carried through for the isotropic guide, we arrive at the following expressions for the interior fields:

$$\begin{aligned} E_\pm &= \pm A_1[(n_e^2/n_o^2)J_{n\pm1}(w\varrho/a) \pm \alpha_1 J_{n\pm1}(u\varrho/a)]e_{n\pm1}, \\ H_\pm &= (ih/\omega\mu)A_1[(k_0^2 n_e^2/h^2)J_{n\pm1}(w\varrho/a) \pm \alpha_1 J_{n\pm1}(u\varrho/a)]e_{n\pm1}, \\ E_z &= -2i(\beta_e/h)A_1 J_n(w\varrho/a)e_n, \\ H_z &= 2(\beta_o/\omega\mu)A_1\alpha_1 J_n(u\varrho/a)e_n \end{aligned} \tag{4.123}$$

where $w = \beta_e a$, $u = \beta_o a$, and $e_p = e^{ip\phi + i\omega t - ihz}$ for $p = n$, $n-1$, and $n+1$, $\alpha_1 = (\omega\mu B_1/ihA_1)wJ_n(w)/uJ_n(u)$. If we now apply Eqs. (4.118) and (4.122) to the isotropic medium outside the core, where $\beta_e = \beta_o = \beta_2$, we obtain the fields

$$\begin{aligned} E_\pm &= \pm A_2(1 \pm \alpha_2)H_{n\pm1}(v\varrho/a)e_{n\pm1} \\ E_z &= -2(i\beta_2/h)A_2 H_n(v\varrho/a)e_n \\ H_\pm &= (ih/\omega\mu)A_2[(k_2^2/h^2) \pm \alpha_2]H_{n\pm1}(v\varrho/a)e_{n\pm1} \\ H_z &= 2(\beta_2/\omega\mu)\alpha_2 A_2 H_n(v\varrho/a)e_n, \end{aligned} \tag{4.124}$$

where $v = \beta_2 a$, $\alpha_2 = \omega\mu B_2/ihA_2$, and $k_2^2 = \omega^2\varepsilon_2\mu_0 = \beta_2^2 + h^2$.

If we now equate the expressions for the longitudinal components of **E** at $\varrho = a$, we find that

$$A_2 = A_1 wJ_n(w)/vH_n(v). \tag{4.125}$$

Equating now the expressions for the longitudinal components of **H**, we find

$$\alpha_1 = \alpha_2 wJ_n(w)/uJ_n(u). \tag{4.126}$$

If we now equate the azimuthal components of **E** and **H** at $\varrho = a$, we can obtain two different expressions for $\alpha_2 = \alpha$. The fact that these two expressions must be equal constitutes the characteristic equation. From the ϕ components of **E**, we have

$$\begin{aligned} \alpha &= [(n_e^2/n_o^2)(\mathscr{J}_{-,w} + \mathscr{J}_{+,w}) - (\mathscr{H}_- + \mathscr{H}_+)]/(\mathscr{J}_- - \mathscr{J}_+ - \mathscr{H}_- + \mathscr{H}_+) \\ &= [(n/u^2) - (n/v^2)]/(\mathscr{J}' - \mathscr{H}') \\ &= (\mathscr{J}_- + \mathscr{J}_+ - \mathscr{H}_- - \mathscr{H}_+)/(\mathscr{J}_- - \mathscr{J}_+ - \mathscr{H}_- + \mathscr{H}_+), \end{aligned} \tag{4.127}$$

where we are using the short-hand notation

$$\mathscr{J}_\pm = J_{n\pm1}(u)/uJ_n(u); \quad \mathscr{H}_\pm = H_{n\pm1}(v)/vH_n(v); \quad \mathscr{J}_{\pm,w} = J_{n\pm1}(w)/wJ_n(w)$$
$$\mathscr{J}' = J_n'(u)/uJ_n(n); \quad \mathscr{H}' = H_n'(v)/vH_n(v); \quad \mathscr{J}_w' = J_n'(w)/wJ_n(w). \tag{4.128}$$

To obtain the last two expressions for α in Eq. (4.127), we have substituted $1/u^2$ for $n_e^2/n_o^2w^2$ [Eq. (4.121)] and used the Bessel recursion relations

$$\mathscr{J}_- + \mathscr{J}_+ = 2n/u^2; \qquad \mathscr{J}_- - \mathscr{J}_+ = 2\mathscr{J}'. \tag{4.129}$$

Identical relations are obeyed by $\mathscr{H}_\pm$. If we now equate the ϕ components of **H** at $\varrho = a$, we obtain an independent expression for $\alpha = \alpha_2$, which is

$$\alpha = [k_0^2n_e^2(\mathscr{J}_{-,w} - \mathscr{J}_{+,w}) - k_0^2n_2^2(\mathscr{H}_- - \mathscr{H}_+)]/h^2(\mathscr{J}_- + \mathscr{J}_+ - \mathscr{H}_- - \mathscr{H}_+)$$
$$= (k_0^2n_e^2\mathscr{J}_w' - k_0^2n_2^2\mathscr{H}')/[nh^2/u^2) - (nh^2/v^2)]. \tag{4.130a}$$

If we now substitute $(k_0^2n_o^2 - \beta_o^2)/\beta_o^2a$ for h^2/u^2, $(k_0^2n_2^2 - \beta_2^2)/\beta_2^2a$ for h^2/v^2, and $k_0^2n_e^2/w^2$ for $k_0^2n_o^2/u^2$, we can reduce Eq. (4.130a) to

$$\alpha = (k_0^2n_e^2\mathscr{J}_w' - k_0^2n_2^2\mathscr{H}')/[(k_0^2n_e^2/w^2) - (k_0^2n_2^2/v^2)]$$
$$= [n_e^2(\mathscr{J}_{-,w} - \mathscr{J}_{+,w}) - n_2^2(\mathscr{H}_- - \mathscr{H}_+)]/[n_e^2(\mathscr{J}_{-,w} + \mathscr{J}_{+,w}) - n_2^2(\mathscr{H}_- + \mathscr{H}_+)] \tag{4.130b}$$

By equating the expressions for α in Eqs. (4.127) and (4.130), we can generate several possible forms of the characteristic equation that defines the proper values of u, v, and w, with n_2 and n_e as parameters. The eigenvalues u, q, and w must also satisfy

$$u^2 - v^2 = (\beta_o^2 + h^2)a^2 - (\beta_2^2 + h^2)a^2 = (2\pi a/\lambda_0)^2(n_o^2 - n_2^2)$$
$$= R^2 = u^2 + q^2 \tag{4.131}$$

and

$$w = n_eu/n_o. \tag{4.132}$$

A useful form of the characteristic equation is obtained by equating the second form of α in Eq. (4.127) with the first form of Eq. (4.130b):

$$\frac{J_n'(u)}{uJ_n(u)} = n\left(\frac{1}{u^2} + \frac{1}{q^2}\right)\frac{[n(n_e^2/w^2) + (n_2^2/q^2)]}{[n_e^2J_n'(w)/wJ_n(w)] + [n_2^2K_n'(q)/qK_n(q)]} - \frac{K_n'(q)}{qK_n(q)}. \tag{4.133}$$

The characteristic equation (4.133) has been written in terms of the real-valued quantities $q^2 = -v^2$ and $K_n'(q)/qK_n(q) = -H_n'(v)/vH_n(v)$ to facilitate its discussion. [Though it is customary to refer to Eq. (4.133) as the characteristic equation, it is more appropriate mathematically to think of the three simultaneous equations (4.131)–(4.133) taken together as the characteristic equation. This point of view will facilitate the understanding of our method of solution, to be discussed below.]

We now note that the physical parameters n_o^2 and R^2 [Eq. (4.131)] do not appear explicitly in Eq. (4.133). We may thus adopt the mathematical point of view that these parameters are dependent variables to be determined from the solutions u, q, and w of Eq. (4.133). We can then regard q and w in Eq. (4.133) as independent variables whose values, for any given values of n, n_e, and n_2, determine appropriate values of the dependent variable u. For any chosen values of n, n_e, n_2, w, and q, Eq. (4.133) is a comparatively simple transcendental equation in u having an infinite number of discrete solutions. Each such solution corresponds to a different physical guide whose remaining parameters n_0 and R are found from Eqs. (4.131) and (4.132). A geometric appreciation of the characteristic equation can then be obtained by examining the first octant of the three-dimensional q, w, u space illustrated in Fig. 4-4.

Let us assume now that n, n_e, and n_2 are given constants. If we then examine Eq. (4.133) in some particular plane $q = \text{const}$ (the range of q is $0 < q < \infty$), we can generate a set of curves in this plane that give u as a function of w, where w is allowed to take on all values in the range $0 < w < \infty$. In particular, we can look at the plane $q = 0$. If we use subscript c to indicate cutoff values, the corresponding curves u_c versus w_c in the plane $q = 0$ will satisfy Eq. (4.133) in the limit as $q \to 0$. (We will examine these limiting forms analytically later. Here, we will simply use the results to provide the desired graphical illustrations.) The case $n = 1$ is particularly simple, so we will begin with it. The analytical result states that $J_1(u_c) = 0$ for all values of w_c, and $J_1(w_c) = 0$ for all values of u_c, independent of the values of n_e and n_2. These conditions are sketched in Fig. 4-5 in the plane $q = 0$. The straight line $u_c = n_o w_c/n_e$ [Eq. (4.132)] locates the successive roots of these cutoff equations for a particular anisotropic guide with indices n_o and n_e. Because $q = 0$, we see from Eq. (4.131) that $u_c = R$ and $w_c = n_e R/n_o$, so that this line and the cutoff curves, taken together, give the cutoff values of all the modes on the corresponding physical guides. The line $w = u$, with unit slope, locates the cutoff values of the modes of order one of the isotropic guide. If we use these to label the roots, we then find that $u_c = w_c = 0$ gives the cutoff value for the $HE_{1,1}$ mode. The

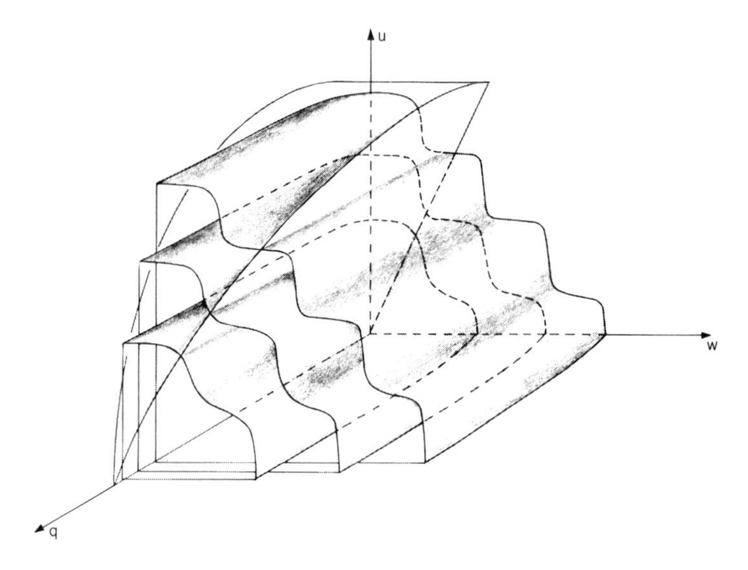

Fig. 4.4. Illustrating a graphical solution of the characteristic equations (4.131)–(4.133). For each mode of order $n = 2$ considered here, there is a corresponding corrugated surface in the first octant of (u, q, w) space described by Eq. (4.133). Each such surface cuts the circular cylinder $u^2 + q^2 = R^2$ [Eq. (4.131)] in a curve. The intersection of this curve with the plane $w = n_e u/n_o$, if one exists for the chosen values of the refractive indices, defines the eigenvalues u, q, and w of the mode. Here, we illustrate the first three mode surfaces, i.e., those for the $HE_{2,1}$, $EH_{2,1}$, and $HE_{2,2}$ modes. Each surface extends to $q = \infty$. Every point on each of these surfaces is a possible solution of the characteristic equations.

next root, in the isotropic guide, is a double root giving the cutoff values of both the $EH_{1,1}$ and $HE_{1,2}$ modes. The third root and all successive roots for the isotropic guide are again double roots. The third root gives the cutoff values for the $EH_{1,2}$ and $HE_{1,3}$ modes. To be consistent with the literature of the isotropic guide, therefore, we should label the roots of the cutoff equation for the anisotropic guide, in the order of their distance from the origin, as $HE_{1,1}$, $EH_{1,1}$, $HE_{1,2}$, $EH_{1,2}$, $HE_{1,3}, \ldots$. We have accordingly shown the HE cutoff curves in Fig. 4-5 as solid lines and the EH cutoff curves as dashed lines. The $HE_{1,1}$ cutoff curve is simply the point at the origin.

We may similarly generate from Eq. (4.133) a set of curves $u(w)$ in every plane $q =$ const. We thus see that the characteristic equation describes, for each chosen set of values of n, n_e, and n_2, an infinite set of surfaces $u(q, w)$ in the first octant of q, w, u space. There is one such surface for each mode. As q varies from 0 to ∞ on a particular mode surface, u and w vary continuously and monotonically over a small range of values. Thus,

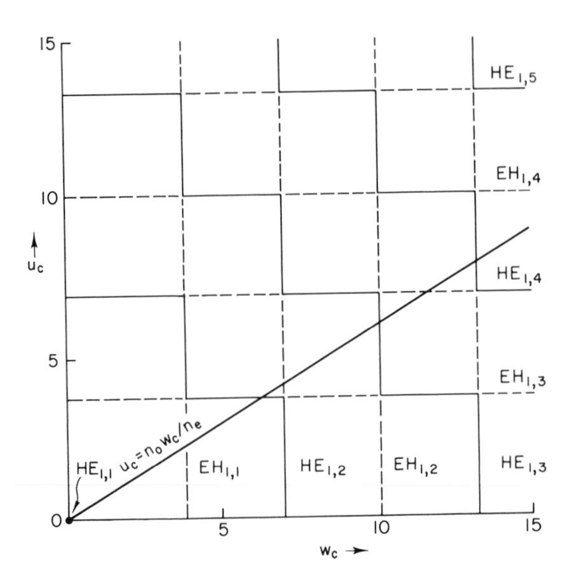

Fig. 4.5. Cutoff curves of the $EH_{1,m}$ (- - - - -) and $HE_{1,m}$ (——) modes The line $u_e = n_o w_e / n_e$ locates the cutoff parameters of a guide with indices n_o and n_e.

as we shall see presently when we study graphically their cross sections, these surfaces have relatively simple shapes. Given the additional physical parameters n_o and R, we may then imagine the simple space curve that is the locus of the intersection point in q, w, u space of the cylindrical surface $R^2 = u^2 + q^2$ with the plane surface $w = n_e u / n_o$ [Eqs. (4.131) and (4.132)]. This space curve intersects each of the first p mode surfaces at one and only one point, where p is an integer that increases with increasing R. If the mode surfaces are ordered according to their distance from the q axis, the successive intersection points of the space curve with the mode surfaces give the parameters of the $HE_{n,1}$, $EH_{n,1}$, $HE_{n,2}$, $EH_{n,2}$, $HE_{n,3}, \ldots$. A particular example of this geometric interpretation of Eqs. (4.131), (4.132), and (4.133) is sketched in Fig. 4-4. The first three mode surfaces for $n_e^2/n_2^2 = 2$ and $n = 2$ are shown. The surfaces looks substantially like corrugated sheets resting diagonally in a right angle wedge formed by the (u, q) and (q, w) planes. Each sheet expands slightly and moves a corresponding distance further from the q axis as q goes from 0 to ∞. Figure 4-4 also shows the space curve described by Eqs. (4.131) and (4.132) as it intersects these three mode surfaces, thus identifying the parameters of the corresponding modes of the chosen guide (R, n_o).

Let us now go back to Eq. (4.133) to better understand the general properties of these modes. We pointed out in Section B that for any mode on the isotropic guide, HE waves have more of the characteristics of TM than TE modes. In the surface wave outside the core, the major axes of their polarization ellipses were accordingly perpendicular to the circles of constant intensity. On the other hand, EH waves, were more like TE modes; the major axes of their polarization ellipses in the surface wave were parallel to the circles of constant intensity. Mathematically, these distinctions arose from the corresponding values of the polarization parameter $\gamma = (1 - \alpha)/(1 + \alpha)$ for the two types of modes. Thus, for $HE_{n,m}$ waves with $n > 0$, we found $0 > \alpha > -1$, while for $EH_{n,m}$ waves, we showed that $\alpha > 1$ for $n > 0$. We will now show that no such distinction can be made for the corresponding modes on the anisotropic guide.

We note, first of all, that the bracketted expression on the right-hand side of Eq. (4.133) is equal to α^{-1} for each and every mode on the anisotropic guide. We further know that for every arbitrarily assigned pair of values of w and q, we can find from Eq. (4.133) an infinite set of values of u, each describing one mode on one of an infinite set of physically realizable (in principle) waveguides, whose parameters R and n_o are determined from the triplets u, q, w with the aid of Eqs. (4.131) and (4.132). Let us now suppose that we have studied some particular isotropic guide of indices n_1 and n_2 and radius a at some particular wavelength λ_0. The guide may be assumed to support the $HE_{1,1}$ mode, and we know all the parameters $u = \beta_1 a$, q, α, and h for this mode. Of course, α is less than unity, as it must be for all HE waves with $0 < n$ on isotropic guides. Let us consider now the $HE_{1,1}$ mode on an anisotropic guide with parameters $n_e = n_1$, $w \equiv \beta_e a = \beta_1 a$, and q equal to the q on the isotropic guide. Because we have not specified R and n_o on the anisotropic guide, we are free to make these assignments of the values of n_e, w, n_2, and q. These, in turn, fix the value of α for the anisotropic guide to be that of the $HE_{1,1}$ mode on the isotropic guide. We can now insert these values into the right-hand side of Eq. (4.133) and solve for u. We thus determine a proper value of u for each and every mode of order one on the anisotropic guide. Each such value pertains to a different anisotropic guide, whose parameters are determined from the triplets u, q, w with the aid of Eqs. (4.131) and (4.132). (A particular choice of w and q for the $HE_{1,1}$ mode represents, in Fig. 4-4, a vertical line rising from the w, q plane and extending to $u = \infty$. It clearly cuts each mode surface once.) We have accordingly shown that for every choice of α for an $HE_{1,1}$ mode on an isotropic guide, there are an infinity of anisotropic guides that support an $EH_{1,m}$ mode with the same value of α, as well as an infinite set of anisotropic guides that support an $HE_{1,m}$ mode with this value of α. We could go through the same argument for any mode on an isotropic guide, with any of its allowed values of α. We conclude, therefore, that there is no essential difference between HE and EH modes on the general anisotropic guide, so that our only reason for using this labeling scheme is to retain continuity with the literature of the isotropic guide.

We conclude that we may not differentiate between EH and HE modes on anisotropic guides in general. All that can be said is that the general mode can have any relative combination of TE and TM components so that the polarization ellipses can have their major axes either radial or tangential. The only exception occurs when $n = 0$. Here, as in the isotropic guide, we may have pure $TE_{0,m}$ and $TM_{0,m}$ modes. Their analysis is entirely straightforward and need not be pursued here. We now wish to study the limiting cases of cutoff and far-from-cutoff propagation so as to be able to specify the range of values of u and w for particular modes.

2. Cutoff and Far-from-Cutoff Behavior of the Modes*

The polarization parameter $\gamma = (1 - \alpha)/(1 + \alpha)$ provides a useful form of the characteristic equation for examining these limits. From the third form for α in Eq. (4.127), we obtain

$$\gamma = -(\mathscr{J}_+ - \mathscr{H}_+)/(\mathscr{J}_- - \mathscr{H}_-), \tag{4.134}$$

whereas from the second form for α in Eq. (4.130b), we obtain

$$\gamma = [(n_e^2 \mathscr{J}_{+,w}/n_2^2) - \mathscr{H}_+]/[(n_e^2 \mathscr{J}_{-,w}/n_2^2) - \mathscr{H}_-]. \tag{4.135}$$

By equating expressions (4.134) and (4.135), we can obtain a relatively simple form of the characteristic equation equivalent to Eq. (4.133). If we substitute $-iq$ for v and the modified Bessel functions $K_n(q) = \frac{1}{2}\pi(-i)^{n+1} H_n^{(2)}(-iq)$ for the Hankel functions, we than have $\mathscr{H}_\pm = \mp\mathscr{K}_\pm \equiv K_{n\pm1}(q)/qK_n(q)$. With these substitutions, the characteristic equation is

$$(\mathscr{J}_+ + \mathscr{K}_+)[(n_e^2 \mathscr{J}_{-,w}/n_2^2) - \mathscr{K}_-] = -(\mathscr{J}_- - \mathscr{K}_-)[(n_e^2 \mathscr{J}_{+,w}/n_2^2) + \mathscr{K}_+] \tag{4.136}$$

For propagation near the critical angle ($u \approx R$, $q \approx 0$), we may use the small-argument approximations for the modified Bessel functions.

* These studies were originally done by one of the authors (J. J. Burke) in collaboration with Professor F. J. Rosenbaum of Washington Univ., St. Louis, Missouri.

Thus, let

$$\begin{aligned} \mathcal{K}_+ &\approx 2n/q^2 && \text{for} \quad n \geq 1, \\ \mathcal{K}_- &\approx 1/2(n-1) && \text{for} \quad n \geq 2, \\ \mathcal{K}_- &\approx \ln(2/\gamma' q) && \text{for} \quad n = 1, \end{aligned} \tag{4.137}$$

where γ' is Euler's constant. Let us now examine Eq. (4.136) for $n \geq 2$ in this approximation. After substituting Eqs. (4.137) into (4.136), we multiply through by $uwq^2 J_n(w) J_n(u)$ to obtain

$$\begin{aligned} &[q^2 J_{n+1}(u) + 2nJ_n(u)]\{\nu J_{n-1}(w) - [wJ_n(w)/2(n-1)]\} \\ &\quad = -\{J_{n-1}(u) - [uJ_n(u)/2(n-1)]\}[q^2 \nu J_{n+1}(w) + 2nwJ_n(w)], \end{aligned} \tag{4.138}$$

where we are temporarily using the short-hand notation $\nu = n_e^2/n_2^2$. We first note that the terms $q^2 J_{n+1}(u)$ and $q^2 J_{n+1}(w)$ will vanish for all values of u and w as q approaches zero. We may, therefore, ignore them. We then note that because $u = n_o w/n_e$, the terms $2nuJ_n(u)$ and $2nwJ_n(w)$ cannot vanish simultaneously unless the guide is isotropic. Thus, the roots of $J_n(x) = 0$ cannot satisfy the cutoff unless $n_o = n_e$.* The cutoff equation is therefore

$$[J_{n-1}(u)/uJ_n(u)] + [n_e^2 J_{n-1}(w)/n_2^2 wJ_n(w)] = 1/(n-1). \tag{4.139}$$

Thus, whereas we had two cutoff equations for the isotropic guide [i.e., $J_n(u) = 0$ for EH modes and $J_{n-1}(u)/uJ_n(u) = n_2^2/(n-1)(n_1^2 + n_2^2)$ for HE modes], we need only one equation to characterize cutoff for the modes of the anisotropic guide. Accordingly, the mathematical distinctions between HE and EH modes have vanished, even in this special case.

To obtain the cutoff equation for $n = 1$, we substitute the approximations (4.137) into (4.136) and proceed as before to the analog of Eq. (4.139), with $1/2(n-1)$ replaced by $\ln(2/\gamma' q)$. As $q \to 0$, this equation can be satisfied for all values of u by $J_1(w) = 0$ and for all values of w by $J_1(u) = 0$. We thus get the cutoff plots shown in Fig. 4-5.

Let us now examine the cutoff equation (4.139) in some detail. From the Bessel recursion relations, we can readily show that $2(n-1)J_{n-1}(u)/uJ_n(u) = J_{n-2}(u)/J_n(u) + 1$, so that Eq. (4.139) can be rewritten

$$[J_{n-2}(w)/J_n(w)] + 1 = -(n_2^2/n_e^2)\{[J_{n-2}(u)/J_n(u)] - 1\}. \tag{4.140}$$

* More precisely, Eqs. (4.138) and (4.139) admit the special solution $J_n(u) = 0$ if and only if $J_n(w) = 0$. This occurs for the isotropic guide, where $w = u$, and for selected anisotropic guides where the ratio of n_o and n_e is equal to the ratio of different roots of $J_n(x) = 0$.

Let us examine Eq. (4.140) for a general n, assuming that $u/w = n_o/n_e < 1$ (The analysis for the case where the inequality is reversed is the same.) For any $n > 1$, the solution of this equation provides the cutoff values of u and w (designated as u_c and w_c) for both the $HE_{n,m}$ and $EH_{n,m}$ modes. In order that the mode labels corresponding to successive roots be consistent with the literature on the isotropic guide, it is necessary to alternate them. Thus, the first root applies to the $HE_{n,1}$ mode, the second to the $EH_{n,1}$, the third to the $HE_{n,2}$, etc. Generally, if $p = 1, 2, 3, \ldots$, the $(2p)$th root applies to the $EH_{n,p}$ mode, while the $(2p - 1)$th root applies to the $HE_{n,p}$ mode.

Although Eq. (4.140) must be solved numerically for $u_c(= R)$ and w_c for particular values of $\varepsilon_{\parallel} = n_e^2/n_2^2$ and $\varepsilon_{\perp} = n_o^2/n_2^2$, it is possible to obtain upper and lower bounds on the solutions. A rough sketch of Eq. (4.140) shows that the isotropic case sets the upper limit on u_c and the lower limit on w_c (and thus b/λ) for any mode with $n > 1$, assuming $w > u$. As $\varepsilon_{\parallel}/\varepsilon_{\perp}$ increases, u_c decreases while w_c increases. For $\varepsilon_{\parallel}/\varepsilon_{\perp}$ large, u_c approaches zero for all modes, while w_c for the $HE_{n,p}$ mode is given by the $(2p - 1)$th nonzero root of $J_n(w_c)$ and the maximum value of w_c for the $EH_{n,p}$ mode is

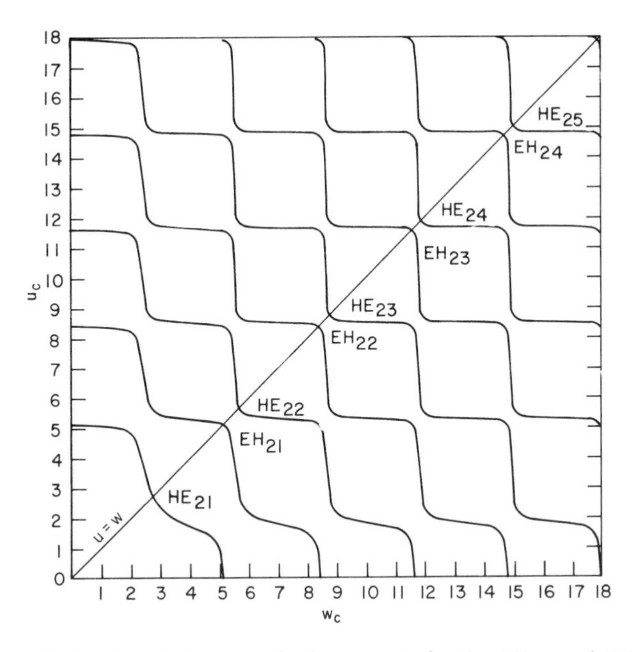

Fig. 4.6. Cutoff values of the normalized arguments for the $HE_{2,m}$ and $EH_{2,m}$ modes with $\varepsilon_{\parallel} = 2$ and $m = 1$–5. (Courtesy of Professor F. J. Rosenbaum, Washington Univ., St. Louis, Mo.).

given by the $(2p)$th nonzero root of $J_n(w_c) = 0$. For example, the range of w_c for the $EH_{2,1}$ mode (with $n_e > n_o$) is given by $5.136 \leq w_c \leq 8.417$.

The cutoff values u_c and w_c found from Eq. (4.140), for $n = 2$, $\varepsilon_{\parallel} = 2$, and for all values of the ratio u/w [$= (\varepsilon_{\perp}/\varepsilon_{\parallel})^{1/2}$] are shown in Fig. 4-6. Each curve pertains to a particular mode, defined to agree with its isotropic ($u = w$) label. Note that the curves are not symmetric in the $u = w$ line, being flatter for $u > w$. It is also apparent that, while the $HE_{n,m}$ and $EH_{n,m-1}$ modes approach degeneracy for large values of m on isotropic guides, they are readily resolvable on anisotropic guides.

The curves of Fig. 4-7 illustrate u_c and w_c for $HE_{n,1}$ and $EH_{n,1}$ modes with $n = 2, 3, 4$. It is apparent that no general statements about mode degeneracy can be made, even in this special case of cutoff. Note that the excursions of the curves decrease as the azimuthal mode number n increases.

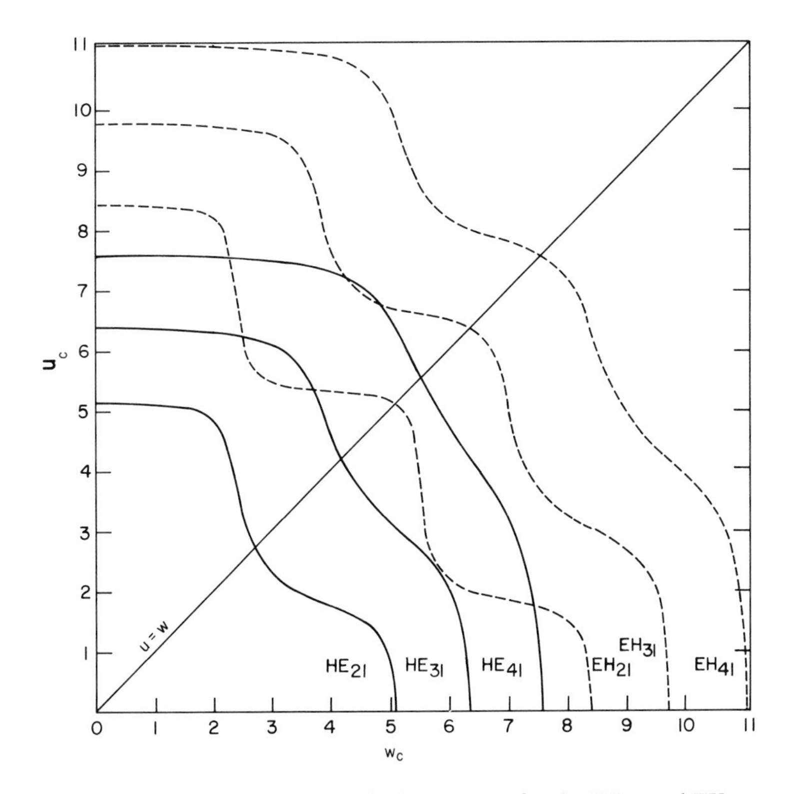

Fig. 4.7. Cutoff values of the normalized arguments for the $HE_{n,1}$ and $EH_{n,1}$ modes with $\varepsilon_{\parallel} = 2$ and $n = 2$–4. (Courtesy of Professor F. J. Rosenbaum, Washington Univ., St. Louis, Mo.)

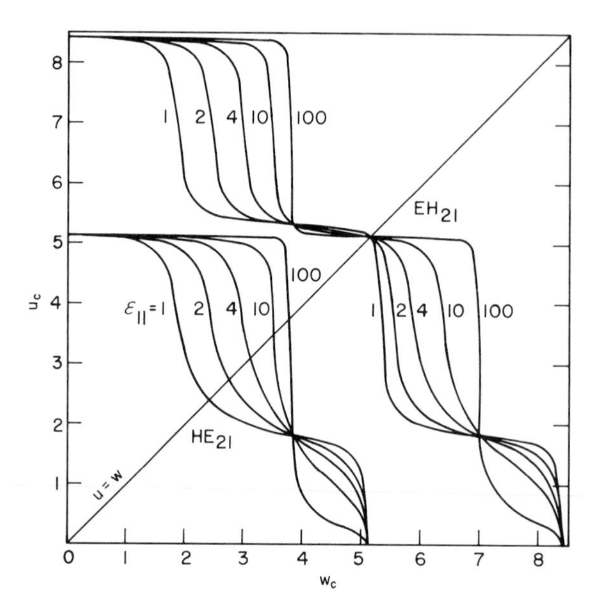

Fig. 4.8. Cutoff values of the normalized arguments for the $HE_{2,1}$ and $EH_{2,1}$ modes with $\varepsilon_{\|}$ as the parameter. (Courtesy of Professor F. J. Rosenbaum, Washington Univ., St. Louis, Mo.)

In Fig. 4-8, u_c and w_c are plotted for the $HE_{2,1}$ and $EH_{2,1}$ modes with $\varepsilon_{\|}$ as parameter. There are several points on these curves that are independent of $\varepsilon_{\|}$. Thus, the coefficient of $\varepsilon_{\|}$ $(= n_e^2/n_2^2)$ in Eq. (4.139) must vanish at these points. This yields a pair of equations

$$J_{n-1}(w_c) = 0 \tag{4.141}$$

and

$$u_c J_n(u_c) = (n-1) J_{n-1}(u_c) \tag{4.142}$$

for the points (u_c, w_c), where the curves intersect.

The curves of Fig. 4-9 show the cutoff values $(u_c = w_c)$ for several low-order modes of the isotropic guide as a function of the relative dielectric constant $\varepsilon_r = \varepsilon_{\|} = \varepsilon_{\perp}$. Many degeneracies can occur. These are most often seen in optical experiments when the relative dielectric constant is near unity. As is apparent from Fig. 4-9, a coherent mixing of the $HE_{n+1,m}$ and the $EH_{n-1,m}$ modes is to be expected near cutoff.

When the quantity $q \to \infty$, the fields outside the rod vanish ($K_{\pm} \to 0$ as $q \to \infty$). This condition is often a good approximation in optical resonators when the propagation is nearly axial. In this limit, Eq. (4.136) becomes

$$J_{n-1}(u)/J_{n+1}(u) = -J_{n-1}(w)/J_{n+1}(w). \tag{4.143}$$

As $\varepsilon_{\parallel} \rightarrow \varepsilon_{\perp}$ from above, Eq. (4.143) reduces to the isotropic case with $J_{n-1}(u) = J_{n-1}(w) = 0$ for the $\mathrm{HE}_{n,m}$ mode and $J_{n+1}(u) = J_{n+1}(w) = 0$ for the $\mathrm{EH}_{n,m}$ modes. These equations define the lower limit of w_{f} and the upper limit u_{f} (subscript f denotes far-from-cutoff values). Again, as $\varepsilon_{\parallel}/\varepsilon_{\perp}$ increase, u_{f} decreases and w_{f} increases. The lower bound of u_{f} for all modes is zero. The upper bound of w_{f} for the $\mathrm{EH}_{n,p}$ mode ($n \geq 1$) is given by the $(2p)$th nonzero root of $J_{n+1}(w_{\mathrm{f}}) = 0$. The upper bound on w_{f} for the $\mathrm{HE}_{n,p}$ mode ($n \geq 1$) is the $(2p - 1)$th nonzero root of this same equation.

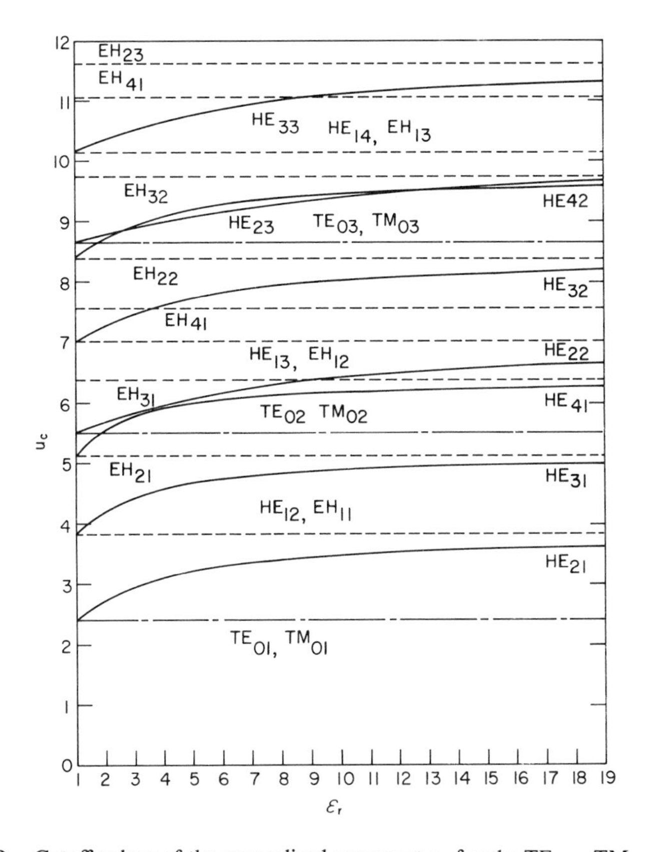

Fig. 4.9. Cutoff values of the normalized argument u_{c} for the $\mathrm{TE}_{0,m}$, $\mathrm{TM}_{0,m}$, $\mathrm{HE}_{n,m}$, and $\mathrm{EH}_{n,m}$ modes on the isotropic dielectric waveguide as a function of relative dielectric constant. (Courtesy of Professor F. J. Rosenbaum, Washington Univ., St. Louis, Mo.)

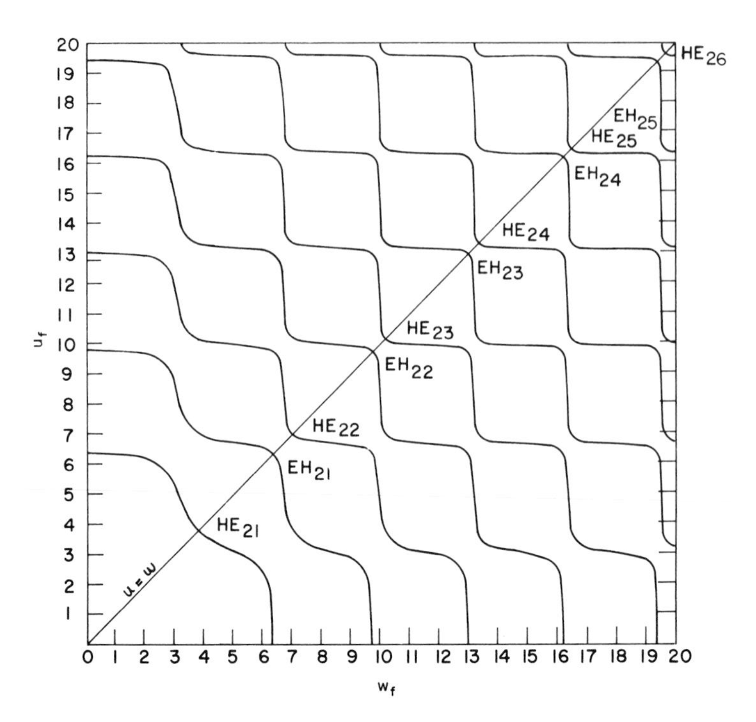

Fig. 4.10. Far-from cutoff values of the normalized arguments for the $HE_{2,m}$ and $EH_{2,m}$ modes with $\varepsilon_{\|} = 2$ and $m = 1$–5. (Courtesy of Professor F. J. Rosenbaum, Washington Univ., St. Louis, Mo.)

In Fig. 4-10, the values of u_f and w_f satisfying Eq. (4.143) are plotted for the $HE_{2,m}$ and $EH_{2,m}$ modes with $m = 1$–5. As is seen from Eq. (4.143), these curves are independent of the relative dielectric constant and are symmetric about the $u = w$ line. The curves are labeled according to their isotropic equivalent. Those modes that are degenerate on the isotropic guide are generally well resolved on the anisotropic guide. However, well-resolved isotropic guide modes can approach degeneracy on anisotropic guides, depending on the degree of anisotropy. On the isotropic guide, the $HE_{n+1,m}$ and $EH_{n-1,m}$ modes are degenerate in the far-from-cutoff limit. This fact is important in correlating these modes with those of optical interferometers, as was discussed in Section D.

The curves of Figs. 4-6 and 4-10 can be combined in a single figure, such as Fig. 4-11, which shows the full range of u and w for particular modes, in this case, $HE_{2,1}$ and $EH_{2,1}$. For each of these modes, we plot the cutoff curve for $\varepsilon_{\|} = n_e^2/n_2^2 = 2$, an intermediate curve with $q = 5$, and the far-from-cutoff curve. Taken together with the curves of Fig. 4-8 these curves show that the range of values of u and w for any particular mode is well defined. The important parameter characterizing a given waveguide is not q, but rather R defined by Eq. (4.131). However, by plotting a family of curves with different values of q, the values of u, q, and w appropriate for any value of R can be determined from Eq. (4.131). Alternatively, one could plot u as a function of q with w as parameter. The intersections of the circle $R^2 = u^2 + q^2$ with the appropriate ($w = n_e u/n_o$) curves would then give u and q for each mode for a particular rod and frequency. Such curves give all the information required to define the modes on any waveguide.

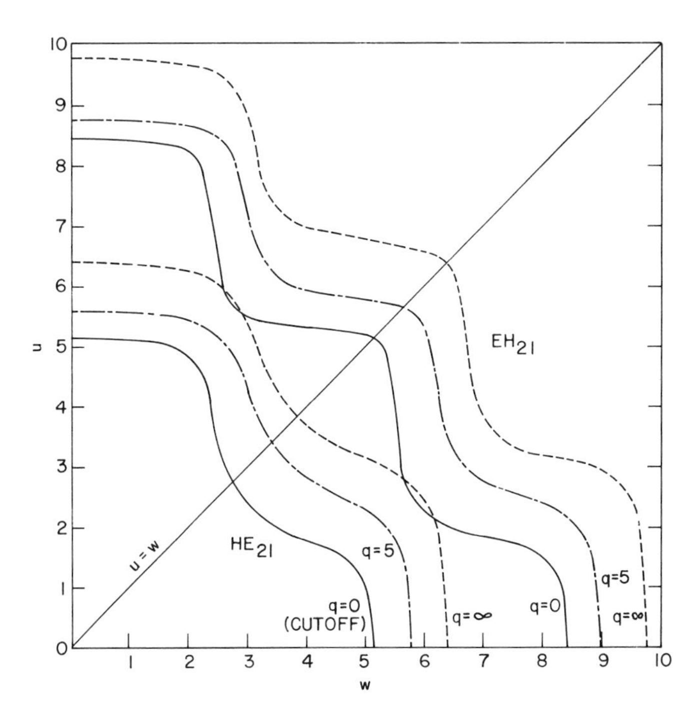

Fig. 4.11. Normalized arguments for the $HE_{2,1}$ and $EH_{2,1}$ modes for $q = 5$, $\varepsilon_{\|} = 2$, bounded by the cutoff values ($q = 0$) and the far-from-cutoff values ($q = \infty$). (Courtesy of Professor F. J. Rosenbaum, Washington Univ., St. Louis, Mo.)

For the general anisotropic guide, α apparently does not take on values that distinguish the HE and EH modes, even in the special cases of cutoff and far-from-cutoff. Near cutoff, for example, we find that, for both modes,

$$\alpha \underset{(q\to 0)}{\to} \frac{[2(n-1)n_e^2 J_{n-1}(w)/n_2^2 w J_n(w)] - 1}{[2(n-1)J_{n-1}(u)/uJ_n(u)] - 1} = -1. \tag{4.144}$$

Far from cutoff, on the other hand, α can take on all the possible values of

$$\alpha_{(q\to\infty)} \to nJ_n(u)/uJ_n'(u) = wJ_n'(w)/nJ_n(w), \tag{4.145}$$

depending on the ratio n_e/n_o. We thus again find no means of distinguishing between the general HE and EH modes.

We conclude this section by noting that the mode fields inside the guide can be decomposed into a superposition of plane waves in a manner paralleling that carried in Section C. The only major difference in the result is that the TM and TE contributions propagate at different angles θ with the guide axis, so that the wave vectors for a single mode form two different cones inside the core of the guide. It is also apparent that our circulating modes of order n carry angular momentum proportional to $-n$ and that all the hybrid modes have a degenerate counterpart described by replacing n by $-n$ in the field expressions (4.123) and (4.124), with the understanding that $\alpha_{-n} = -\alpha_n$. Accordingly, linear combinations of the degenerate modes can occur which have azimuth dependence $\cos n\phi$ and $\sin n\phi$. These describe the noncirculating modes generally preferred by microwave investigators.

REFERENCES

1. J. D. Jackson, "Classical Electrodynamics." Wiley, New York, 1962.
2. W. K. H. Panofsky and M. Phillips, "Classical Electricity and Magnetism." Addison-Wesley, Reading, Massachusetts, 1955.
3. J. A. Stratton, "Electromagnetic Theory." McGraw-Hill, New York, 1941.
4. E. A. J. Marcatili and R. A. Schmeltzer, *Bell Syst. Tech. J.* **43**, 1783 (1964).
5. W. Franz and P. Beckmann, *IRE Trans. Antennas Propagation* **AP-4**, 203 (1956).
6. J. R. Wait, *in* "Advances in Radio Research" (J. A. Saxton, ed.), Vol. 1, pp. 157–217. Academic Press, New York and London, 1964.
7. A. Sommerfeld, "Partial Differential Equations in Physics," p. 88. Academic Press, New York, 1949.
8. J. B. Keller, *J. Opt. Soc. Amer.* **52**, 116 (1962).
9. A. G. Fox and T. Li, *Bell Syst. Tech. J.* **40**, 453 (1961).
10. E. Snitzer, *in* "Advances in Quantum Electronics" (J. R. Singer, ed.), p. 348. Columbia Univ. Press, New York, 1961.
11. E. Snitzer and H. Osterberg, *J. Opt. Soc. Amer.* **51**, 499 (1961).
12. R. E. Collin, "Field Theory of Guided Waves," p. 483. McGraw-Hill, New York, 1960.
13. S. I. Rubinow, *Ann. Phys.* **14**, 305 (1961).
14. F. J. Rosenbaum, *IEEE J. Quant. Electron.* **QE-1**, 367 (1965).

CHAPTER 5

Waveguide Mode Launching

In the last three chapters, we have treated, theoretically, various aspects of wave propagation along planar and circular dielectric waveguides. Accordingly, lossless propagation in an optical waveguide is allowed only in certain field distributions or modes which satisfy appropriate boundary conditions and the homogeneous form of Maxwell's equations. In a large waveguide, the distinctive features of a single mode are lost in a large number of modes that generally become excited. A subject central to various investigations is the problem of how to couple an optical source to a multimode waveguide in such a way as to excite substantially one arbitrarily chosen mode or a combination of desired modes.

In this chapter, we will discuss the various techniques that have been developed in order to excite discrete modes in circular optical waveguides. The first of these methods (*1*) makes use of a monochromatic beam of light incident at the required angle to the axis of the fiber in order to excite appropriate field conditions within the guide.

The second technique for mode launching (*1,2*) consists in imaging an Airy disk pattern on the entrance aperture of the waveguide and moving the Airy disk pattern off-axis at the entrance aperture. Thus, the excitation conditions are varied to propagate various modes along the waveguide.

Third, a spatial filtering technique (*3*) is described that provides a means for launching an arbitrary mode on an optical waveguide. Appropriate amplitude and phase filters are used to discriminate against unwanted modes. These are inserted in the pupil of a launching lens that focuses collimated light onto the end of the guide. A double-path illuminating system is generally needed to obtain the required input polarization. The method is evaluated both theoretically and experimentally, and the agreement is found to be good within the practical limitations of cross-sectional irregularities in the nominally circular optical fibers.

A. EARLY OPTICAL MODE LAUNCHING TECHNIQUES

A given mode can be selectively excited by any technique that establishes at the entrance face of the waveguide (the plane $z = 0$) an electric and magnetic field configuration that matches perfectly the field vectors for that mode. Though no combination of polarizers, spatial filters, and lenses can do this exactly, it has been found that even seemingly crude approximations can provide adequate selection for some purposes. Thus, Snitzer and Osterberg were able to launch desired modes (usually combinations of nearly degenerate modes) by imaging the exit slit of a monochromator onto the end of a fiber with a low-numerical-aperture lens. By tilting the axis of the fiber with respect to the axis of the illuminating cone, they achieved a degree of mode selection.

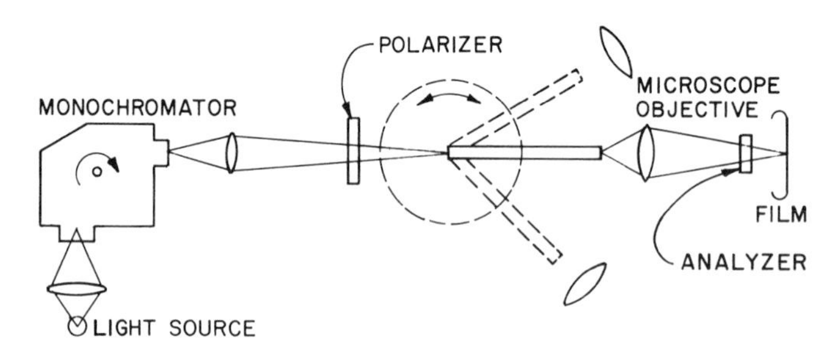

Fig. 5-1. Optical schematic of a system used for exciting discrete modes by changing the angle between the fiber axis and the illumination beam.

Figure 5-1 is an optical schematic of the system used for exciting modes by varying the angle between a low-numerical-aperture light beam and the fiber axis. Illumination from a white-light source is imaged onto the entrance slit of a monochromator, and the exit slit of the monochromator is imaged at a large *f*-number onto the entrance face of the fiber. The fiber is mounted on a rotatory table in order that the angle between the light beam and the fiber axis can be varied accurately at will. The exit end of the fiber is magnified by the use of a microscope system, and the image can be either viewed or photographed at will.

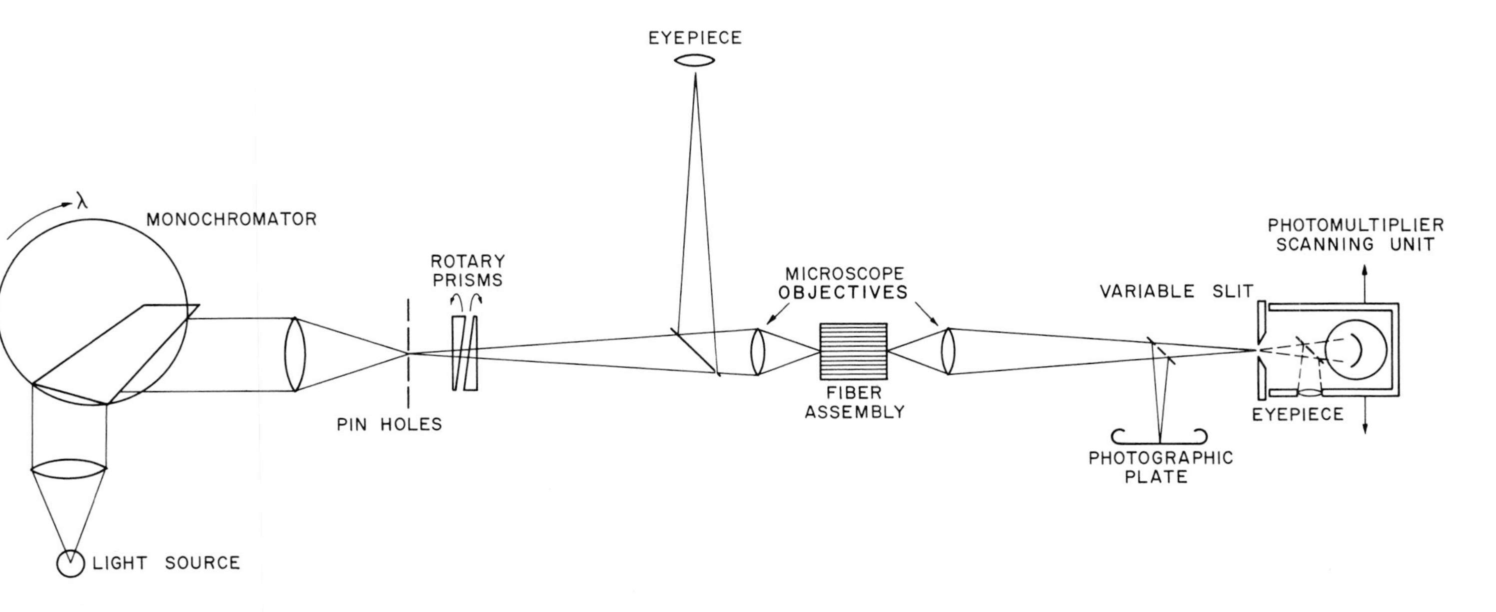

Fig. 5-2. Optical schematic of exciting discrete modes in optical waveguides using an Airy disk.

The system illustrated in Fig. 5-1 provides two variables for changing the excitation conditions, e.g., the wavelength of light (λ) and the angle of excitation (θ). These variables provide sufficient mode selection to enable studies of nearly degenerate mode combinations as well as mode cutoff. However, this system is generally unable to excite very high-purity single modes.

The second method of mode excitation is illustrated in Fig. 5-2. In this system, the illumination from a white-light source is imaged onto the entrance slit of a monochromator, and the exit slit is imaged onto a pinhole. The pinhole is imaged down at the entrance end of the fiber by a highly corrected microscope objective. Two rotatory prisms, which provide for accurate lateral displacements of the pinhole image, and a vertical illumination viewing system are placed between the pinhole and the microscope objective. Thus, the diffraction-limited image of the pinhole is visually placed very accurately with respect to the fiber axis. The other end of the fiber is magnified by the use of a microscope objective and the magnified image can be recorded photographically, viewed, or scanned by a photomultiplier system.

This system provides three variables for changing the field configuration launched onto the entrance aperture of the waveguide; e.g., first, the position of the Airy disk with respect to the fiber axis, second, the size of the Airy disk pattern, which can be changed either by varying the size of the

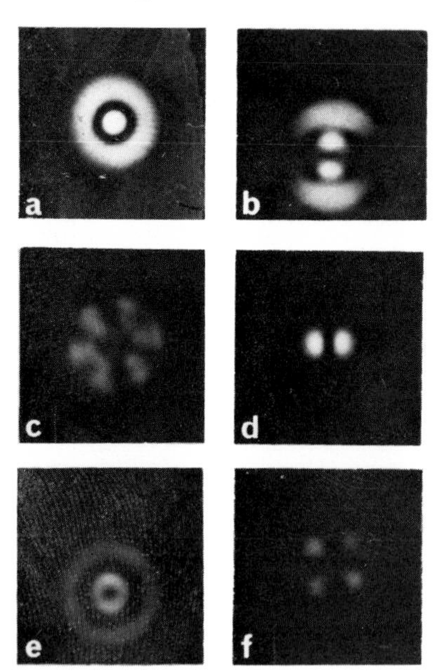

Fig. 5-3. Some typical mode patterns of optical waveguides of slightly noncircular cross section using the optical setup illustrated in Fig. 5-1.

aperture or the numerical aperture of the microscope objective used, and third, the wavelength of incident light.

As we have shown in the previous chapter, pure modes on circular guides are never plane polarized. However, degenerate forms of the $HE_{1,m}$ modes [Chapter 4, Section D, combination (b)] and certain combinations of nearly degenerate modes [Chapter 4, Section D, combinations (c) and (e)] are approximately plane polarized far from cutoff. To excite and analyze these combinations, cross-polarizers can be placed in the optical train at the two ends of the waveguide, as illustrated in Fig. 5-2. Figure 5-3 shows some low-order modes and their combinations. These photomicrographs were produced using linearly polarized light and without an analyzer at the output. Figure 5-3(a) is $HE_{1,2}$ and $EH_{1,1}$ or $HE_{3,1}$; Fig. 5-3(b) is $TE_{0,2}$ or $TM_{0,2} + HE_{2,2}$; Fig. 5-3(c) is perturbed $EH_{2,1} + HE_{4,1}$; Fig. 5-3(d) is $TE_{0,1}$ or $TM_{0,1} + HE_{2,1}$; Fig. 5-3(e) is $TE_{0,2}$ or $HE_{2,2}$; and Fig. 5-3(f) is $EH_{1,1} + HE_{3,1}$.

On the other hand, Fig. 5-4 shows combinations of various modes excited using the system illustrated in Fig. 5-2. Figure 5-4(a) is $TE_{0,2}$ or $TM_{0,2} + HE_{2,2}$; Fig. 5-4(b) is the $HE_{1,2}$ and $EH_{1,2} + HE_{3,2}$ without the use of an analyzer; Fig. 5-4(c) is the $EH_{1,1} + HE_{3,1}$ on an elliptical fiber; and Fig. 5-4(d) is the $HE_{1,2} + EH_{1,2} + HE_{3,2}$ with crossed polarizer and analyzer.

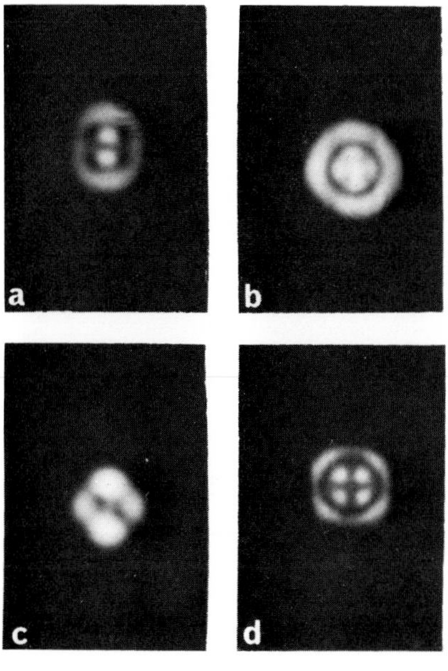

a b c d

Fig. 5-4. Combinations of different modes produced using the optical setup illustrated in Fig. 5-2.

B. MODE LAUNCHING USING SPATIAL FILTERING

Recently, an unusual technique has been developed (*3*) using spatial filtering in order to launch appropriate field conditions at the entrance end of an optical waveguide. The spatial filtering technique to be described here derives from two facts: (1) the waveguide mode fields inside the core are completely describable in terms of the interference of a system of plane waves; and (2) such a system of waves can be generated with a collimated beam, a relatively simple spatial filter, a lens, and, in the general case, a double-path illuminating system in which each of two orthogonally polarized fields can be separately filtered.

This spatial filtering technique is intended to provide an incident electric field in the plane of the entrance face of the waveguide that very closely matches the transverse electric field of a single (arbitrarily chosen) mode. That this is possible is suggested by a particular description of modes in which transverse fields are represented by circularly polarized components (Chapter 4). For convenience, we will repeat the field expressions [Eqs. (4.58)] here. We assume a unit amplitude for the mode and use the symbol A instead of α for the parameter that describes the relative strength of the TM and TE contributions to the hybrid mode. All other notations are the same as in Chapter 4. Accordingly, we have, inside the core,

$$\begin{aligned} E_{\pm} &= E_x \pm iE_y = \pm\,(1 \pm A)\,f_{n\pm 1} \\ H_{\pm} &= H_x \pm iH_y = (ih/\omega\mu)[(k_1{}^2/h^2) \pm A]\,f_{n\pm 1}, \end{aligned} \tag{5.1a}$$

whereas outside the core,

$$\begin{aligned} E_{\pm} &= \pm(1 \pm A)\left[\frac{uJ_n(u)}{vH_n(v)}\right] g_{n\pm 1} \\ H_{\pm} &= \frac{ih}{\omega\mu}\left(\frac{k_2{}^2}{h^2} \pm A\right)\frac{uJ_n(u)}{vH_n(v)}\, g_{n\pm 1}, \end{aligned} \tag{5.1b}$$

where the functions $f_{n\pm1}$ and $g_{n\pm1}$ are given by

$$\begin{aligned} f_{n\pm1} &= J_{n\pm1}(\beta_1 r)\exp[i(n \pm 1)\phi + i\omega t - ihz] \\ g_{n\pm1} &= H_{n\pm1}(\beta_2 r)\exp[i(n \pm 1)\phi + i\omega t - ihz]. \end{aligned} \tag{5.2}$$

$J_{n\pm1}$ and $H_{n\pm1}$ are the Bessel functions of the first and third kinds, respectively, of order $n \pm 1$. The other parameters are as defined in Chapter 4.

Consider now a well-known integral representation of the Bessel function [Eq. (4.93)]

$$J_{n'}(\beta' r)\exp(in'\phi) = (i^{-n'}/2\pi)\int_0^{2\pi} \exp[i\beta' r\cos(\phi - \bar{\phi}) + in'\bar{\phi}]\,d\bar{\phi}. \quad (5.3)$$

This can be recognized as a two-dimensional Fourier transform of the distribution

$$P(\varrho, \bar{\phi}) = [\delta(\varrho - a')/\varrho]\exp(in'\bar{\phi}). \quad (5.4)$$

Evidently, we can obtain a good approximation of the transverse field components inside the core [Eq. (5.1)] by placing a mask with transmittance $P(\varrho, \bar{\phi})$ in the pupil of a lens that focuses collimated, appropriately polarized light onto the end of the waveguide. (Because the diameter of the waveguide core is quite small, we can neglect the quadratic phase factor in the focal plane.) Equation (5.1) shows that the transverse electric field on the guide is generally elliptically polarized, the orientation of the major axis varying with azimuth angle ϕ. To match closely such a field, we require a double-path filtering system (Fig. 5-5), in which each circular component is generated separately. (In many cases of interest, one of the circular components is much stronger than the other. When this is so, as in the experiments to be discussed, only one path is needed.)

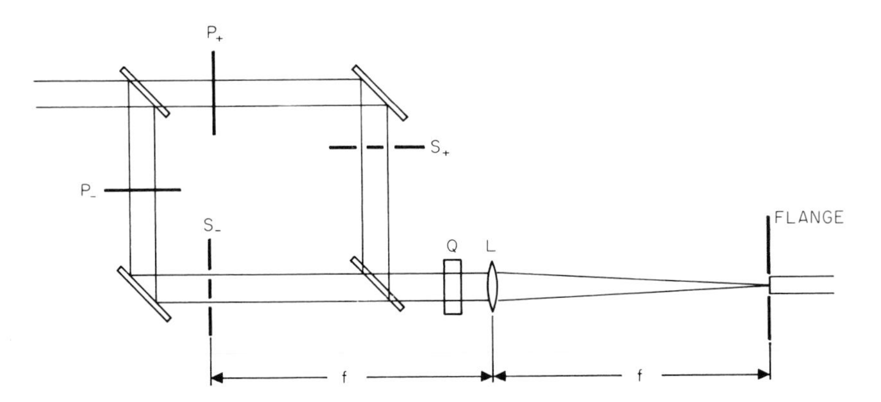

Fig. 5-5. Schematic for launching a particular mode on the waveguide at the right with a collimated beam incident from the left, polarized at an angle of 45° to the plane of the drawing. Polarizers P_+ and P_- select electric field components in and perpendicular to the plane of the drawing for subsequent filtering by spatial filters S_+ and S_-. Quarter-wave plate Q, with retardation axis at 45° to the plane of the drawing, transforms filtered fields into right (+) and left (−) circularly polarized fields that are Fourier-transformed by lens L.

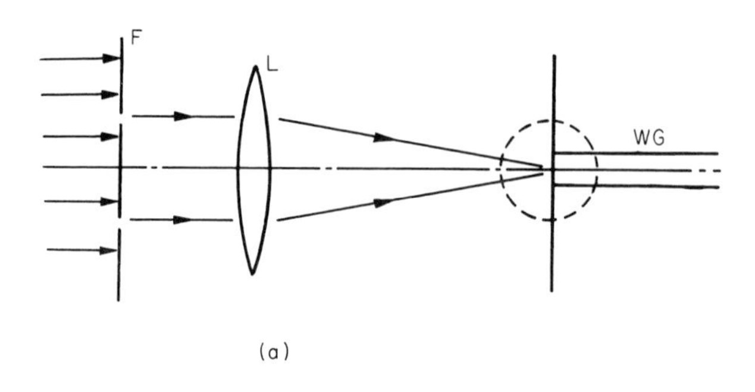

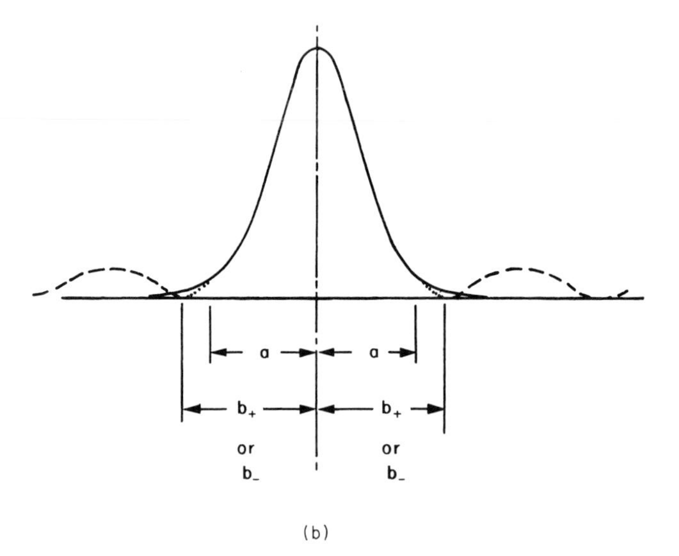

Fig. 5-6. (a) Coupling by a spatial filter F and launching lens L to a waveguide WG; and (b) theoretical mode irradiance (——), effective incident irradiance (. . .) by spatial filter, and eliminated higher-order lobe (– – – – –) of incident irradiance.

The simplified sketch in Fig. 5-6 illustrates the intended operation in each arm of the system in Fig. 5-5. In Fig. 5-6(a), the spatial filter, Fourier-transforming lens, and optical waveguide are shown. A circular aperture in an otherwise opaque mask is placed in the entrance face of the waveguide, the aperture being concentric with the waveguide core. The focal length f of the lens and the radius of the filtering ring mask a' are chosen such that β' in Eq. (5.3) is

$$\beta' = (2\pi/\lambda)(a'/f) = u'/a, \tag{5.5}$$

where $u'r/a'$ is the argument of the desired Bessel function, $J_{n'\pm 1}$.

Figure 5-6(b) shows an enlargement of the portion enclosed by a dashed line in Fig. 5-6(a). The Bessel function distribution from the ring mask is compared with the mode field to be matched to the entrance face of the fiber. The distance on the input face (b_+ or b_-) is u' given in Eq. (5.5) and is such as to correspond with the zero of the Bessel function that is the first to fall outside the core radius a. The higher-order lobes of the Bessel function, those that do not conform to the mode field outside the core, are eliminated by a blocking mask, which is not shown in the figure.

Returning now to Fig. 5-5, we see that the double-path system serves to provide an appropriate polarization in the general case. The collimated beam incident from the left is polarized at 45° to the plane of the drawing. After being divided by a beam splitter, the component polarized in the plane of the drawing is selected by polarizer P_+ and that perpendicular to this plane by polarizer P_-. Filters S_+ and S_- of ring apertures with azimuthly increasing phase transmittances, which may be expressed mathematically as $[\delta(\varrho - a')/\varrho] \exp[i(n' \pm 1)\bar{\phi}]$, respectively, modify the beams. After passing through the second beam splitter, the recombined beams pass through a quarter-wave plate whose retardation axis is oriented at an angle of 45° to the plane of the drawing. The linearly polarized component in the plane of the drawing is thus transformed into a right circularly polarized component, whereas the linear component perpendicular to the drawing becomes a left circularly polarized component. If the path lengths in the two arms are appropriately adjusted, the two beams, after passage through the lens L, will provide the desired elliptically polarized field on the entrance face of the waveguide. [We are implicitly assuming in this discussion that the transverse circular components on the end of the waveguide core are proportional to the corresponding components in the collimated beam incident on the lens. This is consistent with vector Huygens–Kirchhoff theory (*4*) (see Chapter 6). This theory, however, is not a very good approximation for wide angles. We should thus restrict our experiments so that f/a' in Eq. (5.5) is much greater than unity.]

With this understanding of Fig. 5-5, we conclude that it is experimentally possible to generate an incident field in the plane $z = 0$ given by

$$\begin{aligned} E_{\pm}' &= \pm(1 \pm A) f'_{n'\pm1} = \pm\, i(\mu/\varepsilon)^{1/2} H_{\pm}'; \qquad r < b_{\pm} \\ &= 0; \qquad r > b_{\pm}, \end{aligned} \tag{5.6}$$

where

$$f'_{n'\pm1} = J_{n'\pm1}(\beta' r) \exp[i(n' \pm 1)\phi + i\omega t]. \tag{5.7}$$

Assuming this to be the incident field, we can proceed to evaluate theoretically the degree of success to be expected.

1. Theory of Mode Launching by Spatial Filtering

In a recent paper (*5*), Snyder addresses the related problem of determining the relative power in those modes that are excited when a wave is incident on the end of a dielectric cylinder. He considers explicitly two waves: a normally incident infinite plane wave and another such (fictitious) wave, one which is truncated so that it extends only over the cross section of the waveguide core. Our theoretical approach is mathematically equivalent to Snyder's. Thus, we assume that the tangential components of the electric (but not the magnetic) field at the end of the guide are those of the incident wave, the reaction of the guide being substantially ignored. We will state our physical argument somewhat differently, however, in order to obtain self-consistent results. Thus, Snyder assumes sources in the plane $z = 0$, in the form of electric and magnetic current sheets. These currents are specified so as to produce a plane wave that propagates in the z direction for $0 < z$ with the assumption that the backscattered field is negligible. In the half-space $z > 0$, the tangential components of the most general electric field on the waveguide, represented by the general (monochromatic) solution of the homogeneous wave equation in the guide half-space, are matched to the assumed magnetic current sheet. This procedure is mathematically equivalent to equating the tangential components of the incident electric field to those of the general electric field on the guide. It results in a unique set of amplitude coefficients for the modes. However, the magnetic field resulting from this set of coefficients does not match the magnetic field of the incident wave at the boundary and thus cannot correspond to the assumed electric current sheet.

The procedure that we will follow is simply to assume that the tangential components of the incident electric field (but not the magnetic field) must match the general electric field for the guide half-space at the plane $z = 0$. This will provide a unique set of coefficients for the modes, one which corresponds to the approximation that the reflected electric field is zero. To obtain a better approximation, we could proceed by determining the tangential components of the reflected magnetic field, represented by the difference between the tangential components of the incident field and those obtained from the above set of approximate mode coefficients. This reflected magnetic field implies in turn an associated electric field that is not equal to zero. By matching the sum of this reflected electric field and the incident field to the general electric field in the guide half-space, we then could obtain the second approximation. The procedure could then be repeated to yield higher-order approximations. Its convergence properties have not been studied, however, because we expect that the first approximation will provide descriptions of our experiments that are accurate within the experimental errors.

We express the general field in the waveguide half-space as

$$\mathbf{E} = \sum C_n \mathbf{E}_n + \mathbf{E}_{\mathrm{R}} \qquad \text{and} \qquad \mathbf{H} = \sum C_n \mathbf{H}_n + \mathbf{H}_{\mathrm{R}},$$

where $\mathbf{E}_n$ and $\mathbf{H}_n$ represent the field vectors of the nth propagating mode and $\mathbf{E}_{\mathrm{R}}$ and $\mathbf{H}_{\mathrm{R}}$ are the field vectors for the radiating field; C_n is the amplitude of the nth mode. The incident field has electric vector $\mathbf{E}'$. Equating the tangential fields (subscript t) at the plane $z = 0$ gives

$$\mathbf{E}_{\mathrm{t}}' = \sum C_n \mathbf{E}_{n\mathrm{t}} + \mathbf{E}_{\mathrm{Rt}}.$$

By taking the vector cross-product of each side with $\mathbf{H}_{n\mathrm{t}}^*$, integrating over the plane $z = 0$, and invoking the orthogonality condition satisfied by the modes, we obtain

$$C_n = \int \mathbf{E}_{\mathrm{t}}' \times \mathbf{H}_{n\mathrm{t}}^* \, da \Big/ \int \mathbf{E}_{n\mathrm{t}} \times \mathbf{H}_{n\mathrm{t}}^* \, da. \tag{5.8}$$

The power P_n in the nth mode, relative to the incident power, is then given by

$$P_n = \left| \int \mathbf{E}_{\mathrm{t}}' \times \mathbf{H}_{n\mathrm{t}}^* \, da \right|^2 \Big/ \left[\left(\int \mathbf{E}_{\mathrm{t}}' \times \mathbf{H}_{\mathrm{t}}^{*\prime} \, da \right) \left(\int \mathbf{E}_{n\mathrm{t}} \times \mathbf{H}_{n\mathrm{t}}^* \, da \right) \right]. \tag{5.9}$$

We can conclude directly from Eqs. (5.8) and (5.9) that, to this approximation, we can accomplish our purpose by simply providing an incident electric field whose tangential components match those of the desired mode. The incident electric field given by Eqs. (5.6) and (5.7) is a perfect match to the transverse field on the core. We thus expect it to be very selective in all cases where most of the power is in the core. We now proceed to evaluate the amplitude and power coefficients of other mismatched modes (cases I and II below) and of the desired mode (case III). From Eqs. (5.1), (5.2), and (5.6)–(5.9), we have the following.

Case I. $n \neq n'$.

$$C_n = 0 \qquad \text{and} \qquad P_n = 0. \tag{5.10}$$

Case II. $n = n'$ and $\beta_1 \neq \beta_1'$.

$$C_n = \frac{u'J_n(u')(\alpha_+'Q_+' + \alpha_-'Q_-' + \beta_-'N_-' + \beta_-'N_-')}{uJ_n(u)(\alpha_+Q_+ + \alpha_-Q_- + \beta_+R_+ + \beta_-R_-)}$$

$$P_n = \frac{h(\alpha_+'Q_+' + \alpha_-'Q_-' + \beta_+'N_+' + \beta_-'N_-')^2}{k_0(\alpha_+Q_+ + \alpha_-Q_- + \beta_+R_+ + \beta_-R_-)} \times \frac{1}{(\gamma_+'O_+' + \gamma_-'O_-')}, \tag{5.11}$$

where

$$u' = \beta_1' a,$$

$$\alpha_\pm = (1 \pm A)[(k_1^2/h^2) \pm A], \qquad \beta_\pm = (1 \pm A)[(k_2^2/h^2) \pm A],$$

$$\alpha_\pm' = (1 \pm A')[(k_1^2/h^2) \pm A], \qquad \beta_\pm' = (1 \pm A')[(k_2^2/h^2) \pm A],$$

$$\gamma_\pm' = (1 \pm A')^2,$$

$$Q_\pm = \tfrac{1}{2}[\mathscr{J}^2_{n\pm1} - (\mathscr{J}_{n\pm2}/u)],$$

$$Q_\pm' = \{\pm 1/[(u')^2 - u^2]\}(\mathscr{J}'_{n\pm1} - \mathscr{J}_{n\pm1}),$$

$$N_\pm' = \{\pm 1/[(u')^2 - v^2]\}[\mathscr{J}'_{n\pm1} - (b_\pm u'/a)\mathscr{F}'_{n\pm0}\mathscr{G}^*_{n\pm1} + \mathscr{H}^*_{n\pm1}],$$

$$R_\pm = -\tfrac{1}{4}[2\mathscr{H}_{n\pm1}\mathscr{H}^*_{n\pm1} - (\mathscr{H}^*_{n\pm2}/v) - (\mathscr{H}_{n\pm2}/v^*)],$$

$$O_\pm' = (b_\pm^2/2a^2)\mathscr{F}'_{n\pm0}\cdot\mathscr{F}'_{n\pm2},$$

$$\mathscr{J}_{n+i} = J_{n+i}(u)/uJ_n(u), \qquad \mathscr{H}_{n+i} = H_{n+i}(v)/vH_n(v),$$

$$\mathscr{J}'_{n+i} = J_{n+i}(u')/u'J_n(u'), \qquad \mathscr{H}'_{n+i} = H_{n+i}(v')/v'H_n(v'), \qquad (i = -1,\ 1,\ 2)$$

$$\mathscr{H}^*_{n+i} = H^*_{n+i}(v)/v^*H_n^*(v), \qquad (i = -1,\ 1,\ 2),$$

$$\mathscr{F}'_{n\pm i} = J_{n+i}(u'b_\pm/a)/u'J_n(u'), \qquad (i = 0,\ 2),$$

and

$$\mathscr{G}_{n\pm1} = H^*_{n\pm1}(vb_\pm/a)/v^*H_n^*(v).$$

When one circular component is much stronger than the other, we may use the single-path system. This approximation leads to the following relations for $EH_{n,m}$ modes with $n > 0$ and $HE_{n,m}$ modes with $n < 0$:

$$\alpha_+ = \alpha_+' = \beta_+ = \beta_+' = 4$$

$$\alpha_- = \alpha_-' = \beta_- = \beta_-' = 0.$$

Then, Eqs. (5.11) are rewritten as

$$\begin{aligned} C_n &= u'J_n(u')(Q_+{}' + N_+{}')/uJ_n(u)(Q_+ + R_+) \\ P_n &= h(Q_+{}' + N_+{}')^2/k_0(Q_+ + R_+)O_+{}'. \end{aligned} \tag{5.12}$$

For the $EH_{n,m}$ mode ($n < 0$) and $HE_{n,m}$ mode ($n > 0$), the approximate relations are

$$\begin{aligned} \alpha_+ &= \alpha_+{}' = \beta_+ = \beta_+{}' = 0 \\ \alpha_- &= \alpha_-{}' = \beta_- = \beta_-{}' = 4. \end{aligned}$$

Hence, Eqs. (5.11) are also given as

$$\begin{aligned} C_n &= u'J_n(u')(Q_-{}' + N_-{}')/uJ_n(u)(Q_- + R_-) \\ P_n &= h(Q_-{}' + N_-{}')^2/k_0(Q_- + R_-)O_-{}'. \end{aligned} \tag{5.13}$$

Case III. $n = n'$ and $\beta_1 = \beta_1{}'$.

$$\begin{aligned} C_n &= \frac{(\alpha_+Q_+ + \alpha_-Q_- + \beta_+N_+ + \beta_-N_-)}{(\alpha_+Q_+ + \alpha_-Q_- + \beta_+R_+ + \beta_-R_-)} \\ P_n &= \frac{h(\alpha_+Q_+ + \alpha_-Q_- + \beta_+N_+ + \beta_-N_-)^2}{k_0(\alpha_+Q_+ + \alpha_-Q_- + \beta_+R_+ + \beta_-R_-)} \frac{1}{(\gamma_+O_+ + \gamma_-O_-)}, \end{aligned} \tag{5.14}$$

where

$$\begin{aligned} \gamma_\pm &= (1 \pm A)^2, \\ N_\pm &= [\pm 1/(u^2 - v^2)][\mathscr{J}_{n\pm1} - (b_\pm u/a)\, \mathscr{F}_{n\pm0}\mathscr{G}^*_{n\pm1} + \mathscr{H}^*_{n\pm1}], \\ O_\pm &= (b_\pm{}^2/2a^2)\mathscr{F}_{n\pm0}\mathscr{F}_{n\pm2}, \end{aligned}$$

and

$$\mathscr{F}_{n\pm i} = J_{n\pm i}(ub_\pm/a)/uJ_n(u) \qquad (i = 0,\ 2).$$

Other notations are the same as those used in the previous case.

When the single path system is appropriate, these reduce to

$$\begin{aligned} C_n &= (Q_+ + N_+)/(Q_+ + R_+) \\ P_n &= h(Q_+ + N_+)^2/k_0(Q_+ + R_+)O_+, \end{aligned} \tag{5.15}$$

for the $EH_{n,m}$ mode ($n > 0$) and the $HE_{n,m}$ mode ($n < 0$). Furthermore, for the $EH_{n,m}$ mode ($n < 0$) and the $HE_{n,m}$ mode ($n > 0$),

$$\begin{aligned} C_n &= (Q_- + N_-)/(Q_- + R_-) \\ P_n &= h(Q_- + N_-)^2/k_0(Q_- + R_-)O_-. \end{aligned} \tag{5.16}$$

By using Eqs. (5.11) and (5.14), we can numerically evaluate C_n and P_n for several representative cases in which the parameters used are $n_1 = 1.53$, $n_2 = 1.52$, and $dn_2/\lambda = 10.35$. We assume in each case that β' is given by Eq. (5.7) for a particular desired mode. We calculate its power relative to the power in the incident wave. We also determine the relative power of one other mode, one which, apart from the desired mode, is excited most strongly by the incident wave. The results are given in Table I. Insofar as the approximations are appropriate, we expect the method to be quite effective.

Table I

Relative Power of Waveguide Modes Excited by a Diffraction Pattern of a Spatial Filter

Desired mode	Relative power (%)	Unwanted mode	Relative power (%)
$HE_{1,1}$	98.6	$HE_{1,2}$	0.077
$HE_{2,1}$	97.6	$HE_{2,2}$	0.065
$HE_{1,2}$	95.6	$HE_{1,1}$	0.080
$HE_{2,2}$	92.0	$HE_{2,1}$	0.070

2. Experiment in Mode Launching by Spatial Filtering

In the experimental study of the mode launching technique, we focus our attention on the $HE_{1,1}$ and $HE_{1,2}$ modes at $\lambda = 6328$ Å in a waveguide of diameter $d = 15$ μm with core and coating refractive indices of $n_1 = 1.530$ and $n_2 = 1.525$, respectively. For these modes on this waveguide, we obtain the following values for the important parameters:

$$R = (\pi d/\lambda)(n_1{}^2 - n_2{}^2)^{1/2} \approx 6.5,$$
$$u_{1,1} \approx 2.25; \qquad A_{1,1} = -A_{-1,1} \approx -0.998$$
$$u_{1,2} \approx 5.1; \qquad A_{1,2} = -A_{-1,2} \approx -0.988.$$

From Eqs. (5.1a) and (5.1b), we see that the E_- component (or E_+ component) is much stronger than the E_+ component (or E_- component) in both cases, so that we may obtain a good match to the mode field by using only one path of Fig. 5-5. Furthermore, in this case, we should note that it is not necessary to use a quarter-wave plate and polarizers to divide the radiant field into each component. The reason for this is that the field obtained without such optics is a combination of each dominant field of the two $HE_{1,m}$ modes (one is the E_+-component dominant mode and the other is the E_--component dominant mode). The combination of the two $HE_{1,m}$ modes gives us a conventional $HE_{1,m}$ mode [whose dominant component is plane polarized—see Chapter 4, Section D, combination (b)].

By using the values of u defined above, we can find the appropriate ring filter radius a' and focal length f from Eq. (5.5).

As is apparent from Eqs. (5.1) and (5.2), we have chosen the $HE_{1,m}$ modes to eliminate the technical difficulties attendant on preparing variable-phase pupil filters. For all other modes, such variable-phase filters would be required. It appears to us that the synthetic hologram technique of Lohmann (*6*) would be well-suited to meeting such a requirement. However, we have made no attempt to verify this because of the low irradiance available in the holographic real image. (In Appendix B, Sawatari demonstrates the use of an amplitude filter to achieve the exponential phase variation of the fields.)

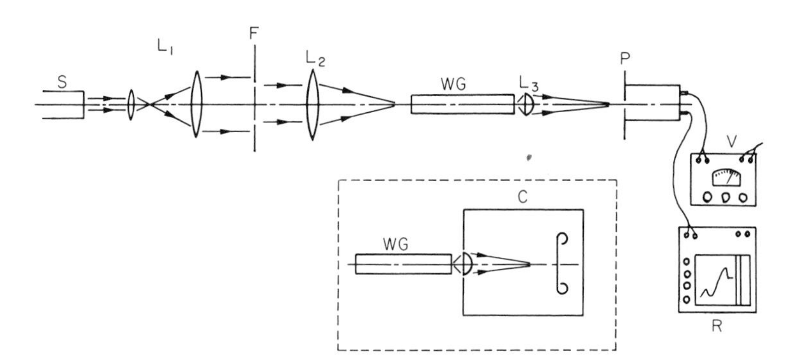

Fig. 5-7. Experimental arrangement for exciting a desired mode in a multimode waveguide. *S*, He–Ne laser; L_1, collimator lens; *F*, spatial filter (ring aperture); L_2, launching lens; *WG*, tested fiber; L_3, microscope objective; *P*, pinhole with photodetector; *V*, high-voltage supply; *R*, recorder; and *C*, camera.

A schematic of the experimental arrangement is shown in Fig. 5-7. The source is an OTI Model 170 He–Ne laser providing a 0.25-mW output in the TEM_{00} mode. Two simple lenses expand and recollimate the beam, which is then incident on the annular ring spatial filters. The filters were made photographically. The ratio of ring width to mask diameter was approximately 0.01 for both masks. Masks with appropriately uniform phase transmittance over the rings were chosen from a large group of photographic negatives (Kodak Ortho type 3).

The focal length of the lens, which follows the ring filter, was 150 mm for the $HE_{1,1}$ mode and 30 mm for the $HE_{1,2}$ mode. The diameters of the ring in the corresponding masks were 9 and 4 mm. The use of a different launching lens for the $HE_{1,2}$ mode was necessitated by the difficulty of constructing a mask whose phase transmittance was uniform over a 20-mm ring.

The waveguide was a 5-cm-long optical fiber made of UK-50 glass with a soda-lime-glass coating. The coating diameter (225 μm) was 15 times that of the core. Before final drawing, the outside surface of the coating was covered with black paint. This provides for the absorption of stray light in the guide. The fiber was embedded in thick-walled capillary tubing, to provide mechanical support, and the ends were finely polished.

A 50× oil-immersion objective was used to image the output end of the fiber in the plane of the photographic film (Kodak Pan X, housed in an Asahi Pentax camera) or a scanning pinhole of 20 μm diameter attached to a photomultiplier housing. A photomultiplier tube, EMI 9558B, together with a suitable amplifier and a strip-chart recorder, provided a photometric trace through the center of the image of the waveguide exit face.

Alignment of the axis of the waveguide and the launching lens had to be done by observation of the waveguide output end. This was necessary because the beam splitters or other optical components in the launching beam (which might otherwise have provided for direct observation of the relative position of the incident diffraction pattern and the waveguide axis) would have disturbed the phase relations in the beam. Therefore, the waveguide mount was equipped with fine-motion translation and rotation screws to provide maximum output at the axis of the waveguide, this being a common characteristic of all $HE_{1,m}$ modes. A 10× eyepiece placed in the microscope tube body, which supports the 50× oil-immersion objective, was used for these adjustments. This same microscope objective, with or without the eyepiece, was used to observe visually, or photographically, or to record photometrically, the irradiance distribution at the plane of the waveguide entrance face. For such measurements, the waveguide itself was removed from the bench.

An appropriate ring filter and launching lens were first placed in the expanded laser beam. The diffraction pattern at the focal plane of the lens was then recorded photographically and photometrically. The entrance end of the waveguide was then placed in this plane, and the lateral position and angular orientation of the waveguide were adjusted to give maximum output power on the axis of the guide. The image of the output end of the guide was then recorded photographically and photometrically.

Table II

VARIOUS QUANTITIES USED IN THE MODE LAUNCHING EXPERIMENTS

	HE_{11}		HE_{12}	
	Computed (mm)	Measured (mm)	Computed (mm)	Measured (mm)
Diameter of ring aperture	—	9.0	—	4.0
Focal length of a lens	—	150.0	—	30.0
Diameter of the input pattern	0.016	0.016	0.017	0.017
Diameter of the output pattern	—	0.017	—	0.019
Diameter of the fiber	—	0.015	—	0.015

In Table II, we list the important linear dimensions in these experiments. For each mode, the diameter of the ring filter and the focal length of the launching lens are given. These are followed by the observed and calculated diameters of the truncated diffraction patterns which describe the input irradiance distribution (the first and second zeros of the patterns for the $HE_{1,1}$ and $HE_{1,2}$ modes, respectively). The diameters of the observed output patterns at the exit face of the guide are then given, followed by the diameter of the core of the waveguide. As indicated by the diameter of the $HE_{1,2}$ output, this mode carries significant power in the surface wave outside the core.

The photographs in Fig. 5-8 show one of the ring masks and its diffraction pattern. Except for the scale, the patterns are essentially the same for both modes. Only the central spot is used for $HE_{1,1}$ excitation, the rest being blocked by a mask on the fiber face. The central spot and first ring are used for $HE_{1,2}$ excitation. Figure 5-8 also shows photometric traces across the center of the diffraction patterns for the two excitation conditions. The solid curve is for $HE_{1,1}$ and the broken curve is for $HE_{1,2}$. Both patterns were truncated at a radius of 7.5 μm by the fiber mask and fit the desired Bessel function within about 10%, except near the zeros. (The dashed curves in Figs. 5-11 and 5-12, which will be discussed later, indicate the exact discrepancies between these patterns and the desired Bessel function distributions.) The apparent asymmetries of the traces in Fig. 5-8 are due primarily to residual phase nonuniformity in the ring masks.

Figure 5-9 shows a photograph and photometric trace of the irradiance distribution at the output face of the waveguide for the $HE_{1,1}$ input corre-

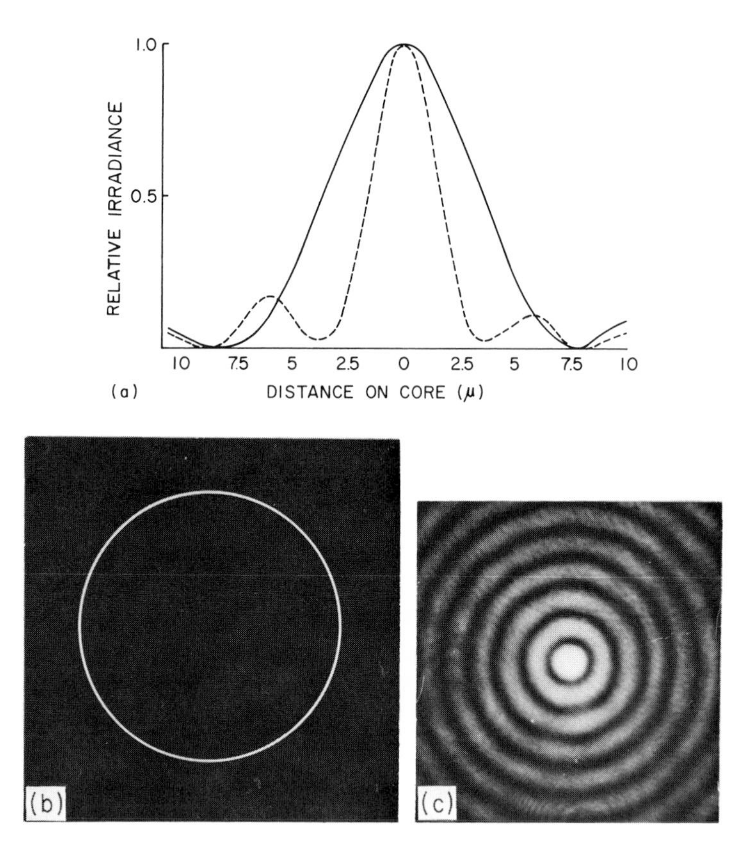

Fig. 5.8. Photometric trace (a) and photograph (c) of the diffraction pattern of a ring aperture (b). These diffraction patterns excite the $HE_{1,1}$ or $HE_{1,2}$ mode in a fiber.

sponding to the solid curve in Fig. 5-8. This distribution fits the theoretical $HE_{1,1}$ distribution to within 20% over most of its range. It displays, however, a noticeable asymmetry, the apparent consequence of a slight flattening of the fiber core noted visually in the image of its output face. (The detailed discrepancies between this distribution and the $HE_{1,1}$ distribution of a perfectly circular guide are shown by the solid curve in Fig. 5-11.)

Figure 5-10 shows a photograph and photometric trace of the output intensity distribution corresponding to the $HE_{1,2}$ excitation (the broken curve of Fig. 5-8). Asymmetry in the pattern is again apparent. The diameter of the scanning aperture corresponds to approximately 0.5 mm on this scale and is thus not capable of resolving well the zero of the pattern at approximately 3.7 mm. However, the weaker of the circular transverse electric field components can have an appreciable value for this mode. This can also explain in part the nonzero minimum.

The curves in Figs. 5-11 and 5-12 of the $HE_{1,1}$ and $HE_{1,2}$ modes, respectively, show the ratio of the observed input (dashed) and output (solid) irradiance distributions to those predicted theoretically. In the case of the

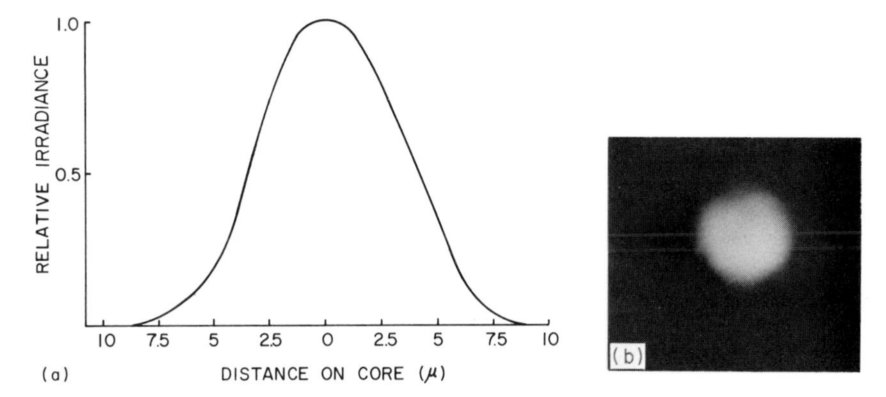

Fig. 5-9. Photometric trace and photograph of the $HE_{1,1}$ mode excited in a fiber by a ring aperture diffraction pattern.

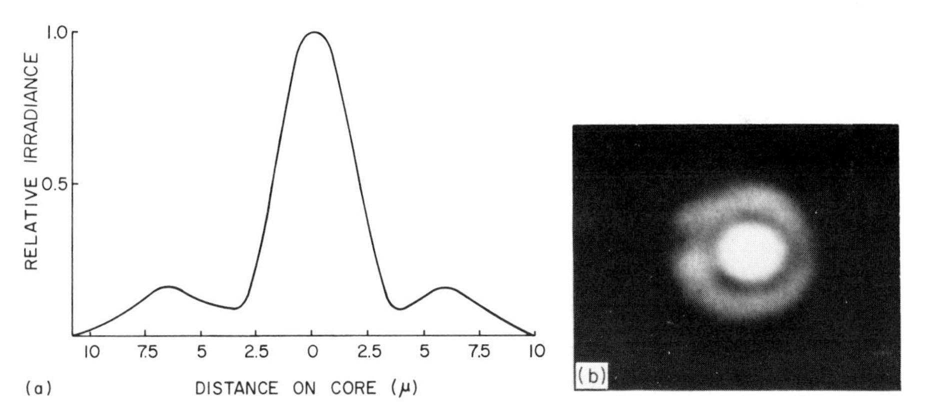

Fig. 5-10. Photometric trace and photograph of the $HE_{1,2}$ mode excited in a fiber by a ring aperture diffraction pattern.

output distributions, only the Bessel function describing the dominant component is considered. For the $HE_{1,1}$ mode (Fig. 5-11), the errors are apparently explicable in terms of the noncircularity in the fiber cross section. The structure of the error curve does not give any indication of the presence of other modes, indicating that any such modes present do not carry enough power to be detectable in our experiment. Figure 5-12, for the $HE_{1,2}$ mode, however, shows an unexpected structure near the axis in both the input and output distributions; the output follows, to some extent, the input. The latter fact indicates the probable presence of a small amount of power in some other mode or modes. It is apparent from Fig. 5-10, however, that the relative power of the unwanted mode to the $HE_{1,2}$ mode is small, since only a little structure appears in the output distribution. Filtering out other modes in the presence of so much relative power in the $HE_{1,2}$ mode would be quite difficult.

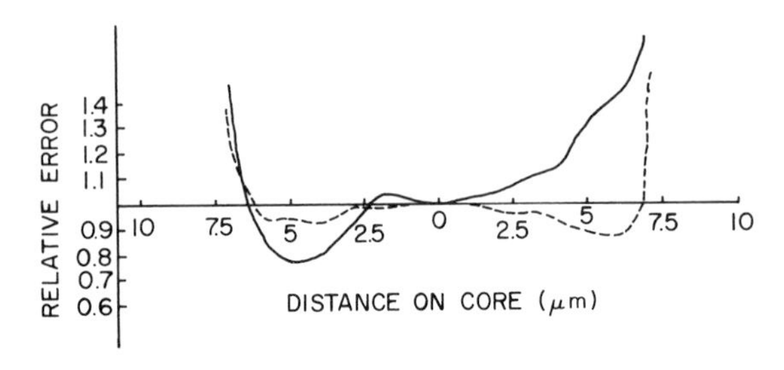

Fig. 5.11. Comparison of the observed pattern (——), input distribution (- - - - -), and theoretical value for the $HE_{1,1}$ mode (horizontal axis).

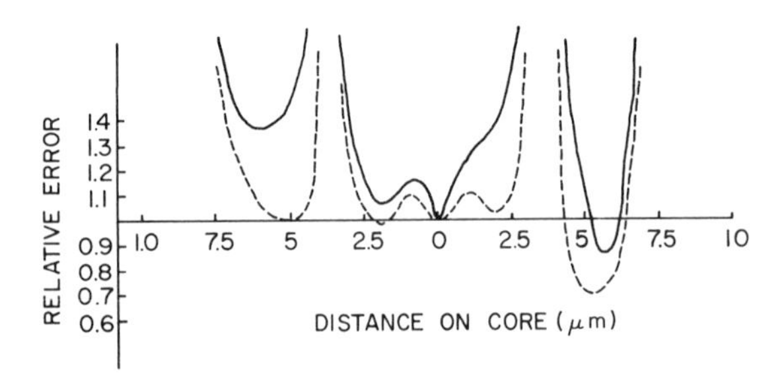

Fig. 5.12. Comparison of the observed pattern (——), input distribution (- - - - -), and theoretical value for the $HE_{1,2}$ mode (horizontal axis).

Thus, an arbitrary mode can be launched on a multimode dielectric waveguide with a double-path spatial filtering system composed of a single lens, beam splitters, plane mirrors, and relatively simple filters. The theoretical evaluation indicates that the discrimination is very high for all modes that confine most of their power to the waveguide core. Experiments with the $HE_{1,1}$ and $HE_{1,2}$ modes substantiate the theoretical predictions. For the general mode, more complicated filters than the simple uniform phase ring masks used here are needed. Sawatari discusses one realization of such filters in Appendix B.

REFERENCES

1. E. Snitzer and H. Osterberg, *J. Opt. Soc. Amer.* **51**, 499 (1961).
2. N. S. Kapany and J. J. Burke, *J. Opt. Soc. Amer.* **51**, 1067 (1961).
3. N. S. Kapany, J. J. Burke, and T. Sawatari *J. Opt. Soc. Amer.* **60**, 1178 (1970).
4. N. S. Kapany, J. J. Burke, and K. Frame, *Appl. Opt.* **4**, 1534 (1965).
5. A. W. Snyder, *J. Opt. Soc. Amer.* **56**, 601 (1966).
6. A. W. Lohmann and D. P. Paris, *Appl. Opt.* **6**, 1739 (1967).

CHAPTER 6

Near and Far-Field Observation of Circular Waveguide Modes

In this chapter, we will be concerned with the directly observable properties of the modes of fiber optical waveguides. We will begin with a qualitative description of the intensity distributions and polarizations observed when the exit end of a fiber is viewed with the aid of a microscope. Then, we will discuss the observations of the far-field (Fraunhofer region) radiation patterns of low-order modes. Finally, we will describe and illustrate a spatial filtering technique that yields a quantitative measure of the relative power in several modes propagating simultaneously in an optical waveguide.

A. QUALITATIVE DESCRIPTION OF OBSERVED MODE PATTERNS

It is obvious that the intensity distributions and polarizations observed at the output end of a fiber waveguide depend primarily on the excitation conditions. If, as in the earliest detailed optical experimental studies of the modes (*1*), the input fields are plane polarized, then we may expect the output radiation to be approximately plane polarized as well. However, because the $HE_{1,m}$ modes are the only ones that have a degenerate form that is nearly plane polarized, a combination of nondegenerate (different phase velocities) modes will usually be required to match the input field. As these modes propagate toward the exit end of the fiber, their phase differences lead to continuously changing intensity distributions and polarizations. When the combinations are relatively simple and the spectral bandwidth of the launching radiation is sufficiently narrow, we can observe such interesting phenomena as rotation of the plane of polarization and/or the nodal lines of a pattern, as discussed in Chapter 4, Section D.

Let us begin by restating the general expressions for the fields inside the fiber core, which we will then specialize to the interesting cases of near-cutoff ($\theta_1 \approx \theta_c$, $h \approx k_2$) and near-axial ($\theta_1 \approx \pi/2$, $h \approx k_1$) propagation. From Eqs. (4.58), we have

$$\begin{aligned} E_\pm &= E_x \pm iE_y = \pm(1 \pm A) f_{n\pm1}, \\ H_\pm &= H_x \pm iH_y = (ih/\omega\mu)[(k_1^2/h^2) \pm A] f_{n\pm1}, \\ E_z &= -(i\beta_1/h) f_n, \qquad H_z = (ih/\omega\mu) A E_z, \end{aligned} \tag{6.1}$$

where the functions

$$f_p = J_p(\beta_1 r) \exp(ip\phi + i\omega t - ihz); \qquad p = n-1,\ n,\ n+1,$$

describe all temporal and spatial variations of the fields. J_p is the Bessel function of the first kind of order p and

$$\beta_1^2 + h^2 = k_1^2 = \omega^2 \varepsilon_1 \mu = (2\pi/\lambda_0)^2 n_1^2.$$

For convenience, we are using the symbol A, instead of α, for the TM/TE ratio of the hybrid modes.

We now use these expressions to obtain coarse (see Chapter 4 for a more precise treatment) approximations for the far-from-cutoff conditions of the four basic mode types. Table I summarizes the results. It is clear from this table that the intensity distributions of the basic modes, in this approximation, are very simple. A mode of order m has m radial maxima in the fiber cross section and is circularly symmetric in intensity. The intensity vanishes at the core–coating interface. The polarization is circular, its phase at any given time in a plane $z =$ const varying with azimuth through $p + 1$ cycles for EH modes and $p - 1$ cycles for HE modes. To illustrate these basic modes, we sketch the fields in the plane $z = 0$ at time $t = 0$ for the $EH_{1,1}$ and $HE_{3,1}$ modes. The upper four diagrams of Fig. 6-1 show the electric vectors at eight azimuth locations around the ring of maximum intensity ($m = 1$) for these modes. As is clear from the table, the field varies radially as $J_2(u\varrho/a)$, with $J_2(u) \approx 0$ for all these modes. One-sided arrowheads in the illustrations indicate that the subject electric field rotates with time in the sense indicated by the arrowhead. Two-sided arrowheads indicate that the electric vector is linearly polarized.

By adding the fields of the basic modes in various combinations, as illustrated in the other parts of Fig. 6-1, we can obtain linearly polarized degenerate forms of the $EH_{1,1}$ mode (combination 1) or $HE_{3,1}$ mode (combination 2), circularly polarized fields with azimuth modulation propor-

Table I

Summary of Important Properties of Basic Modes of a Circular Waveguide in the Far-from-Cutoff Approximation

General mode designation	$EH_{n,m}$		$HE_{n,m}$	
Specific mode designation ($p = \lvert n \rvert$)	$EH^{+}_{p,m}$	$EH^{-}_{\bar{p},m}$	$HE^{+}_{-p,m}$	$HE^{-}_{\bar{p},m}$
A	1	−1	1	−1
Significant field components $f_n = J_n(u\varrho/a)e^{i\omega t + in\phi - ihz}$	$E_+, H_+ \Rightarrow f_{p+1}$ $E_- \approx 0$	$E_-, H_- \Rightarrow f_{-p-1}$ $E_+ \approx 0$	$E_+, H_+ \Rightarrow f_{-p+1}$ $E_- \approx 0$	$E_-, H_- \Rightarrow f_{p-1}$ $E_+ \approx 0$
u	$J_{p+1}(u) = 0$		$J_{p-1}(u) = 0$	
$h = [k_1{}^2 - (u^2/a^2]^{1/2}$	$h \approx k_1$ [see Eqs. (4.79) and (4.86)]		$h \approx k_1$ [see Eqs. (4.79) and (4.86)]	
Intensity distribution	$J^2_{p+1}(u\varrho/a)$		$J^2_{p-1}(u\varrho/a)$	

tional to cos 2ϕ as in combination 4, circularly symmetric, plane polarized distributions (combination 3), and plane polarized fields with azimuth modulation cos 2ϕ as in combination 5. The roughened lines in combinations 4 and 5 indicate nodal lines. As discussed in Chapter 4, there is a z-dependent modulation of the fields in combinations 3–5. On the other hand, slight perturbations in the shape of the guide may cause combinations 3–5 to be stable and thus the true modes of the perturbed guide.

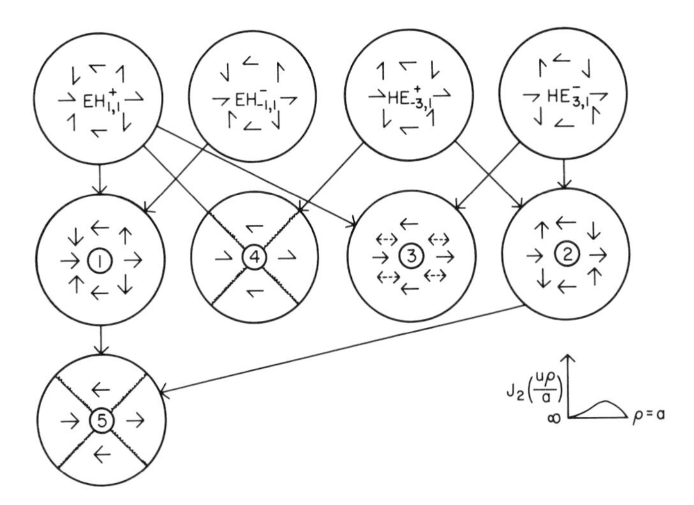

Fig. 6-1. Schematic illustration of the basic $EH_{1,1}$ and $HE_{3,1}$ modes far from cutoff (top) and some of the patterns arising from their mixture. The electric vectors are illustrated at eight azimuth positions in the plane $z = 0$ at time $t = 0$, around the radial ring of maximum intensity. Linearly polarized field vectors are shown with two-sided arrowheads. Circularly polarized fields are shown with one-sided arrowheads indicating the sense of rotation with increasing time.

In Fig. 6-2, we illustrate similar combinations of the $TE_{0,1}$, $TM_{0,1}$, and $HE_{2,1}$ modes in the approximation of near-axial propagation. For the $TE_{0,1}$ modes, we have $A = \infty$ in Eqs. (6.1), while for the $TM_{0,1}$ modes, $A = 0$. Accordingly, these two basic modes have no circularly polarized degenerate forms, unlike the hybrid modes. Apart from this difference in polarization, however, their properties are described adequately by Table I in the far-from-cutoff approximation.

The conditions for modes near cutoff are also important, because these modes, as a consequence of their significant power in the evanescent surface wave, present a larger effective aperture to the launching radiation and are often preferentially excited. Table II, which is based on the developments of Chapter 4, summarizes the properties of the modes near cutoff. We see from the table that the basic modes are again circularly symmetric in intensity, with m radial maxima. The last maximum is quite close to the edge of the core because of the cutoff conditions that must be satisfied by the scale parameter u. Figure 4-1 (Chapter 4) can be used to visualize the radial geometry of the general pattern. The equations defining the cutoff parameter u in Table II show that in fiber optical guides with $n_1 \approx n_2$, the $HE_{p+1,m}$ and $EH_{p-1,m}$ modes are nearly degenerate, so that combinations like those

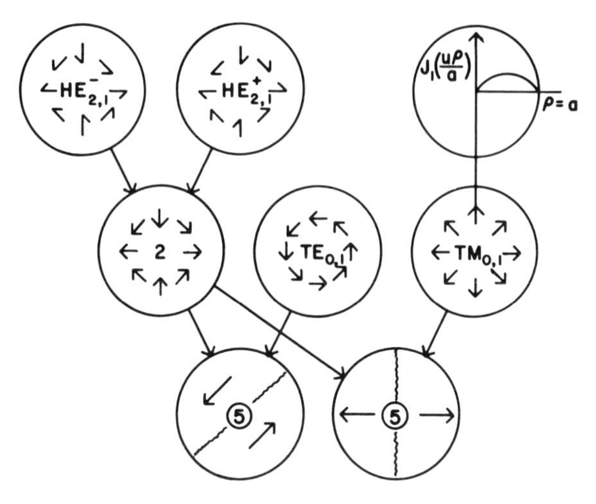

Fig. 6-2. Schematic illustration of the basic $TE_{0,1}$, $TM_{0,1}$, and $HE_{2,1}$ modes (top) and patterns arising from their mixture. Rough lines indicate nodes. The electric vectors are illustrated at eight azimuth positions in the plane $z = 0$ at time $t = 0$, around the radial ring of maximum intensity. Linearly polarized field vectors are shown with two-sided arrowheads. Circularly polarized fields are shown with one-sided arrow heads indicating the sense of rotation with increasing time.

of Figs. 6-1 and 6-2 are again observable with modes near cutoff. Additionally, it can be noted that the $HE_{1,m+1}$ mode cuts off at the same value of u as the $EH_{1,m}$ mode, so that mixtures of the $HE_{1,m+1}$, $EH_{1,m}$, and $HE_{3,m}$ modes are common in fiber guides. Those that are observed with plane polarized launching radiation have been studied by Snitzer and Osterberg (*1*). Many other combinations of these six (two degenerate forms of each) modes are observable under different launching conditions. The interested reader can readily predict them from Table II, following the graphical method of Figs. 6-1 and 6-2 or the analytical approach of Chapter 4.

Table II

SUMMARY OF THE IMPORTANT PROPERTIES OF BASIC MODES OF CIRCULAR WAVEGUIDE AT CUTOFF

General mode designation	$EH_{n,m}$		$HE_{n,m}$	
Specific mode designation ($p = \|n\|$)	$EH^{+}_{p,m}$	$EH^{-}_{-p,m}$	$HE^{+}_{-p,m}$	$HE^{-}_{p,m}$
A	n_1^2/n_2^2	$-n_1^2/n_2^2$	1	-1
Significant field components $f_n = J_n(u\varrho/a)e^{in\phi+i\omega t-ihz}$	$E_+, H_+ \Rightarrow f_{p+1}$ $H_- \approx 0$	$E_-, H_- \Rightarrow f_{-p-1}$ $H_+ \approx 0$	$E_+, H_+ \Rightarrow f_{-p+1}$ $E_- \approx 0$	$E_-, H_- \Rightarrow f_{p-1}$ $E_+ \approx 0$
u	$J_p(u) = 0$ $J_0(u) = 0$ for TE and TM		$\dfrac{J_{p-2}(u)}{J_p(u)} = -\dfrac{n_1^2 - n_2^2}{n_1^2 + n_2^2}; \; p > 1$ $J_1(u) = 0; \; p = 1$	
$h = [k_2^2 + (q^2/a^2)]^{1/2}$	$\approx k_2$		$\approx k_2$	
Intensity distributions	$J^2_{p+1}(u\varrho/a)$		$J^2_{p-1}(u\varrho/a)$	

B. RADIATION PATTERNS OF CIRCULAR WAVEGUIDE MODES

In this section, we will obtain theoretically, and verify experimentally, predictions of the intensity distribution in the far radiation field of a waveguide excited in a single mode. The theoretical method is a classical one described in detail by Schelkunoff (*2*). As applied to our problem, it is the vector field equivalent of the scalar Huygens–Kirchhoff theory of classical optics. Accordingly, the mode fields incident on the exit plane of the fiber waveguide are replaced by equivalent electric and magnetic current sheets of densities proportional to the tangential components of the electric and magnetic fields of the mode. These current sheets constitute continuously distributed arrays of electric and magnetic dipoles that act as sources for the fields observed in the uniform half-space beyond the exit plane of the fiber. It is well known that, in the presence of sources, Maxwell's equations are most easily solved in terms of vector and scalar potentials. The solutions for the potentials are expressible as sums or integrals over the source distribution. The corresponding fields are then found from the potentials by differentiation. For the dielectric-rod antenna at microwave frequencies, the equivalent sources are located all along and around the surface of the rod, as well as at its end. However, in the fiber waveguide, whose thick, low-index coating is surrounded by an absorbing material, it is sufficient to consider only the end of the guide.

1. Derivation of Expressions for the Radiation Patterns

The radiation patterns of circular, dielectric-rod waveguides have been the subject of many theoretical and experimental studies at microwave frequencies (*3–5*). The results of several of these are reviewed in a monograph by Kiely (*3*). Horton and Watson (*4*) were apparently the first to apply Schelkunoff's method to the dielectric waveguide. They found it necessary to include equivalent sources all along the surface of the rod. We shall consider only the end plane. To avoid extensive derivation, we will employ Schelkunoff's formalism (*2*, Chapter IX) translating his formulas, as necessary, to the circular vector component forms used throughout this book.

Consider a dielectric waveguide of radius a, refractive index n_1, and axis coincident with the z axis of a rectangular coordinate system. The waveguide is embedded in a medium of index n_2 that fills the space $z < 0$ not occupied by the waveguide. The space $z > 0$ is occupied by a homogeneous medium of refractive index n_0. We assume that a waveguide mode (labeled by subscripts n, m) is launched on the rod at $z = -\infty$ and radiates into the homogeneous half-space $z > 0$. The field components for the incident wave inside the waveguide are then given by Eqs. (6.1). The fields outside the rod can be obtained from Eqs. (6.1) by multiplying all components by the factor $uJ_n(u)/vH_n(v)$ and by substituting the functions

$$g_p = H_p(\beta_2 r) \exp(ip\phi + i\omega t - ihz), \qquad p = n,\ n-1,\ n+1,$$

for f_p and the parameters k_2 and β_2 for k_1 and β_1. Here, H_p is the Hankel function of the second kind of order p and

$$\beta_2{}^2 + h^2 = k_2{}^2 = \omega^2 \varepsilon_2 \mu = (2\pi/\lambda_0)^2 n_2{}^2,$$
$$u = \beta_1 a, \qquad v = \beta_2 a.$$

The parameters h, A, β_1, β_2, u, and v are all characteristic of the (n, m) mode and should be understood to be so subscripted. The subscripts have been omitted for convenience. The parameter A, through the ratio

$$\gamma = \frac{1-A}{1+A} = -\frac{[J_{n+1}(u)/uJ_n] - [H_{n+1}(v)/vH_n]}{[J_{n-1}(u)/uJ_n] - [H_{n-1}(v)/vH_n]}, \tag{6.2}$$

describes the polarization of the mode in the transverse plane, as is clear from Eqs. (6.1). For $n > 0$, $A_n > 1$ for $EH_{n,m}$ modes, and $-1 < A_n < 0$ for $HE_{n,m}$ modes. For $n < 0$, $A_n = -A_{-n} < -1$ for $EH_{n,m}$ modes, and $0 < A_n < 1$ for $HE_{n,m}$ modes. For $n = 0$, $A = 0$ for $TM_{0,m}$ modes and $A \to \infty$ for $TE_{0,m}$ modes. Equations (6.1) thus describe all of the modes, including degeneracies. Since, in the plane $z = 0$, the boundary conditions are independent of azimuth, we can retain complete generality by restricting our attention to the fields given explicitly by Eqs. (6.1), assuming $n \geq 0$. The mode pattern as well as the diffracted field will accordingly be circularly symmetric. To obtain the diffraction pattern associated with mode combinations yielding an azimuth-dependent mode pattern, it is sufficient to impose the same azimuthal modulation on the diffracted field as is observed in the direct image of the output end of the fiber. With this understanding of Eqs. (6.1), we can proceed with the application of Schelkunoff's theory.

The incident field, Eqs. (6.1), is replaced in the plane $z = 0$ by magnetic and electric current sheets of density $\mathbf{M} = \mathbf{E} \times \mathbf{z}$ and $\mathbf{J} = \mathbf{z} \times \mathbf{H}$, where $\mathbf{z}$ is the unit vector in the z direction (*2*, Chapter IX). We have, accordingly,

$$M_\pm = M_x \pm iM_y = \mp E_\pm,$$
$$J_\pm = J_x \pm iJ_y = \pm H_\pm.$$

The components of the radiation vectors (*2*, p. 332) are then given by

$$\left.\begin{matrix} L_\pm \\ N_\pm \end{matrix}\right\} = \int_0^{2\pi}\int_0^\infty \left\{\begin{matrix} M_\pm(r,\phi) \\ J_\pm(r,\phi) \end{matrix}\right\}\{\exp[ik_0 r(\sin\theta)\cos(\phi' - \phi)]\} r\, dr\, d\phi,$$

where θ is the angle of diffraction with reference to the z axis ($\theta = 0$) and ϕ' is the azimuth angle about the z axis.

The intensity in the far field is then given by (*2*, p. 333)

$$\begin{aligned} I &= I_{11} + 2I_{12} + I_{22}, \\ I_{11} &= \frac{(\mu_0/\varepsilon_0)^{1/2}}{8\lambda_0^2}\left\{\frac{1+\cos^2\theta}{4}(|N_+|^2 + |N_-|^2)\right. \\ &\qquad \left. - \frac{1-\cos^2\theta}{2}\operatorname{Re}[N_+N_-^*\exp(-2i\phi')]\right\}, \\ I_{22} &= \frac{(\varepsilon_0/\mu_0)^{1/2}}{8\lambda_0^2}\left\{\frac{1+\cos^2\theta}{4}(|L_+|^2 + |L_-|^2)\right. \\ &\qquad \left. - \frac{1-\cos^2\theta}{2}\operatorname{Re}(L_+L_-^*\exp(-2i\phi')]\right\}, \\ I_{12} &= \frac{-1}{8\lambda_0^2}\operatorname{Re}\left\{\frac{\cos\theta}{2i}[N_+L_-^*\exp(-2i\phi') - N_-L_+^*\exp(2i\phi')]\right\}. \end{aligned} \tag{6.3}$$

The evaluation of Eqs. (6.3) is straightforward, since the integrals defining the components of the radiation vectors are readily reduced to one-dimensional integrals involving only the squares of Bessel functions. The result is given by

$$I = CF(x, u, q), \tag{6.4}$$

where

$$\begin{aligned} x &= k_0 a\sin\theta, \qquad u = k_1 a\sin\theta_1, \\ q &= v/i = k_0 a(n_1^2\cos^2\theta_1 - n_2^2)^{1/2} \\ C &= (\varepsilon_0/\mu_0)^{1/2}(k_0^2 a^4)(1 + A)^2 u^2 J_n^2(u)/32, \\ F(x, u, q) &= [(G_+ - \gamma G_-)\cos\theta_1\cos\theta \\ &\qquad -(F_+ - \gamma F_-)]^2 + [(G_+ + \gamma G_-)\cos\theta_1 + (F_+ + \gamma F_-)\cos\theta]^2, \end{aligned}$$

and

$$G_+ = \left(\frac{\delta_1}{u^2 - x^2} + \frac{\delta_2}{q^2 + x^2}\right) J_{n+1}(x) - xJ_n(x)\left(\frac{\delta_1 \mathscr{J}}{u^2 - x^2} + \frac{\delta_2 H}{q^2 + x^2}\right),$$

$$G_- = \left(\frac{\zeta_1}{u^2 - x^2} + \frac{\zeta_2}{q^2 + x^2}\right) J_{n-1}(x) - xJ_n(x)\left(\frac{\zeta_1 \mathscr{J}'}{u^2 - x^2} + \frac{\delta_2 H'}{q^2 + x^2}\right),$$

$$F_+ = \left(\frac{1}{u^2 - x^2} + \frac{1}{q^2 + x^2}\right) J_{n+1}(x) - xJ_n(x)\left(\frac{\mathscr{J}}{u^2 - x^2} + \frac{H}{q^2 + x^2}\right),$$

$$F_- = \left(\frac{1}{u^2 - x^2} + \frac{1}{q^2 + x^2}\right) J_{n-1}(x) - xJ_n(x)\left(\frac{\mathscr{J}'}{u^2 - x^2} + \frac{H'}{q^2 + x^2}\right),$$

and

$$\delta_{1,2} = \frac{(k_{1,2}^2/h^2) + A}{1 + A}, \qquad \zeta_{1,2} = \frac{(k_{1,2}^2/h^2) - A}{1 - A},$$

$$\mathscr{J} = \frac{J_{n+1}(u)}{uJ_n(u)}, \qquad \mathscr{J}' = \frac{J_{n-1}(u)}{uJ_n(u)};$$

$$H = \frac{H_{n+1}(v)}{vH_n(v)}, \qquad H' = \frac{H_{n-1}(v)}{vH(v)}.$$

The angle θ_1 is the angle between the z axis and the normals to the plane waves that comprise the waveguide mode.

For the special case $n_2 \simeq n_1$, it can be shown that $\gamma \gg 1$ for $\mathrm{EH}_{n,m}$ modes (n assumed > 0). We then find that I is given approximately by

$$I \simeq C({F_+}^2 + {G_+}^2 \cos^2 \theta_1)(1 + \cos^2 \theta). \tag{6.5}$$

With the knowledge that $\mathscr{J} \simeq H$ and ${k_1}^2 \simeq {k_2}^2$ for this case, we see that F_+ and G_+ are of essentially the same form. Equation (6.5) thus reduces to

$$I \simeq C' \left\{ \frac{J_{n+1}(x) - xJ_n(x)[J_{n+1}(u)/uJ_n(u)]}{(u^2 - x^2)(q^2 + x^2)} \right\}^2 \tag{6.6}$$

where

$$C' = C(1 + \cos^2 \theta)(1 + \cos^2 \theta_1)(u^2 + q^2)^2,$$
$$x = k_0 a \sin \theta, \qquad u = k_1 a \sin \theta_1,$$
$$q = k_0 a({n_1}^2 \cos^2 \theta_1 - {n_2}^2)^{1/2}.$$

To obtain the radiation pattern (apart from constant factors) for $\mathrm{HE}_{n,m}$ modes in the same approximation, one need only replace J_{n+1} by J_{n-1} in Eq. (6.6). Though this equation is not generally a highly accurate approximation, it can be seen from the exact equations (6.3) that it nevertheless describes the general characteristics of the radiation field quite well. We will, therefore, use it to discuss the general diffraction pattern of an (n, m) mode. The corresponding mode pattern is proportional to $J_{n+1}^2(\beta_1 r)$.

2. General Discussion of the Radiation Pattern

For $q \gg u > 1$ (the far-from-cutoff case), the pattern has its principal maximum when $x \simeq u$, i.e., $\theta = \theta_{pr} \simeq \sin^{-1}[(k_1/k_0) \sin \theta_1]$. This is the angle of refraction for plane waves incident on the plane $z = 0$ at the characteristic angle θ_1 of the mode. For $m = 1$ (only one radial maximum in the mode pattern), there are no subsidiary maxima at angles θ less than θ_{pr}. For $m > 1$, there are $m - 1$ subsidiary maxima between $\theta = 0$ and $\theta = \theta_{pr}$. For angles greater than θ_{pr}, an infinite number of subsidiary maxima are predicted, their strengths decreasing monotonically as θ increases and all being much weaker than the principal maximum. The solid curve in Fig. 6-3 illustrates the far-from-cutoff case for the $TM_{0,2}$ mode ($n_1 = 1.75$, $n_2 = 1.52$). The slash mark at the peak indicates the angle $\theta_0 = \sin^{-1}[(k_1/k_0) \sin \theta_1]$. [It should be noted that Eq. (6.6) remains a good approximation for $TM_{0,m}$ modes, even though $\gamma = 1$.]

The effect of the factor $1/(q^2 + x^2)$ is to increase the strength of the forward lobes at the expense of the principal lobe while at the same time shifting θ_{pr} to an angle considerably less than $\sin^{-1}[(k_1/k_0) \sin \theta_1]$. For $q \ll u$ (the near-cutoff condition), most of the radiation is in the first lobe.

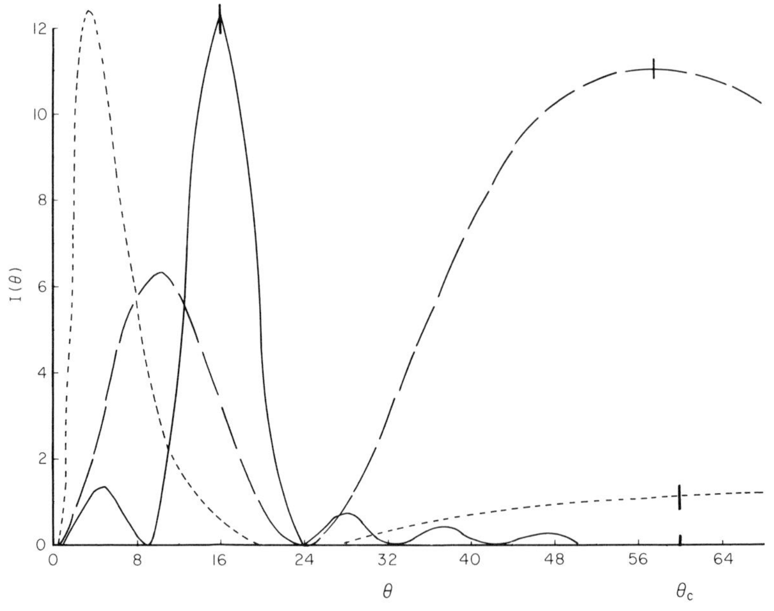

Fig. 6-3. Theoretical far-field pattern of $TM_{0,2}$ mode near cutoff [(- - - - -) $ka = 6.437$, $q \ll u$], an intermediate case [(——) $ka = 7.376$, $q \simeq u$], and far from cutoff [(————) $ka = 24.34$, $q \gg u$]. Core index $n_1 = 1.75$. Coating index $n_2 = 1.52$.

As $q \to 0$, the angle at which the intensity is strongest approaches $\theta = 0$. The short-dashed curve of Fig. 6-3 exemplifies the near-cutoff case for the $TM_{0,2}$ mode.

For the intermediate case, q is of the order of u, the mode angle θ_1 is close to the critical angle, and a substantial fraction of the radiated power has been taken from the principal lobe and shifted to the first lobe. This is illustrated by the long-dashed curve of Fig. 6-3. The slash mark again indicates the position of the angle θ_0 for the $TM_{0,2}$ mode.

The $HE_{1,m}$ modes, which have a maximum in the forward direction, are somewhat exceptional. For these modes, θ_{pr} is significantly smaller than $\theta_0 = \sin^{-1}[(k_1/k_0) \sin \theta_1]$ and the central lobe is generally relatively strong. For the $HE_{1,1}$ mode, the maximum intensity is always in the forward direction. For modes other than the $HE_{1,m}$, the forward direction is always a null, except in the limit as $q \to 0$, when the near-forward, ring-shaped lobe coalesces to a point. Since the characteristic angle θ_1 of the mode can take on any value in the range $0 < \theta_1 < \sin^{-1}[1 - (n_2{}^2/n_1{}^2)]^{1/2}$, the angle θ_{pr} of the principal maximum can vary between 0 and $\sin^{-1}(n_1{}^2 - n_2{}^2)^{1/2}$.

The curves of Fig. 6-3 were obtained by evaluating the complete equation (6.4). We will present further results of such calculations in the following section.

3. Theoretical Calculations of Radiation Pattern

In this section, we treat, theoretically, several particular cases of interest for the purpose of gaining a fuller understanding of the radiation characteristics of optical dielectric waveguides.

Figure 6-4 indicates the changes in the radiation pattern of the $TM_{0,1}$ mode from the intermediate case ($u = 2.953$, $q = 2.4$, $ka = 30.84$, lower curve) to the near-cutoff case ($u = 2.434$, $q = 0.2$, $ka = 19.79$, upper curve) for a waveguide with core index 1.525 and coating index 1.52. A second intermediate case ($u = 2.652$, $q = 1.0$, $ka = 22.97$, middle curve) is also shown. The slash marks on the curves indicate the angles $\theta_0 = \sin^{-1}[(k_1/k_0) \sin \theta_1]$. The slash mark on the abscissa indicates the angle in the field that corresponds to critical angle propagation in the fiber core. The $TE_{0,1}$ and $HE_{2,1}$ modes are described, except for polarization and very fine detail, by these same curves.

Three patterns for the $TM_{0,2}$ mode near cutoff for $n_1 = 1.525$ and $n_2 = 1.520$ are plotted to a linear scale in Fig. 6-5. The slash marks again indicate the angle θ_0 and the complement of the critical angle. The upper curve pertains to values of ka, u, and q equal, respectively, to 45.52, 5.584,

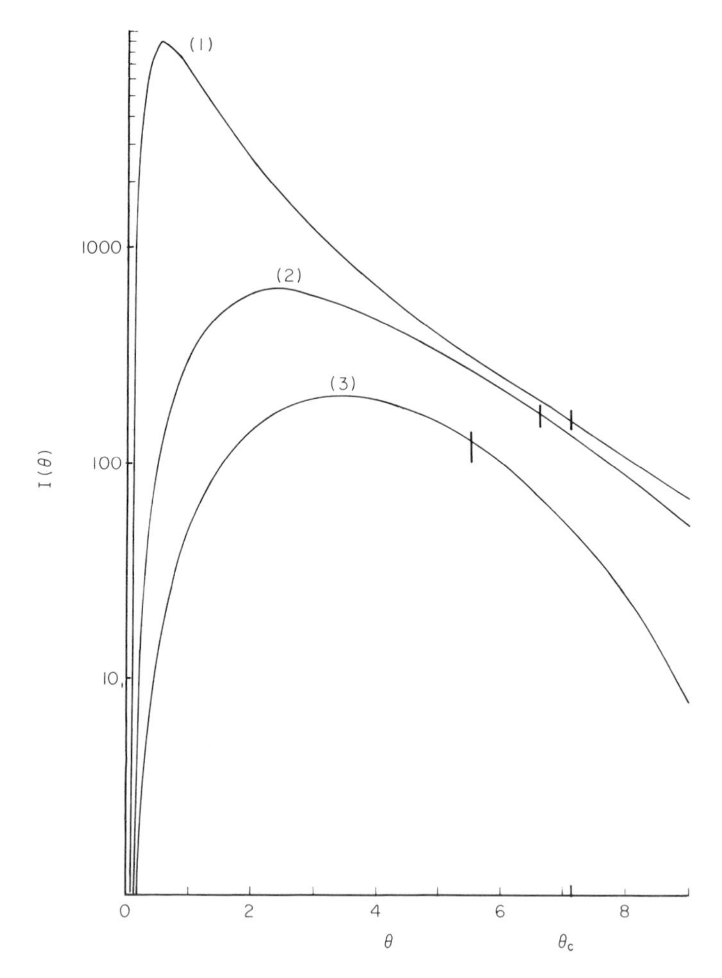

Fig. 6-4. Theoretical far-field patterns of $TM_{0,1}$ mode near cutoff: (1) $ka = 19.79$, $q = 0.2$, $u = 2.434$; and two intermediate cases: (2) ka, q, $u = 22.97$, 1.0, 2.652 and (3) 30.84, 2.4, 2.953, respectively, $n_1 = 1.525$, $n_2 = 1.52$. Slash marks indicate angle θ_0.

and 0.6; the middle curve 46.45, 5.643, and 1.0; the lower curve 48.23, 5.732, and 1.6. The outer ring shows no significant variation over this range. These curves substantially describe the $TE_{0,2}$ and $HE_{2,2}$ modes as well.

Figures 6-6(a) and 6-6(b) show $HE_{2,5}$ radiaton patterns for waveguides with $n_1 = 1.62$, $n_2 = 1.52$. Figure 6-4(a) is the near-cutoff case ($ka = 26.68$, $u = 14.95$, $q = 0.2$), where the innermost ring is very strong and almost in the forward direction. The fourth ring is too weak to appear on this scale, being an order of magnitude less than the second ring. The third ring is just strong enough to be represented. At 46° (not shown), the first of the subsidiary maxima beyond θ_{pr} (33° in this case) appears, its value two orders of magnitude below that at θ_{pr}. For all practical purposes, therefore, the radiation pattern has the same number of rings (five in this case) as the mode pattern.

Figure 6-6(b) shows the $HE_{2,5}$ radiation pattern for an intermediate case ($ka = 42.18$, $u = 15.78$, $q = 17.6$). It is seen that the angle of the principal maximum θ_{pr} is just slightly less than $\theta_0 = \sin^{-1}[(k_1/k_0)\sin\theta_1]$, indicated by the slash mark. The first subsidiary maximum beyond θ_{pr} occurs at about 28.5° and is just barely in evidence on the figure. The pattern was not calculated beyond this angle. As was noted in the earlier discussion, the far-from-cutoff condition ($q \gg u$) gives a pattern with a strong maximum at θ_{pr} and subsidiary maxima on each side, the maxima beyond θ_{pr} taking on values of the same order of magnitude as those less than θ_{pr}. In this far-from-cutoff condition, the waveguide radiation pattern appears similar to the far-field diffraction pattern of a system of coherent plane waves incident from all azimuths at an angle θ_0 on an aperture of radius a in an opaque screen.

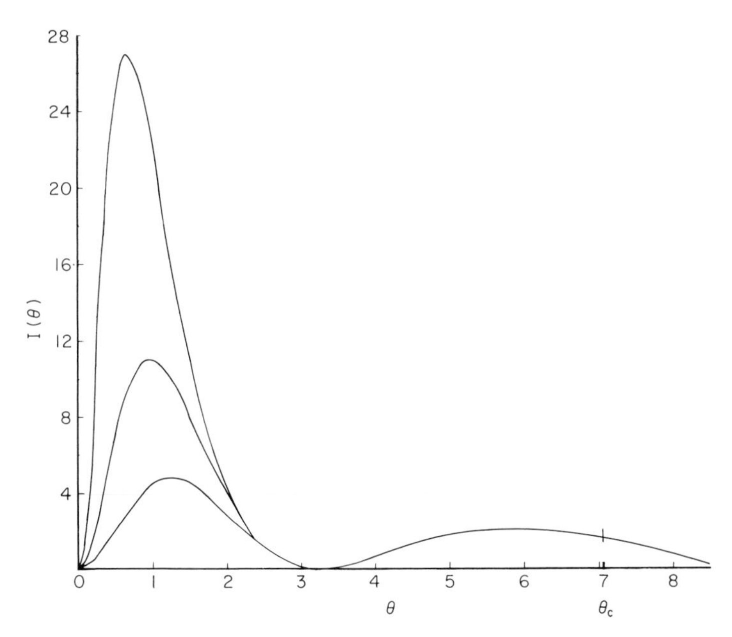

Fig. 6-5. Theoretical far-field patterns of $TM_{0,2}$ mode as it approaches cutoff ($q = 0$). Lower curve: $ka = 48.23$, $q = 1.6$, $u = 5.732$; middle curve: $ka = 46.45$, $q = 1.0$, $u = 5.643$; upper curve: $ka = 45.52$, $q = 0.6$, $u = 5.584$. Slash mark indicates angle θ_0. $n_1 = 1.525$, $n_2 = 1.520$.

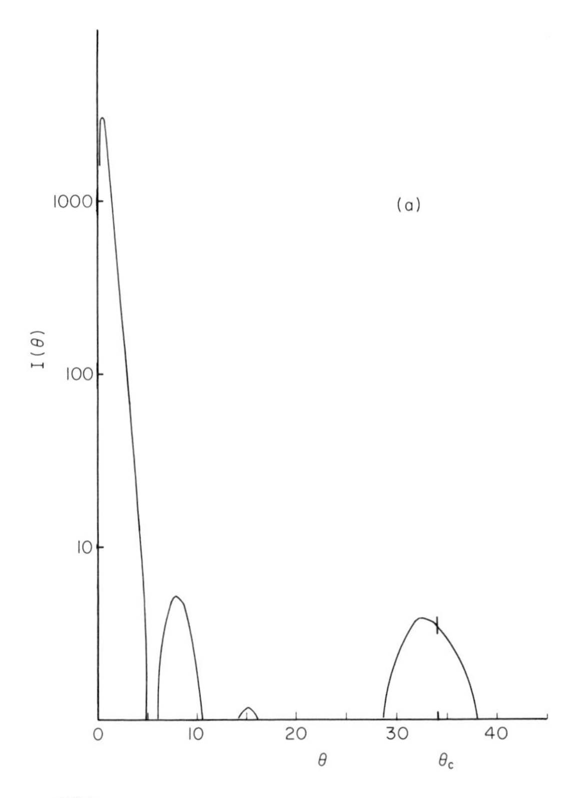

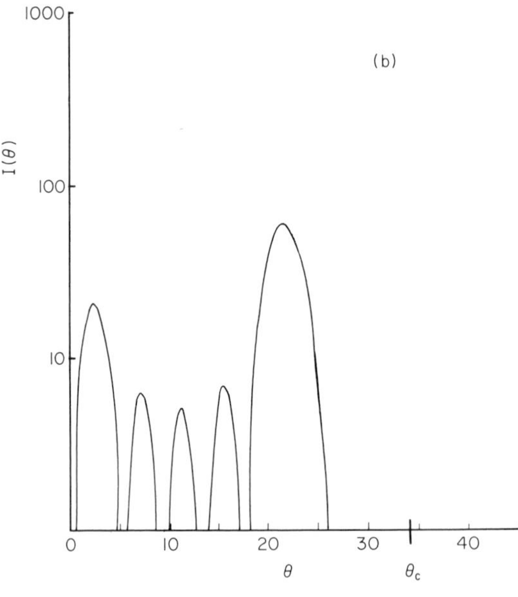

Fig. 6-6. $HE_{2,5}$ radiation pattern near cutoff: (a) $ka = 26.68$, $q = 0.2$, $u = 14.95$, and intermediate (b) $ka = 42.18$, $q = 17.6$, $u = 15.78$. $n_1 = 1.62$, $n_2 = 1.52$.

4. Experimental Observation of Waveguide Radiation Patterns

To verify the theory experimentally, we chose optical fibers made of Corning Signal-Yellow filter glass No. 3-77 ($n_1 = 1.525$) coated in soda lime glass ($n_2 = 1.5188$). Fibers of these materials are of good optical quality. Their low numerical aperture permits the study of waveguide modes with fibers of reasonable size (1–10 μm) and their light-absorbing property provides an added degree of mode selection.

The fibers were drawn in a three-step process. The filter glass was first drawn to yield rods of diameter 0.5 mm. These were inserted in thick-walled soda lime capillary tubing (o.d. 6 mm, i.d. 0.5 mm) and drawn to yield coated fibers of outside diameter 0.5 mm. These fibers were then inserted in a second length of capillary tubing for the final drawing, which provided fibers of the required core sizes 9.8 cm long. A film of black paint was deposited on the outer surface of the capillary tubing before drawing. This serves to damp out stray radiation in the dielectric coating, a serious source of noise in these experiments.

Waveguide modes are launched on these fibers by focusing the image of a back-lighted pinhole (diameter 500 μm) onto the entrance end of the fibers with a 10× microscope objective (N.A. = 0.25) as shown in Fig. 6-7. The light is provided by a mercury arc source, dispersed by an Engis Equipment Co. Model S 05-02 monochromator. The mode patterns (im-

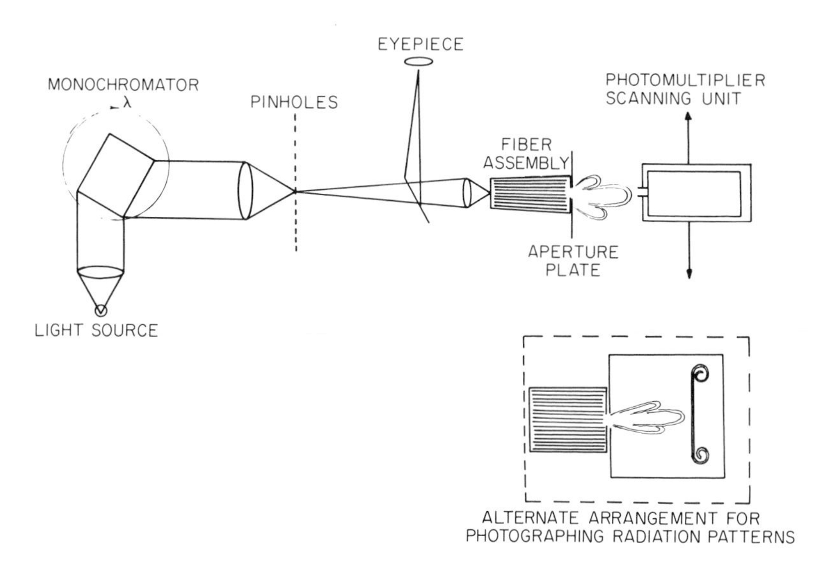

Fig. 6-7. Schematic of experimental arrangement for far-field pattern measurements.

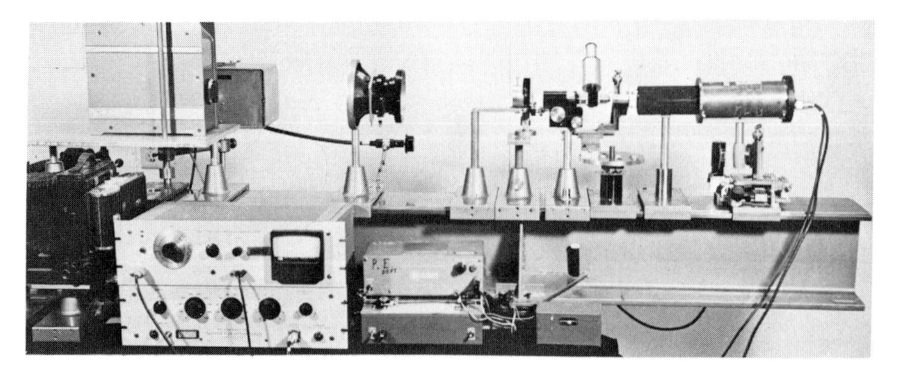

Fig. 6-8. Photograph of apparatus.

ages of the output ends of the fibers) are viewed with a high-power microscope. We used two methods to view the radiation pattern. In the first, suggested by Snitzer and Osterberg (*1*), a $10\times$ telescope eyepiece is used to view the back focal plane of the microscope objective employed to observe the mode pattern. However, it was generally more satisfactory to observe the pattern directly by removing the microscope and placing the eye directly behind the output end of the fiber.

As indicated schematically in Fig. 6-7, photographs of the far-field radiation pattern were made with the fibers radiating directly into the camera housing to expose the film, located 5.4 cm from the fiber. A photomultiplier unit, displaced the same distance from the output plane, was used to scan the radiation field, its pickup aperture (diameter 1 mm) traveling in a line perpendicular to the fiber axis. The linear, rather than circular, scan was adequate for these low-N.A. fibers that do not radiate significantly at angles greater than 8° from the forward direction.

A photograph of the apparatus, with the photomultiplier in position, is shown in Fig. 6-8. The amplified signal from the photomultiplier was used to drive a strip-chart recorder. The large photographic objective shown served as a condenser, focusing light from the monochromator onto the pinhole source, which is mounted in front of the launching microscope. Also apparent is an auxiliary microscope used to position the image of the pinhole at the desired position.

5. Identification of Observed Modes

The cross sections of the fibers studied in this experiment were not perfectly circular; all appear to be slightly flattened along the circumference of one small sector. As a direct consequence of this, the mode patterns observed were not precisely those of the circular dielectric waveguide. Nevertheless, they could be described in terms of particular combinations of circular waveguide modes. These same combinations are also observable in circular waveguides (*1*), where, however, their characteristics are somewhat different. These differences are most readily described with the aid of Fig. 6-9, which indicates schematically the observed mode patterns. The numbers at the top of each schematic designate the modes of our slightly noncircular waveguides, while those at the bottom designate the corresponding circular modes. The former are labeled in the same way as are laser interferometer modes (*6*), the first number indicating the number of azimuthal nodal lines in the transverse fields, the second the number of radial nodal rings. The circular modes, on the other hand, have been labeled, historically, according to their longitudinal fields. The schematics have been drawn to indicate that the circular modes, when excited individually in their simplest forms, will be manifested by circularly symmetric mode patterns. When they appear as coherent combinations of the designated modes, however, the

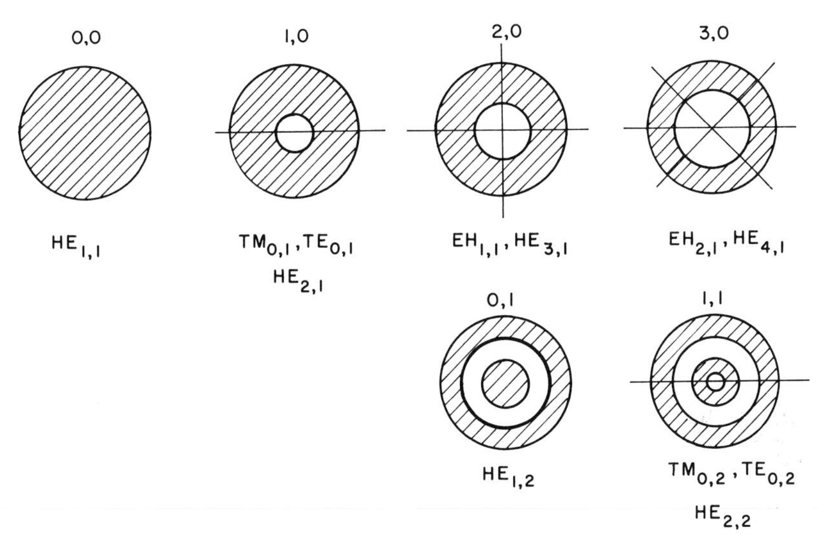

Fig. 6-9. Schematic of cross-section energy distributions of fiber core for the modes studied. (Cross-hatched areas indicate high flux density.) Upper label describes experimental modes, the first digit giving the number of azimuthal nodal lines, the second the number of radial nodes. Lower labels refer to circular waveguide modes which combine coherently to describe the observed modes. The illustrated circularly symmetric distributions pertain to the simplest form of individual circular modes. Azimuthal lines dictate nodes arising from coherent combinations. Each mode in second row can combine coherently with the mode pictured above it.

patterns may be azimuth-modulated, as indicated by the lines; the degree of modulation depends on the ratio of the length of the guide to the free-space wavelength. Thus, the mode combination $EH_{2,1} + HE_{4,1}$, if observed as a six-lobed, nearly linearly polarized pattern with unit modulation at one frequency, will manifest a continuously varying degree of polarization and modulation as the wavelength is varied, as well as a rotation of the pattern. The analogous (3, 0) modes, however, have a six-lobed pattern that does not manifest these variations with changing wavelength.

Similarly, the (1, 0) modes appear as stable two-lobed patterns that may take four possible forms. Two of these have the patterns indicated in Fig. 6-6 and are observed to be approximately linearly polarized in perpendicular directions. The other two are rotated 90° with respect to the pattern shown.

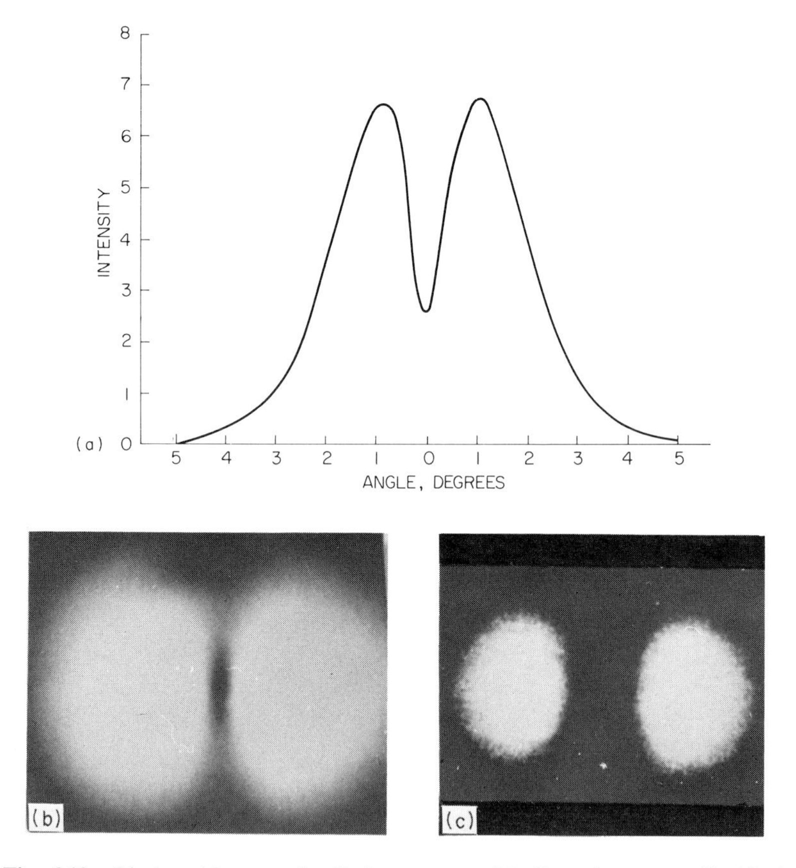

Fig. 6-10. Photometric trace of radiation pattern of (1, 0) mode near cutoff with photographs of mode pattern (right) and radiation pattern.

The nodal line is either perpendicular or parallel to the flattened portion of the fiber cross section. The (1, 0) modes also appear in coherent combinations, some of which take the form, at particular wavelengths, of pure circular modes.

Figure 6-10 shows photographs of the (1, 0) mode near cutoff and a photometric scan through the center of its radiation pattern. The mode pattern is shown in the photo at the right, the radiation pattern at the left. The latter does not show any of the side lobes predicted by theory, because they are too weak to be seen at this scale. The patterns are unpolarized. It can be seen that the mode is not perfectly symmetric about the nodal line and that the lobes are somewhat broader than expected. Both effects are apparently direct consequences of the asymmetry in the fiber cross section and preclude the possibility of comparing the details of the pattern with theoretical predictions. There is another way to effect a meaningful comparison,

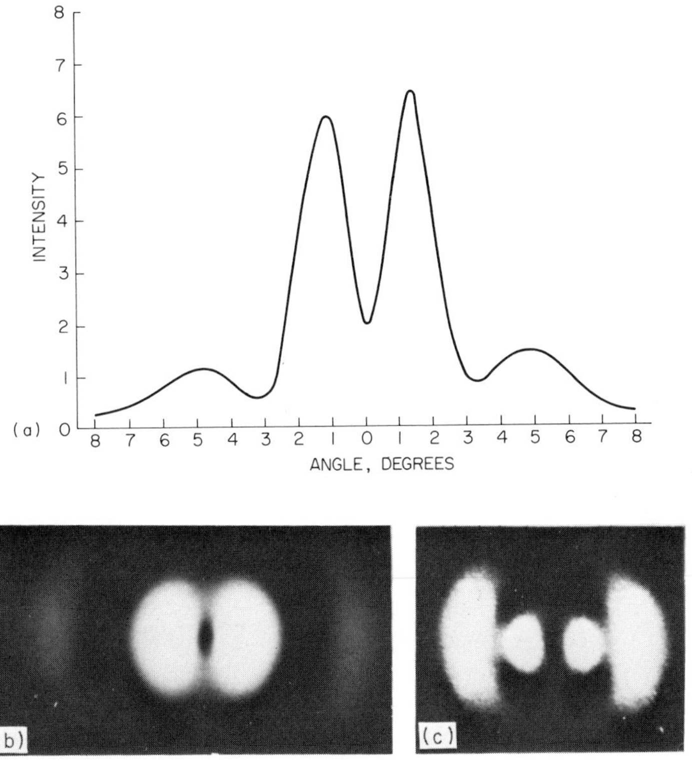

Fig. 6-11. Photometric trace of (1, 1) mode near cutoff with photographs of mode pattern (right) and radiation pattern (left).

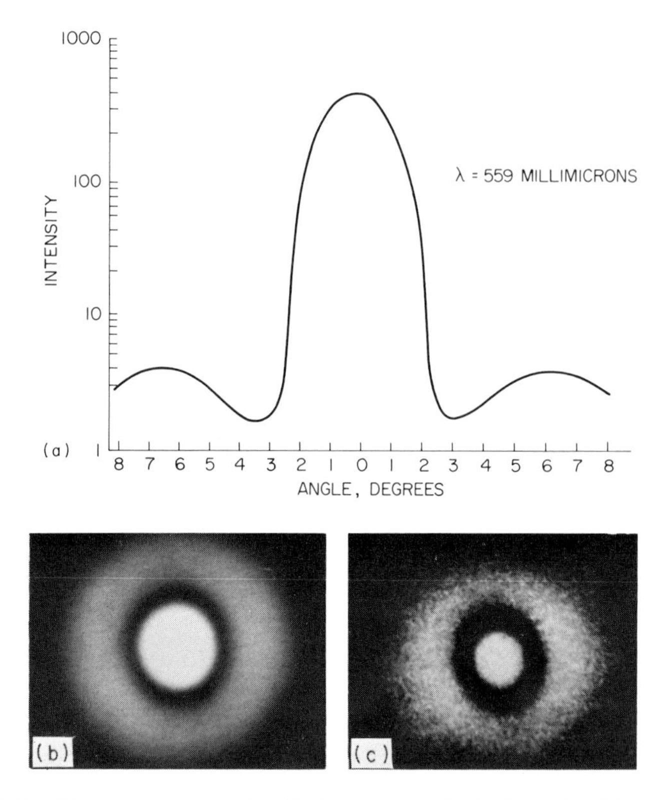

Fig. 6-12. Photometric trace of radiation pattern of (0, 1) mode with photographs of mode pattern (right) and radiation pattern (left). $\lambda = 559$ nm.

however, which we will discuss in the next section. The lobe broadening is also a consequence of the finite bandwidth of the source (approximately 100 Å).

Figure 6-11 shows the mode pattern (right) and the radiation pattern (left) of the (1, 1) mode in the near-cutoff region. It is unpolarized, being an incoherent combination of the two linearly polarized forms of the (1, 1) mode.

Figure 6-12 shows the mode pattern (right) and radiation pattern (left) of the (0, 1) mode. A weak background of the (2, 0) modes is also present. The (0, 1) mode is excited here in the near-cutoff condition. It is propagating in the core at an angle near the critical angle, with a large fraction of its energy in the coating. The central lobe is thus quite strong with respect to the outer ring. The lack of circular symmetry is again a consequence of the asymmetry in the fiber, as is the breadth of the central lobe. An observed

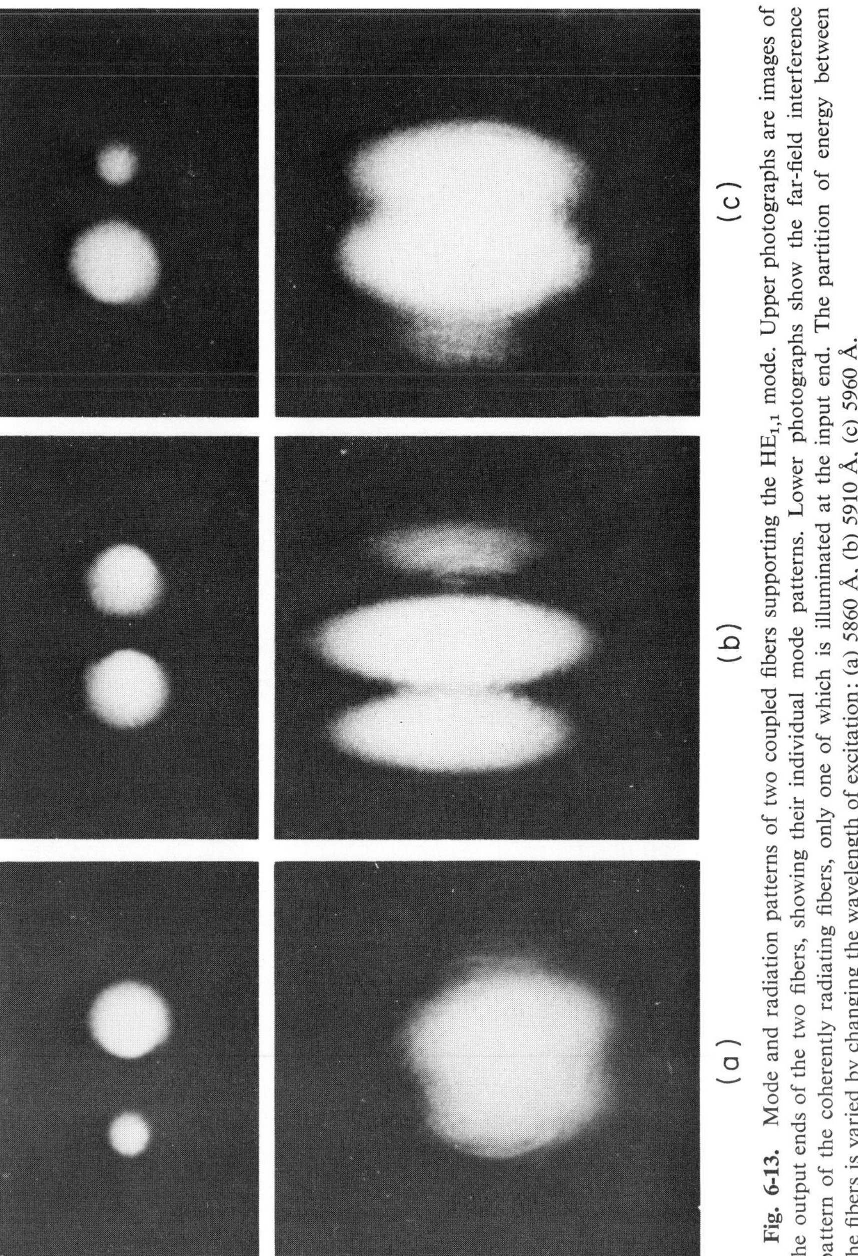

Fig. 6-13. Mode and radiation patterns of two coupled fibers supporting the $HE_{1,1}$ mode. Upper photographs are images of the output ends of the two fibers, showing their individual mode patterns. Lower photographs show the far-field interference pattern of the coherently radiating fibers, only one of which is illuminated at the input end. The partition of energy between the fibers is varied by changing the wavelength of excitation: (a) 5860 Å, (b) 5910 Å, (c) 5960 Å.

variation of the relative heights of the central and side lobes of this mode with changing wavelength offers a means of comparing theory and experiment in greater detail. This is pursued in the next section.

Figure 6-13 shows the mode and far-field-interference pattern of two coupled fibers supporting the (0, 0) ($HE_{1,1}$) mode. Only one fiber is illuminated at the input end. The flux transfers periodically with distance from the input plane and any desired distribution between the two fibers can be selected by varying the wavelength of excitation.

We conclude from the observations reported in this section that the surface perturbations in the fibers studied give rise to true modes which can be described by linear combinations of those circular waveguide modes that have almost equal propagation constants. Though the observed modes are not generally degenerate, we label with the same numbers the four forms that result from combining two doubly degenerate circular modes. This is because these four modes have, except for polarization, almost identical field variations in the fiber cross section and are, therefore, equivalent in the Kirchhoff approximation for the radiation pattern.

6. Comparison of Radiation Pattern Theory and Experiment

It was noted in the discussion of the theory that as a mode approached cutoff, the energy in the radiation pattern shifts from the outer major lobe in the vicinity of $\theta_0 = \sin^{-1}[(k_1/k_0) \sin \theta_1]$ to the innermost lobe. This phenomenon is illustrated by the theoretical curves of Fig. 6-14(a), which give the predicted radiation patterns of the $HE_{1,2}$ mode for three values of ka near cutoff in a waveguide with core index 1.525 and coating index 1.52. This mode is cut off at $ka = 32.1$. At $ka = 32.22$ (not shown in the figure), the intensity at $\theta = 0$ is 1600 times as great as that at the maximum of the outer ring. For this case, $u = 3.971$ and $q = 0.2$. For $ka = 34.80$ ($u = 4.176$, $q = 1.0$), the ratio is 25, as shown in the figure.

In Fig. 6-14(b) are shown two measured diffraction patterns of the (0, 1) mode in our slightly noncircular fibers, excited by linearly polarized, narrow-band radiation at 559 and 544 nm. Photographs of this mode are shown in Fig. 6-12. It is observed to cut off at $\lambda \simeq 580$ nm. If we assume that this cutoff wavelength corresponds to $ka = 32.1$, the cutoff value for the theoretical $HE_{1,2}$ mode, then the experimental curves correspond to $ka = 33.24$ and 34.14, values in the range of the plotted theoretical patterns with which they may be compared. The theoretical values are absolute, as calculated from Eq. (6.4), apart from a factor $a^2(\varepsilon_0/\mu_0)^{1/2}$. The experimental curves are plotted relative to the value of the central maximum at $\lambda = 559$ nm,

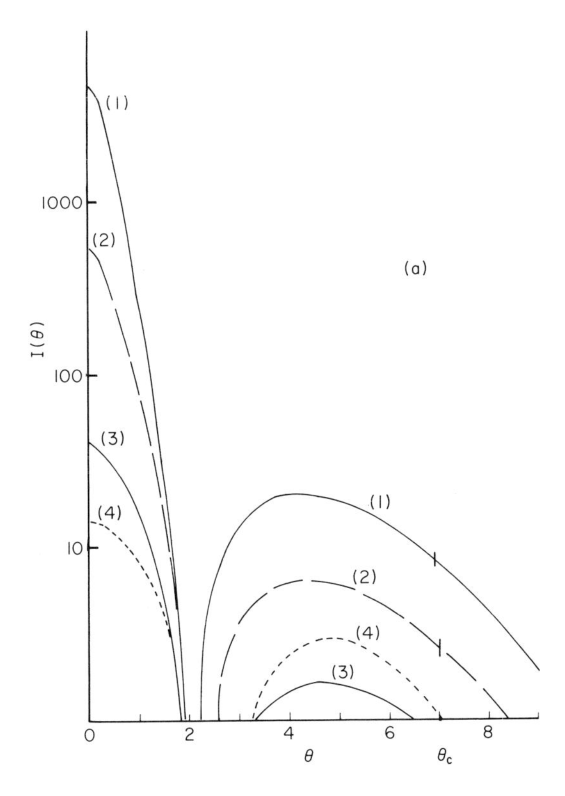

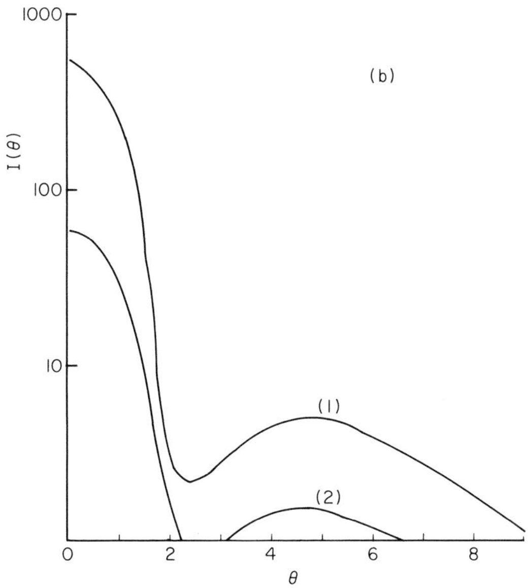

Fig. 6-14. (a) Theoretical radiation patterns of $HE_{1,2}$ mode and (b) experimental patterns of (0, 1) mode, showing changes in the relative intensity of the central and side lobes with changing wavelength near cutoff. In (a), $ka = 32.82$ (1); 33.47 (2); 34.80 (3); 40.92 (4). In (b), $\lambda = 5590$ Å (1) and 5440 Å (2) (see text).

which is adjusted to be equal to the theoretical maximum for $ka = 33.47$. The experimental lobes of the (0, 1) mode are seen to be considerably broader than the theoretical $HE_{1,2}$ lobes, and the minimum and second maximum are shifted toward slightly greater angles. The former effect appears to be a consequence of the slight noncircularity of the fiber cross section as well as the finite bandwidth of the exciting radiation. The outward shift of the maximum and minimum, on the other hand, are probably caused as well by the difference between the effective radius of the experimental fiber ($n_1 = 1.525$, $n_2 = 1.519$) and the theoretical waveguide ($n_1 = 1.525$, $n_2 = 1.520$). In the vicinity of cutoff, slight changes in these parameters effect large variations in the waveguide mode and, consequently, in the radiation pattern. In view of these sources of error, the agreement is quite reasonable.

To compare theory and experiment further, we have examined the variation with wavelength in the relative values of the two maxima of the (0, 1) mode and the corresponding variation in the theoretical $HE_{1,2}$ mode. The result is shown in Fig. 6-15. The solid curve represents the theory for the same values of the parameters as in Fig. 6-14(a). The circles are experimental values as in Fig. 6-14(b). The theoretical cutoff is at 580 nm. Below 545 nm, a background of the (2, 0) mode is observed experimentally (see

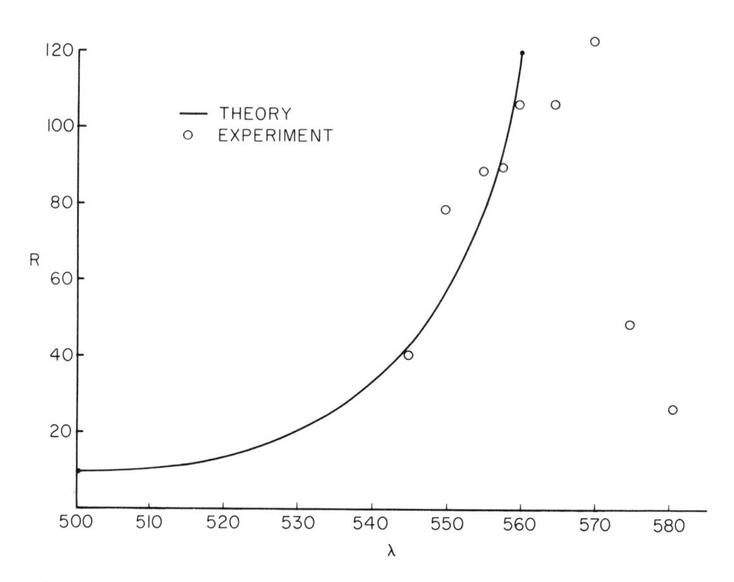

Fig. 6-15. Ratio of peak intensities in central and side lobes versus wavelength for the theoretical $HE_{1,2}$ mode and experimental (0, 1) mode near cutoff, as in Fig. 6-14. Experimental points beyond 570 nm are strongly influenced by the presence of (2, 0) modes, as discussed in the text.

Fig. 6-9) and the ratio could not be measured. At 560 nm, the brightness of the pattern reaches a maximum. Beyond this, the power in the mode spreads further into the coating and is attenuated by the film of black paint outside the coating. At 575 nm, the (0, 1) mode is considerably weakened and the single ring pattern of the (1, 0) modes (see Fig. 6-9) can also be observed. The latter is probably also detected by the photomultiplier at 565 nm. Because the theoretical cutoff is at 580 nm and the experimental (0, 1) mode is detectable out to 582 nm, the theoretical curve should perhaps be shifted slightly to the right.

C. QUANTITATIVE MODE DISCRIMINATION IN OPTICAL WAVEGUIDES AND RESONATORS

Experimental studies of optical waveguides and resonators are made difficult by the multimode nature of these structures. Investigations of optical waveguides and laser resonators (*1*, *7–9*), for example, have been limited by the lack of accurate means of determining the relative power in several modes propagating simultaneously. It is often difficult, in fact, to identify all the modes present. Measurements of mode coupling induced by surface irregularities in optical fibers (*10*) are further frustrated by the absence of accurate methods of mode launching. Definitive work thus requires the development of techniques that assure that almost all the energy focused into a waveguide is coupled into the desired mode. Equally important are techniques that permit the discrimination between signals carried simultaneously by different modes in the same waveguide. The two problems, i.e., mode launching and mode selection, are intimately connected, of course, in that solutions of one are usually applicable in some way to the other. In this section, we describe a technique for detecting a signal in a desired mode in the presence of other modes. In the last chapter, a definitive method of mode launching was described.

Because of the considerable analytical detail required in the quantitative description of this mode selection technique, it is appropriate to introduce the subject with a qualitative discussion. This is done with reference to the schematic shown in Fig. 6-16. This Mach–Zehnder interferometer-like arrangement is intended to accomplish spatial filtering of a multimode vector field in such a way as to provide a photocurrent proportional to the power in the mode selected by the filters S_+ and S_-.

It is assumed, first of all, that there are several modes propagating simultaneously on the fiber at the left and radiating into the system. The lens L serves to collimate the radiation field. As shown in the theoretical section, the electric field behind it is essentially transverse to the optical axis. This field is decomposed into two special orthogonally polarized components by the quarter-wave plate. The beam is then split, and one of these components is selected by each of the two polarizers P_+ and P_- for subsequent processing by the spatial filters S_+ and S_-. After filtering, common components of the fields are selected by analyzers A_+ and A_-. These are recombined at the second beam splitter to yield the desired signal at the pinhole in the focal plane of lenses L_+ and L_-. This signal is an approximate optical analog of the orthogonality condition satisfied by the waveguide modes.

The filters S_+ and S_- are chosen to be as simple as possible while still providing a high degree of discrimination between all modes, including the two degenerate forms of the hybrid modes. As shown later, the orthogonality condition is approximately satisfied in the radiation field as well as for modes which are not near the cutoff condition. Optimum simplicity in the filters is achieved by making their amplitude transmittances proportional to the conjugates of the circular components $E_\pm' = E_x' \pm iE_y'$ of the radiation field of the desired mode. In the Huygens–Kirchhoff approximation which we assume here, these components are specified in terms of the Fourier transforms of the corresponding components, i.e., E_+ and H_+ or E_- and H_-, in the plane of the exit end of the waveguide.

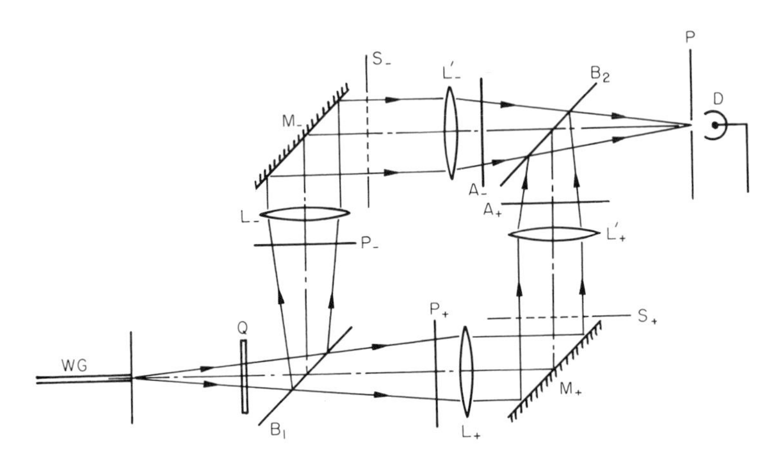

Fig. 6-16. An experimental schematic for discrimination among low-order modes of multimode waveguides in the form of an optical analog of the orthogonality condition satisfied by the modes. *WG*, waveguide; *Q*, quarter-wave plate; B_1 and B_2, beam splitters; $P_\pm$ and $A_\pm$, polarizers and analyzers; $L_\pm$ and $L_\pm'$, lenses; $M_\pm$, mirrors; $S_\pm$, spatial filters; *P*, pinhole; and *D*, detector.

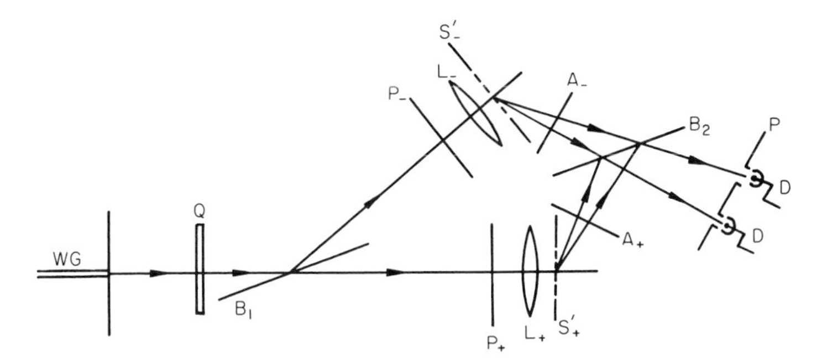

Fig. 6-17. Holographic analog of Fig. 6-16 providing for simultaneous detection of several modes.

The purpose of the quarter-wave plate in Fig. 6-16 is to transform the circular components of the radiation field into orthogonal linear vibrations. Thus, if the optical axis of the plate is coincident with the x axis, the phase of the y component of the field is advanced by $\pi/2$ with respect to the x component. Consequently, the right circular component of the incident field, i.e., $E_x + iE_y$, is transformed into a linear vibration in a plane at an angle of 45° to the x, z plane. The left circular component becomes a linear vibration perpendicular to this. Then, with the aid of polarizers P_+ and P_-, the right and left circular components of the radiation fields of all modes can be processed separately and later recombined.

If holography is used to make the spatial filters, a more convenient parallel processing method (*14*, *15*) can be considered. An illustrative schematic is shown in Fig. 6-17, where the quarter-wave plate, beam splitters, polarizers, lenses, and analyzers operate in the same way as shown in Fig. 6-16. The only difference is that the spatial filters are off-axis holograms. By this technique, two photographic plates S_+' and S_-' can simultaneously record the amplitude and phase distributions corresponding to several different modes, each on its own unique carrier wave. Here, the carrier wave for a given mode should be of the same frequency for both filters but have opposite sign in S_+' and S_-'. If such filters are placed as shown in Fig. 6-17, the desired signal for a particular mode is obtained at the unique pinhole corresponding to its carrier frequency.

1. Theoretical Description of the Technique

In this section, we will evaluate theoretically the degree of success to be expected from the mode selection technique described in the previous section. To do this, we need expressions for the fields at the exit end of the waveguide, as well as in the Fraunhofer region. In the Huygens–Kirchhoff approximation, the former are given the mode fields themselves. The latter are then expressible in terms of Fourier transforms of the former. After deriving them, we will use them to evaluate the photocurrent to be expected at the pinhole in Fig. 6-16.

a. Transverse Fields on the Waveguide

It is convenient for later analysis to combine the expressions for the components of the mode fields on the interior (superscript i) and exterior (superscript e) of the waveguide core into single expressions. This is accomplished with the use of the simple multiplicative functions $W^i(\varrho)$ and $W^e(\varrho)$, defined by

$$W^{\mathrm{i}} = \begin{cases} 1; & 0 \leq \varrho \leq a \\ 0; & a < \varrho < \infty \end{cases}$$

$$W^{\mathrm{e}} = \begin{cases} 0; & 0 \leq \varrho \leq a \\ 1; & a < \varrho < \infty. \end{cases}$$

It then follows directly from Eqs. (6.1) that for any mode of a dielectric waveguide of radius a, core index of refraction n_1, and coating index n_2, the transverse electric and magnetic fields can be described by

$$\begin{aligned} E_+ &= E_x + iE_y = F_+{}^{\mathrm{i}}W^{\mathrm{i}} - F_+{}^{\mathrm{e}}W^{\mathrm{e}}, \\ E_- &= E_x - iE_y = -\gamma(F_-{}^{\mathrm{i}}W^{\mathrm{i}} + F_-{}^{\mathrm{e}}W^{\mathrm{e}}), \\ H_+ &= H_x + iH_y = (i/Z_0)(\sigma_+{}^{\mathrm{i}}F_+{}^{\mathrm{i}}W^{\mathrm{i}} - \sigma_+{}^{\mathrm{e}}F_+{}^{\mathrm{e}}W^{\mathrm{e}}), \\ H_- &= H_x - iH_y = (i\gamma/Z_0)(\sigma_-{}^{\mathrm{i}}F_-{}^{\mathrm{i}}W^{\mathrm{i}} + \sigma_-{}^{\mathrm{e}}F_-{}^{\mathrm{e}}W^{\mathrm{e}}), \end{aligned} \tag{6.7}$$

where

$$\begin{aligned} F_\pm{}^{\mathrm{i}} &= [J_{n\pm1}(u\varrho/a)e^{i(n\pm1)\phi+ihz-i\omega t}][1/uJ_n(u)], \\ F_\pm{}^{\mathrm{e}} &= [K_{n\pm1}(q\varrho/a)e^{i(n\pm1)\phi+ihz-i\omega t}][1/qK_n(q)], \end{aligned}$$

and

$$Z_0 = (\mu_0/\varepsilon_0)^{1/2}.$$

J_n is the Bessel function of the first kind of order n, and K_n is the modified Bessel function of the second kind. The values of the constants u, q, h, γ, $\sigma_+{}^{\mathrm{i}}$, $\sigma_+{}^{\mathrm{e}}$, $\sigma_-{}^{\mathrm{i}}$, and $\sigma_-{}^{\mathrm{e}}$ are different for different modes. If we characterize a particular waveguide according to the values of the physical parameters δ and R defined by

$$\delta = 1 - (n_2{}^2/n_1{}^2), \qquad R^2 = k_1{}^2a^2\delta, \tag{6.8}$$

where $k_1 = 2\pi n_1/\lambda_0$, then

$$q = (R^2 - u^2)^{1/2},$$

and u is a solution of the characteristic transcendental equation

$$(k_1{}^2J_+ + k_2{}^2K_+)/(k_1{}^2J_- - k_2{}^2K_-) = -(J_+ + K_+)/(J_- - K_-) = \gamma. \quad (6.9)$$

This equation also defines γ. The $J_\pm$ and $K_\pm$ are given by

$$J_\pm = J_{n\pm 1}(u)/uJ_n(u),$$
$$K_\pm = K_{n\pm 1}(q)/qK_n(q),$$

and

$$k_2 = 2\pi n_2/\lambda_0.$$

The remaining characteristic parameters are given by

$$h^2 = k_1{}^2 - \frac{u^2}{a^2},$$
$$\sigma_\pm{}^{\mathrm{i}} = \frac{h}{k_0}\,\frac{(k_1/h)^2 \pm A}{1 \pm A}, \quad (6.10)$$
$$\sigma_\pm{}^{\mathrm{e}} = \frac{h}{k_0}\,\frac{(k_2/h)^2 \pm A}{1 \pm A},$$

where $k_0 = 2\pi/\lambda_0$ and $A = (1 - \gamma)/(1 + \gamma)$. We list in Table III the $\mathrm{EH}_{n,m}$ and $\mathrm{HE}_{n,m}$ modes by name, including the two degenerate forms of each. The plus and minus superscripts in the labels distinguish between the degenerate modes by indicating the fact that the corresponding circular field

Table III

SUMMARY OF MODE DESIGNATIONS FOR CIRCULAR CYLINDER DIELECTRIC WAVEGUIDES

Region of n	Region of A	Name of mode
$n > 0$	$A_{nm} > 1$	$\mathrm{EH}^{\oplus}_{n,m}$
	$-1 < A_{nm} < 0$	$\mathrm{HE}^{\ominus}_{n,m}$
$n < 0$	$A_{nm} = -A_{-nm} < -1$	$\mathrm{EH}^{\ominus}_{n,m}$
	$0 < A_{nm} = -A_{-nm} < 1$	$\mathrm{HE}^{\oplus}_{n,m}$
$n = 0$	$A_{nm} = 0$	$\mathrm{TM}_{0,m}$
	$A_{nm} \to \infty$	$\mathrm{TE}_{0,m}$

components are the major components of the electric and magnetic field vectors. The range of values of the parameter A (and therefore γ) is also given.

b. Field Vectors in the Fraunhofer Region

As in Section B, we follow Schelkunoff's approach (*2*, Chapter IX) to obtain expressions for the electric and magnetic fields in the Fraunhofer region. A Cartesian coordinate system (x, y, z) with its origin at the center of the exit end of the waveguide is employed. The negative-z axis is coincident with the axis of the guide. Spherical (r, θ, ϕ) and polar (ϱ, ϕ, z) coordinates with the same origin are also employed as required. The conventional relations (*2*, Chapter IX) among the three representations are maintained throughout the discussion.

In the Fraunhofer region, E_r and H_r vanish and we have (*2*, p. 333)

$$E_\theta = Z_0 H_\phi = -(ik_0/4\pi r)(Z_0 N_\theta + L_\phi)e^{-ik_0 r}$$
$$E_\phi = -Z_0 H_\theta = (ik_0/4\pi r)(-Z_0 N_\phi + L_\theta)e^{-ik_0 r}.$$

The vectors N and L are Schelkunoff's radiation vectors. Consider now a lens placed in the far field with its optical axis coincident with the extended axis of the waveguide. Its first principal point lies in the plane $z = f$, where f is the focal length of the lens. We assume that the lens is coated so as to obviate depolarization effects. The passage of light through the lens may be accurately described by vectorial geometric optics (*13*). The θ components of the field vectors are transformed by the lens into ϱ components, that is, they become components perpendicular to the optical axis. The ϕ components remain unchanged. Because the aperture of the lens is large and the structure of the radiation patterns is coarse, the fields at considerable distances beyond the lens continue to be accurately described by vector geometric optics. We thus have, in the region beyond the lens,

$$E_\varrho = Z_0 H_\phi = -iC(Z_0 N_\varrho + L_\phi)e^{-ik_0(z-f)}$$
$$E_\phi = -Z_0 H_\varrho = -iC(Z_0 N_\phi - L_\varrho)e^{-ik_0(z-f)},$$

where

$$C = (k_0/4\pi f)e^{-ik_0 f}$$

In order to proceed, we must transform this representation of the fields into one involving circular components. We begin by forming the components

$$E_\varrho \pm iE_\phi = \mp iZ_0(H_\varrho \pm iH_\phi) = -iC[Z_0(N_\varrho + iN_\phi) \mp i(L_\varrho \pm iL_\phi)]e^{-ik_0(z-f)}.$$

Since, for any vector field B,

$$B_{\pm} = B_x \pm iB_y = (B_\varrho \pm iB_\phi)e^{\pm i\phi},$$

we can rewrite the above equation as

$$E_{\pm} = \mp iZ_0H_{\pm} = -iC(Z_0N_{\pm} \mp iL_{\pm})e^{-ik_0(z-f)}. \tag{6.11}$$

We now obtain $N_{\pm}$ and $L_{\pm}$ [as in (*2*, Chapter IX) and Section B]

$$N_{\pm} = \pm ih_{\pm} \qquad \text{and} \qquad L_{\pm} = \mp ie_{\pm}, \tag{6.12}$$

where $h_{\pm}$ and $e_{\pm}$ are the Fourier transforms of the corresponding components of the field vectors at the exit end of the waveguide. For example,

$$h_{\pm} = \int_0^{2\pi} \int_0^{\infty} H_{\pm}(r_0, \phi_0)e^{ik_0(\varrho/f)r_0\cos(\phi-\phi_0)}r_0\, dr_0\, d\phi_0.$$

The subscript zero is used here to distinguish between the source points and field points, which are not subscripted. Substituting Eq. (6.12) into (6.11), we then have the final form of the fields behind the lens:

$$E_{\pm} = \mp iZ_0H_{\pm} = C(\pm Z_0h_{\pm} + ie_{\pm})e^{-ik_0(z-f)}. \tag{6.13}$$

c. Quantitative Evaluation of the Mode Selection Technique

The orthogonality condition satisfied by the waveguide modes can be written as

$$\int_0^{2\pi} \int_0^{\infty} [E_+(u)H_+^*(u') - E_-(u)H_-^*(u')]\varrho_0\, d\varrho_0\, d\phi_0 = N(u')\, \delta_{u,u'}.$$

The integration is over any cross section of the waveguide perpendicular to its axis. The $E_{\pm}(u)$ are the transverse components of the electric field in the waveguide or at its exit end (in the Huygens–Kirchhoff approximation) for a general mode, designated by the parameter u. The $H_{\pm}^*(u')$ are the conjugates of the transverse components of the magnetic field for a particular mode, designated by u'. Explicit expressions for these field components are given by Eq. (6.7). Here, $N(u')$ is a normalizing factor and $\delta_{u,u'}$ is the Kronecker delta.

The analogous integral in the Fraunhofer region is given by

$$\int_0^{2\pi}\int_0^{\infty} [E_+(u)H_+{}^*(u') - E_-(u)H_-{}^*(u')]\varrho \, d\varrho \, d\phi. \qquad (6.14)$$

Here, the integration is over a plane behind the collimating lens L in Fig. 6-16, and the Fraunhofer fields of the general and particular modes are given by Eq. (6.13). If the pupil filters S_+ and S_- in Fig. 6-16 have amplitude transmittances proportional to $H_+{}^*(u')$ and $H_-{}^*(u')$, respectively, then the amplitude of the electric field at the pinhole in Fig. 6-16 will be proportional to this integral. The detected signal, on the other hand, will be proportional to its square. Therefore, to the extent to which the radiation fields of the modes are orthogonal, the detected signal will be proportional to the power in the u' mode. We therefore examine Eq. (6.14) in detail, treating in particular the problem of low-order modes on multimode waveguides.

If we use Eqs. (6.7) to carry out the Fourier operations implied in Eq. (6.13), then the field vectors in the Fraunhofer region become

$$\begin{aligned} E_+ &= i(1+\sigma_+{}^{\mathrm{i}})\,\mathrm{conv}(f_+{}^{\mathrm{i}}, w^{\mathrm{i}}) - i(1+\sigma_+{}^{\mathrm{e}})\,\mathrm{conv}(f_+{}^{\mathrm{e}}, w^{\mathrm{e}}) = -Z_0H_+ \\ E_- &= i\gamma(1+\sigma_-{}^{i})\,\mathrm{conv}(f_-{}^{\mathrm{i}}, w^{\mathrm{i}}) - i\gamma(1+\sigma_-{}^{\mathrm{e}})\,\mathrm{conv}(f_-{}^{\mathrm{e}}, w^{\mathrm{e}}) = Z_0H_- \end{aligned} \qquad (6.15)$$

The notation conv(f, w) indicates the convolution of two functions f and w. The subscripted and superscripted functions f and w are the Fourier transforms of the corresponding functions F and W of Eqs. (6.7).

Consider now the operation of the filter in the lower half of the system in Fig. 6-16. Suppose the right circular component of the (Fraunhofer) electric field (E_+) induced by a general mode passes through a filter S_+ with amplitude transmittance proportional to the conjugate of the right circular component of the (Fraunhofer) magnetic field ($\bar{H}_+{}^*$) due to another mode. (In this and subsequent developments, we use unbarred and barred quantitites, respectively, to distinguish between a general mode and the mode to be selected by the filter.) From Eqs. (6.15), we have

$$\begin{aligned} E_+\bar{H}_+{}^* = {} & (1+\sigma_+{}^{\mathrm{i}})(1+\bar{\sigma}_+{}^{\mathrm{i}})\,\mathrm{conv}(f_+{}^{\mathrm{i}}, w^{\mathrm{i}})\,\mathrm{conv}(\bar{f}_+{}^{\mathrm{i}}, w^{\mathrm{i}})^* \\ & + (1+\sigma_+{}^{\mathrm{e}})(1+\bar{\sigma}_+{}^{\mathrm{e}})\,\mathrm{conv}(f_+{}^{\mathrm{e}}, w^{\mathrm{e}})\,\mathrm{conv}(\bar{f}_+{}^{\mathrm{e}}, w^{\mathrm{e}})^* \\ & - (1+\sigma_+{}^{\mathrm{i}})(1+\bar{\sigma}_+{}^{\mathrm{e}})\,\mathrm{conv}(f_+{}^{\mathrm{i}}, w^{\mathrm{i}})\,\mathrm{conv}(\bar{f}_+{}^{\mathrm{e}}, w^{\mathrm{e}})^* \\ & - (1+\bar{\sigma}_+{}^{\mathrm{i}})(1+\sigma_+{}^{\mathrm{e}})\,\mathrm{conv}(\bar{f}_+{}^{\mathrm{i}}, w^{\mathrm{i}})\,\mathrm{conv}(f_+{}^{\mathrm{e}}, w^{\mathrm{e}})^*. \end{aligned}$$

This is the electric field just behind the filter S_+. In the focal plane of the lens L_+, the electric field has a component proportional to the Fourier transform of this field. Applying the convolution theorem, we find that this component makes a contribution to the field at the axial pinhole given by

$$\int_0^{2\pi}\int_0^{\infty}(E_+\bar{H}_+{}^*)\varrho\,d\varrho\,d\phi = -2\pi(i/Z_0)$$
$$\times\Big[(1+\sigma_+{}^{\mathrm{i}})(1+\bar{\sigma}_+{}^{\mathrm{i}})\int_0^a F_+{}^{\mathrm{i}}(\bar{F}_+{}^{\mathrm{i}})^*\varrho_0\,d\varrho_0$$
$$+(1+\bar{\sigma}_+{}^{\mathrm{e}})(1+\sigma_+{}^{\mathrm{e}})\int_a^{\infty}F_+{}^{\mathrm{e}}(\bar{F}_+{}^{\mathrm{e}})^*\varrho_0\,d\varrho_0\Big]. \tag{6.16}$$

In a similar manner, we find that the field behind the filter S_- in the upper branch of the system makes a contribution to the field at the axial pinhole given by

$$\int_0^{2\pi}\int_0^{\infty}(E_-\bar{H}_-{}^*)\varrho\,d\varrho\,d\phi = 2\pi\gamma\bar{\gamma}(i/Z_0)$$
$$\times\Big[(1+\bar{\sigma}_-{}^{\mathrm{i}})(1+\sigma_-{}^{\mathrm{i}})\int_0^a F_-{}^{\mathrm{i}}(\bar{F}_-{}^{\mathrm{i}})^*\varrho_0\,d\varrho_0$$
$$+(1+\sigma_-{}^{\mathrm{e}})(1+\bar{\sigma}_-{}^{\mathrm{e}})\int_a^{\infty}F_-{}^{\mathrm{e}}(\bar{F}_-{}^{\mathrm{e}})^*\varrho_0\,d\varrho_0\Big]. \tag{6.17}$$

The field at the axial pinhole is proportional to the difference between Eqs. (6.16) and (6.17) provided that the path lengths in the two arms of the system are appropriately adjusted.

We are interested in evaluating Eqs. (6.16) and (6.17) to order $1/R^3$. We note that if the filter S_+ and S_- do not contain the terms conv($f_\pm{}^{\mathrm{e}}$, w^{e})* corresponding to the evanescent surface wave on the guide, the second terms of both expressions vanish.

From the characteristic equation, we obtain for u and γ in this approximation,

$$J_+ \approx -\frac{\varepsilon_1+\varepsilon_2}{2\varepsilon_1}\left(\frac{1}{q}+\frac{3n+\frac{1}{2}}{R^2}\right), \qquad \gamma_{\mathrm{EH}} \approx -\frac{u^2\delta}{4nq}. \tag{6.18}$$

These equations apply to EH modes. For HE modes, we have

$$J_- \approx \frac{\varepsilon_1+\varepsilon_2}{2\varepsilon_1}\left(\frac{1}{q}-\frac{3n+\frac{1}{2}}{R^2}\right), \qquad \gamma_{\mathrm{HE}} \approx -\frac{4nq}{u^2\delta}. \tag{6.19}$$

In these equations, $\varepsilon_1 = n_1{}^2\varepsilon_0$, etc. It should be noted that γ_{HE} is of the order of $1/\gamma_{\rm EH}$, which is much greater than unity in this approximation. Thus, the power in an HE mode as expressed by Eqs. (6.7) is $\gamma_{\rm HE}^2$ greater than the power in an EH mode. We should, therefore, multiply all components of HE modes represented by Eqs. (6.7) by the factor $1/\gamma_{\rm HE}$. This will normalize the expressions in such a way as to assure that $B^2/\bar{B}^2$ represents the relative power in the modes, where B and $\bar{B}$ are arbitrary amplitude factors for two different modes.

Proceeding with the evaluation of Eqs. (6.16) and (6.17), we have

$$\sigma_\pm{}^{\rm i} \approx n_1[1 + (\gamma^{\pm 1}u^2\delta/2R^2)],$$

where u and γ are given by Eqs. (6.18) and (6.19). Neglecting terms which are less than or equal to $u^2\delta/R^2$, we obtain

$$\sigma_+{}^{\rm i} \approx n_1 \qquad \text{and} \qquad \sigma_-{}^{\rm i} \approx n_1[1 - (2nq/R^2)] \tag{6.20}$$

for EH modes, and

$$\sigma_-{}^{\rm i} \approx n_1 \qquad \text{and} \qquad \sigma_+{}^{\rm i} \approx n_1[1 + (2nq/R^2)] \tag{6.21}$$

for HE modes. Finally, we need the relations

$$\begin{aligned} \int_0^a \bar{F}_\pm{}^{\rm i}(\bar{F}_\pm{}^{\rm i})^*\varrho_0\, d\varrho_0 &= (a^2/2\bar{u}^2)[1 - 2(n \pm 1)J_\pm(\bar{u}) + \bar{u}^2 J_\pm{}^2(\bar{u})] \\ \int_0^a F_\pm{}^{\rm i}(\bar{F}_\pm{}^{\rm i})^*\varrho_0\, d\varrho_0 &= \pm[a^2/(u^2 - \bar{u}^2)][J_\pm(u) - J_\pm(\bar{u})]. \end{aligned} \tag{6.22}$$

Direct substitution of Eqs. (6.18)–(6.22) into Eqs. (6.16) and (6.17) will now yield expressions for the field at the pinhole accurate to order $1/R^3$. We consider first the case when the filters in Fig. 6-16 match the mode. The branch of the system that matches the dominant component of the electric field will contribute $(4nR/u^2\delta)^4$ times more power at the pinhole than the other branch. Accordingly, the amplitude at the pinhole when the mode and filter are matched is substantially given by

$$\int\int \bar{E}_+\bar{H}_+{}^*\varrho\, d\varrho\, d\phi \approx \frac{i}{Z_0}\,\frac{\pi a^2(n_1+1)^2}{\bar{u}^2} \times \left[1 + 2(n+1)\,\frac{\varepsilon_1+\varepsilon_2}{2\varepsilon_1\bar{q}} + \frac{\bar{u}^2}{R^2}\left(\frac{\varepsilon_1+\varepsilon_2}{2\varepsilon_1}\right)^2\right]$$

for EH modes, and

$$\frac{1}{\gamma^2{}_{\mathrm{HE}}} \int\int \bar{E}_{-}\bar{H}_{-}{}^{*}\varrho \, d\varrho \, d\phi \approx \frac{i}{Z_0} \frac{\pi a^2(n_1+1)^2}{\bar{u}^2} \times \left[1 - 2(n-1)\frac{\varepsilon_1+\varepsilon_2}{2\varepsilon_1 \bar{q}} + \frac{\bar{u}^2}{R^2}\left(\frac{\varepsilon_1+\varepsilon_2}{2\varepsilon_1}\right)^2\right]$$

for HE modes. If we divided this and subsequent expressions by the factors outside the square brackets, the order of magnitude of all contributions to the signal can be compared to the unit (substantially) contribution of the matched mode.

We now consider the contributions from modes that are not matched. Substituting as earlier and normalizing, we obtain

$$\int\int E_{+}\bar{H}_{+}{}^{*}\varrho \, d\varrho \, d\phi \approx -\frac{\varepsilon_1+\varepsilon_2}{\varepsilon_1} \frac{\bar{u}^2}{q^2-\bar{q}^2} \frac{\bar{q}-q}{q\bar{q}} = O\left(\frac{\bar{u}^2}{R^3}\right)$$

for an EH mode and mismatched EH filter,

$$\approx \frac{\delta \bar{u}^2}{(u^2-\bar{u}^2)\bar{q}}\left[1 + \frac{2n\bar{q}n_1}{(n_1+1)R^2}\right] = O(4n\bar{\gamma}) \tag{6.23a}$$

for an EH mode and HE filter, and

$$\approx \frac{\delta^2 \bar{u}^2}{4nR^2}\left[1 + \frac{2nn_1(q+\bar{q})}{(n_1+1)R^2}\right] = O\left(\frac{\delta\bar{\gamma}}{R}\right) \tag{6.23b}$$

for an HE mode and mismatched HE filter. Similarly,

$$\int\int E_{-}\bar{H}_{-}{}^{*}\varrho \, d\varrho \, d\phi \approx -\frac{\delta^2 \bar{u}^2}{4nR^2}\left[1 - \frac{2nn_1(q+\bar{q})}{(n_1+1)R^2}\right] = O\left(\frac{\delta\bar{\gamma}}{R}\right) \tag{6.23c}$$

for an EH mode and mismatched EH filter,

$$\approx \frac{\delta \bar{u}^2}{(u^2-\bar{u}^2)q}\left[1 - \frac{2nn_1(q+\bar{q})}{(n_1+1)R^2}\right] = O(4n\bar{\gamma}) \tag{6.23d}$$

for an EH mode and HE filter, and

$$\approx \frac{\varepsilon_1+\varepsilon_2}{\varepsilon_1} \frac{\bar{u}^2}{(q^2-\bar{q}^2)} \frac{\bar{q}-q}{q\bar{q}} = O\left(\frac{\bar{u}^2}{R^3}\right) \tag{6.23e}$$

for an HE mode and mismatched HE filter.

Equations (6.23a)–(6-23e) show that either branch of the system in Fig. 6-16 is adequate if all the radiating modes are of the same type. Thus, a mode of amplitude B makes a contribution to the detected signal, relative to that of the desired mode (amplitude $\bar{B}$), no greater than $(B/\bar{B})^2(\bar{u}^4/R^6)$ or $(B/\bar{B})^2(\delta^4\bar{u}^4/16n^2R^4)$, whichever is larger. Modes of different type, on the other hand, give relative contributions $(B/\bar{B})^2\delta^2\bar{u}^4/(u^2 - \bar{u}^2)^2R^2$ in each branch. It can be shown that $(u^2 - \bar{u}^2)$ is at least of order unity, so that these contributions are often negligible. If not, the two branches of Fig. 6-16 can be combined coherently to reduce the net amplitude to $4n\delta u^2 n_1/(u^2 - \bar{u}^2)R^2$. The noise contribution is accordingly reduced by a factor $1/R^2$.

2. Experimental Tests of the Technique

Several experiments have been performed to establish the applicability of the technique for dielectric optical waveguides. An He–Ne laser source was alternately coupled to the $HE_{1,1}$ and $HE_{1,2}$ modes of a relatively large waveguide by the mode launching technique described in Chapter 5. Each mode was alternately filtered by an $HE_{1,1}$ and an $HE_{1,2}$ mask. A schematic diagram of this experiment is shown in Fig. 6-18. The portion enclosed by the dashed line in the figure is the mode launching arrangement. The beam of the laser source is first expanded and collimated by means of the inverted telescope arrangement. The diffraction pattern of an annular ring mask formed by the condenser lens on the entrance face of the waveguide excites the desired $HE_{1,m}$ mode. By changing the diameter of the ring, a different $HE_{1,m}$ mode can be excited.

The fiber waveguide had the following properties: the core glass was UK-50 (refractive index 1.53). The diameter of the core was approximately

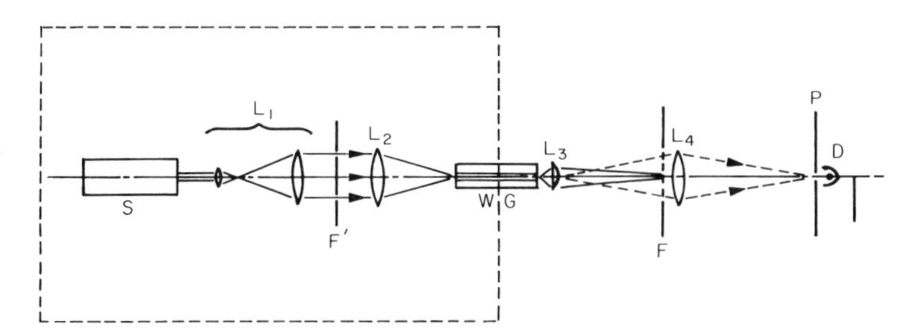

Fig. 6-18. Experimental arrangement for discrimination between $HE_{1,m}$ modes. S, He–Ne laser; L_1, collimator lens; F', spatial filter to launch a mode; L_2, launching lens; WG, waveguide; L_3, microscope objective; F, spatial filter to discriminate modes; L_4, imaging lens; P, pinhole; and D, detector.

25 μm, and the ratio between the core and coating diameter was about 1/15. The length was 5 cm, and the number of modes which this fiber could support was more than 15.

Theory indicates that single-path filtering should be entirely adequate on this waveguide. Thus, from Eqs. (6.23a)–(6.23e), we expect that the errors caused by the presence of $EH_{1,m}$ modes in either branch would be no greater than $\delta\bar{u}^2/R$ multiplied by the relative power in the EH mode and the desired HE mode. Here, $\delta = 0.0065$, $R = 15.5$, and $\bar{u} \approx 2.4$ for the $HE_{1,1}$ mode and 5.5 for the $HE_{1,2}$ mode. Thus, $\delta\bar{u}^2/R < 0.013$. Furthermore, the relative power of EH modes was negligible because of the circularly symmetric launching arrangement (*8*).

From Eqs. (6.23a)–(6.23e), we can also estimate errors due to mismatched $HE_{1,m}$ mode and filter. These will be of order $\bar{u}^4/R^6$ times the relative power. Such errors are also seen to be entirely negligible.

A further simplification results from the fact that a single-path experiment such as this need not be made polarization-sensitive. Thus, from Eqs. (6.18) and (6.19), we see that the dominant transverse field components carry approximately $(4R/\delta\bar{u}^2)^2$ more power than the weaker components. For the $HE_{1,2}$ mode, this is about equal to 10^5 on our waveguide. Thus, the transverse fields of the $HE_{1,m}^-$ mode are essentially given by the components E_- and H_- while those of the $HE_{1,m}^+$ mode are given by E_+ and H_+. With the double-path arrangement shown in Fig. 6-16, we would be able to measure the power in either of these degenerate modes simply by blocking the path corresponding to the opposite components of the fields, using identical filters in each path. Here (Fig. 6-18), we choose not to discriminate between the two degenerate forms (or any linear combination of them). This is effected by omitting both the quarter-wave plate and the polarizers in the single-path experiment.

To simplify the experiment further, the Fourier plane filtering arrangement of Fig. 6-16 can be replaced by a double-diffraction method (*14*) in which the spatial filter is the so-called matched filter in coherent optical systems (*14–16*). This has the added practical advantage of reducing errors in the Fraunhofer region that result from spurious radiation traveling in the fiber coating. The equivalence of the double-diffraction arrangement of Fig. 6-18 and either path in Fig. 6-16 is a consequence of Parseval's theorem.

The microscope objective employed was 50×, and the focal length of the condenser lens was 15 cm. For this double-diffraction method, the form of the matched filter is a magnified version of the transverse magnetic field at the exit end of the fiber, rather than its Fourier transform, as in Fig. 6-16. The filters were made to have amplitude transmittance proportional to

$J_0(\bar{u}_{1,m}\varrho/Ma)$, where M is a magnification factor and $m = 1$ or 2. No attempt was made to incorporate into the filter the Hankel-function dependence of the fields for $Ma < \varrho$. (This would be physically incorrect, in fact, since the image intensity cannot be discontinuous at $\varrho = Ma$. The Bessel function dependence is thus more appropriate.) Because less than 2% of the mode power is carried in the evanescent wave for these modes on this guide, it is expected that the error resulting from this approximation will be no greater than 2%.

The absolute values of the amplitude transmittances of the filter were generated on a photographic film from the diffraction pattern of a ring mask. Kodak Panatomic X 35 mm film was used to obtain the negative of the pattern and Kodak commercial sheet film for the positive. The gammas of these films were controlled to be unity by choosing suitable times and temperatures of development. An example of the intensity transmittance of one of the filters is shown in Fig. 6-19(a) together with a calculated curve of $J_0^2(x)$. Figure 6-19(b) is a photograph of the filter.

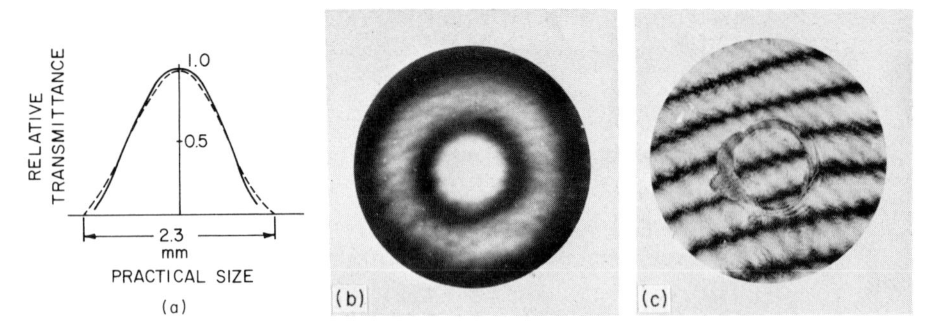

Fig. 6-19. (a) Measured and ideal transmittance of spatial filter for selection of $HE_{1,1}$ mode; (b) photograph of $HE_{1,2}$ filter; and (c) interferometric measurement of phase plate used with $HE_{1,2}$ amplitude mask.

In order to make the negative portion of the Bessel function on the filter, phase plates were made by cutting suitable holes in thin glass plates. These phase plates retard the phase of the negative portion of the mask by an amount π with respect to the positive portion.

The combination of these plates and the photographic filters mentioned above gives the desired spatial filters whose amplitude transmittances are proportional not to the absolute value of the Bessel function, but to the function itself. An example of the phase retardation of one of these filters is shown in Fig. 6-19(c) in the form of Mach–Zehnder interferometer fringes. In order to eliminate the spurious radiation from the fiber coating, a black painted metal plate which has a hole corresponding to the size of the spatial filter was placed at the spatial filter plane (that is, the image plane of the microscope objective).

The intensity on the optical axis in the image of the pupil of the microscope objective formed by the second condenser lens was measured through a pinhole (20 μm diameter) by a photodetector, EMI 9558B, an amplifier, and a recorder. This intensity is proportional to the square of the mode amplitude.

a. Experimental Results

For the first set of measurements with the arrangement shown in Fig. 6-18, most of the laser energy coupled into the waveguide was in the $HE_{1,1}$ mode. A photograph of the intensity distribution in the image of the output end of the waveguide for this case is shown in Fig. 6-20(a). Spatial filters with amplitude transmittances matching the magnetic field of $HE_{1,1}$ and $HE_{1,2}$ modes were alternatively placed in this plane. Figure 6-20(b) is a photograph of the intensity distribution in the image plane of the condenser lens (the plane of the pinhole in Fig. 6-18) when the $HE_{1,1}$ filter was in place. Figure 6-20(b′) is a photometric trace across the center of this pattern. Figures 6-20(c) and 6-20(c′) are the corresponding photograph and trace when the $HE_{1,2}$ amplitude filter was used.

For the second set of data, the $HE_{1,2}$ mode was launched on the waveguide. Figure 6-21(a) is a photograph of the image of the output end of the waveguide for this case. Figures 6-21(b) and 6-21(b′) are photographic and photometric records of the distribution in the image plane of the condenser lens when the $HE_{1,1}$ filter was used. Figures 6-21(c) and 6-21(c′) are the corresponding records for the $HE_{1,2}$ filters.

In each case, the photometric traces have been normalized by setting the intensity of the central peak for the matched filter equal to unity. The other record was thus referenced to this one. Photoelectric measurements of the relative transmittances of the central regions of the filters were made to provide correction factors yielding the appropriate normalization. Thus, the ordinate at the center of the photometric trace is a direct measure of the relative power in each of the two modes for the two sets of measurements. The results are summarized in Table IV.

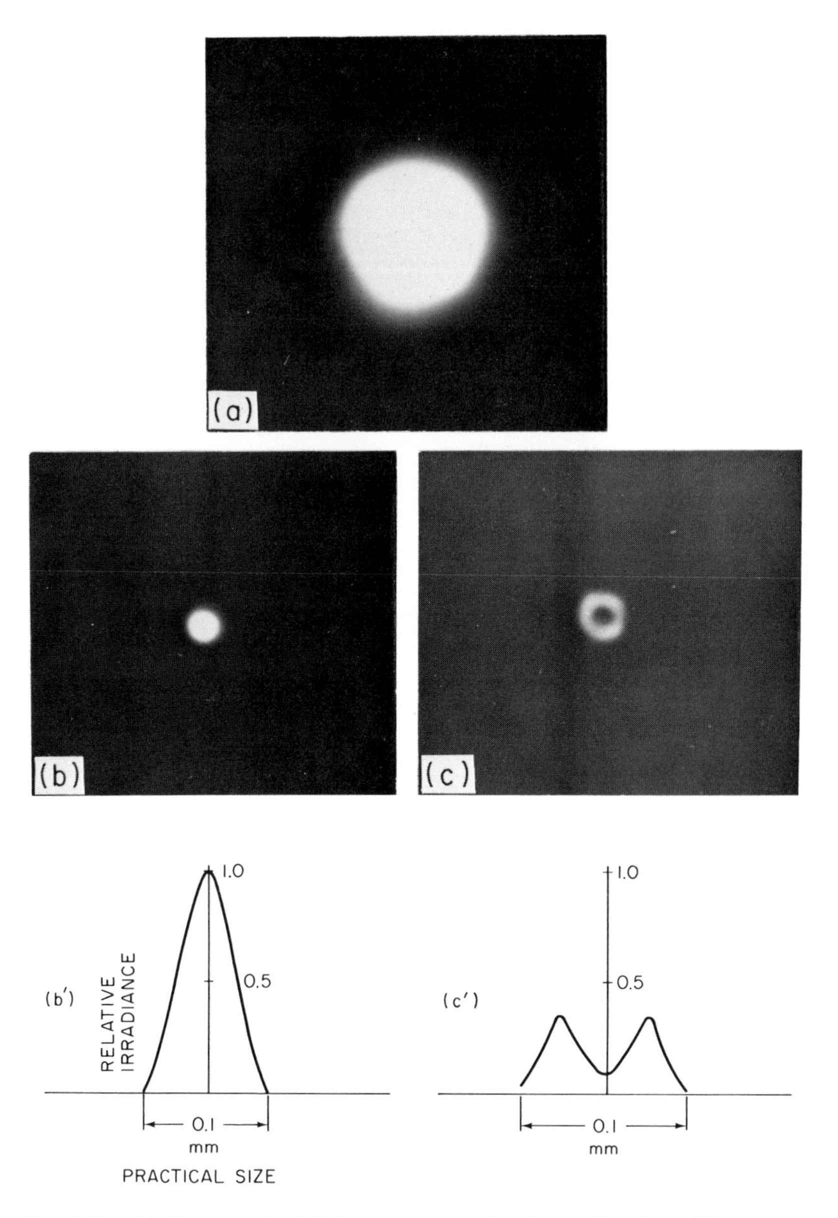

Fig. 6-20. (a) Photograph of $HE_{1,1}$ mode excited in 25-μm fiber by a diffraction pattern of a ring mask; (b) photographic and (b′) photometric recordings in the image plane of a spatial filtering lens for $HE_{1,1}$ field with $HE_{1,1}$ filter; (c) photographic and (c′) photometric recordings for $HE_{1,1}$ field with $HE_{1,2}$ filter.

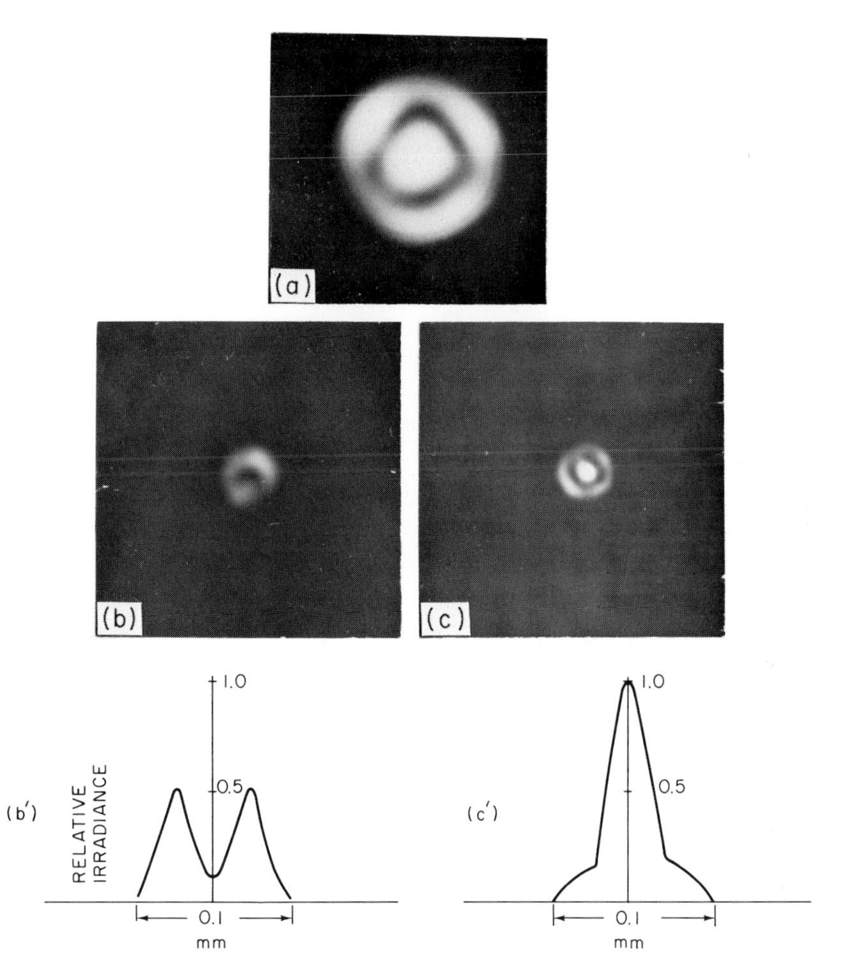

Fig. 6-21. (a) Photograph of $HE_{1,2}$ mode excited in a 25-μm fiber for a diffraction pattern of a ring mask; (b) photographic and (b′) photometric recordings in the image plane of a spatial filtering lens for $HE_{1,2}$ field with $HE_{1,1}$ filter; (c) photographic and (c′) photometric recordings for $HE_{1,2}$ field with $HE_{1,2}$ filter.

Table IV. SUMMARY OF EXPERIMENTAL RESULTS

Mode excited in a fiber	Form of the filter	Relative amplitude	Relative power
$HE_{1,1}$	$HE_{1,1}$	1.0	1.0
$HE_{1,1}$	$HE_{1,2}$	0.26	0.07
$HE_{1,2}$	$HE_{1,2}$	1.0	1.0
$HE_{1,2}$	$HE_{1,1}$	0.36	0.13

b. Experimental Errors

It is apparent from Table IV that the degree of discrimination achieved in these experiments is less than that estimated from theory. The differences between theory and experiment are within experimental error, however. The principal causes of the experimental errors are the lack of precision in fabricating the fiber waveguide and the filters, the finite size of the pinhole detector, and lens imperfections. The error caused by the lenses is in fact very small, since the divergence of the beam from the end of the fiber to the observation plane is small enough to justify the use of Gaussian geometric optics in its description. The error caused by the deviations in the amplitude transmittance of the filters from their theoretical value is approximately 15%. This is the calculated ratio of the flux that passes through the experimental filter to that which would pass through the ideal filter. The finite size of the pinhole contributes an error of about 5%; this estimate is based on the assumption that the photometric traces display all the significant structure in the images which were scanned. Finally, the waveguide cross section was not perfectly circular. This error has been estimated as between 10% and 15%. Within the limits of these experimental errors, the applicability of this technique of mode discrimination is adequately demonstrated.

REFERENCES

1. E. Snitzer and H. Osterberg, *J. Opt. Soc. Amer.* **51**, 499 (1961).
2. S. A. Schelkunoff, "Electromagnetic Waves." Van Nostrand–Reinhold, Princeton, New Jersey, 1943.
3. D. G. Kiely, "Dielectric Aerials." Methuen, London, 1953.
4. C. W. Horton and C. W. Watson, *J. Appl. Phys.* **19**, 836 (1948).
5. S. P. Schlesinger, P. Diament, and V. L. Granatstein, Scientific Rep. No. 80, Contract No. AFCRL-63-522, Columbia Univ. Dept. of Elec. Eng. (March 1963).
6. A. G. Fox and T. Li, *Bell Syst. Tech. J.* **40**, 453 (1961).
7. N. S. Kapany, "Fiber Optics: Principles and Applications." Academic Press, New York, 1967; N. S. Kapany and J. J. Burke, *J. Opt. Soc. Amer.* **51**, 1067 (1961).
8. A. W. Snyder, *J. Opt. Soc. Amer.* **56**, 601 (1966).
9. N. S. Kapany, J. J. Burke, K. L. Frame, and R. E. Wilcox, *J. Opt. Soc. Amer.* **58**, 1176 (1968).
10. A. L. Jones, Ph. D. Thesis, Purdue Univ. (1964).
11. E. Leith and J. Upatnieks, *J. Opt. Soc. Amer.* **52**, 1123 (1962).
12. E. Leith *et al.*, Laser Focus **1**, 15 (November 1965).
13. M. Kline and I. W. Kay, "Electromagnetic Theory and Geometrical Optics." Wiley (Interscience), New York, 1965.
14. J. Tsujiuchi, *in* "Progress in Optics" (E. Wolf, ed.), Vol. II, p. 133. Wiley (Interscience), New York, 1963.
15. E. Leith and J. Upatnieks, *J. Opt. Soc. Amer.* **53**, 1377 (1963).
16. A. Vander Lugt, *IEEE Trans.* **IT-10**, 140 (1964).

CHAPTER 7

Evanescent Surface Waves and Waveguide Interactions

In the previous chapters, we have treated the theoretical and experimental aspects of mode launching, identification, and radiation characteristics. In this chapter, the accent is on the surface wave outside the core of the guide. We will examine, both theoretically and experimentally, its field configuration, propagation characteristics, and coherent interactions between guides which it induces when modes are launched on aggregates of dielectric optical waveguides (active or passive).

It is well known (*1–3*) that as the diameter of a dielectric waveguide decreases, more energy of the supported modes is conducted outside the guide. Thus, the absorption coefficient of the core wavelength material becomes only partially effective, and the "effective absorption coefficient" is a function of the surrounding medium also. The effective absorption coefficient can be determined by calculating the fractions of energy conducted inside and outside the guide and applying the absorption coefficients to the two media. We will calculate the effective absorption coefficient, as a function of the waveguide's optical and physical parameters, for several low-order modes. Also, we will discuss the results of experiments conducted to verify the theory.

Furthermore, in this chapter, we will present theoretical and experimental studies of the electromagnetic coupling between parallel, passive, glass-fiber waveguides and active (Nd^{3+}-doped laser) resonators. Predicted evanescent-wave coupling strengths are verified to within a factor of two. It will be shown that the far-field interference patterns of these phase-locked, optical waveguides manifest a high degree of coherence. Coupling of two fiber lasers is also demonstrated through common-cavity end-mirror reflectors.

A. ENHANCED TRANSMISSION WITH SURFACE WAVES

The conventional techniques of light conduction along large-diameter fibers have yielded observable transmission in the visible region of the spectrum in 100–200-ft lengths. This is possible because of the availability of highly transparent materials in the visible region. On the other hand, in the ultraviolet and infrared regions, comparatively higher absorption coefficients are encountered, and high transmission in long fibers is not easily obtainable. If a substantial fraction of the mode power is in the evanescent wave, however, the effective absorption coefficient can be reduced considerably. Thus, it is possible to conduct energy along greater lengths than may be achievable in larger fibers of like materials for the fiber core and the surround.

The dielectric waveguide theory shows that both the number of modes supported and the depth of penetration of the evanescent-boundary wave in the rarer surrounding medium are determined by the fiber characteristic term $R = [(\pi d/\lambda)(n_1^2 - n_2^2)^{1/2}]$, where d is the diameter and n_1 and n_2 the refractive indices of the fiber core and coating, respectively. The fraction of the total energy guided by the fiber coating is determined by the fiber physical characteristics and the particular mode under consideration. By integrating the time-average axial component of the Poynting vector over appropriate limits, it is possible to calculate the fractions of energy supported in the fiber core (P_1) and the coating (P_2). Once the ratio P_2/P_1 is calculated, it is relatively simple to calculate the "effective absorption coefficient" α of the fiber from the absorption coefficients of the core α_1 and the coating material α_2. In this section, the expressions for P_2/P_1 and α are derived, and calculations for various modes in fibers of different R values are made. It is seen that a significant reduction in the effective absorption coefficient is achievable with the appropriate choice of various parameters.

In an experimental demonstration of this phenomenon, fibers with filter-glass core material were used. Visual observation of fibers of different diameters and lengths showed that high transmission is obtained from smaller-diameter fibers. Quantitative measurements at two different wavelengths showed a reduction of the absorption coefficient by a factor of two on the reduction of the fiber-core diameter.

1. Calculation of Effective Absorption Coefficient

We have already determined that the energy guided by a straight, circularly cylindrical, dielectric waveguide is not totally confined within the core of the guide. A fraction of the total energy guided by a fiber is thus conducted in the surrounding material. It travels parallel to the fiber axis in the form of a surface wave whose amplitude falls off approximately as $\exp(-\beta r)/(\beta r)^{1/2}$, where r measures the distance from the fiber axis. Since some of the energy is conducted in the coating, the effective absorption coefficient α of a given length of coated fiber will be less than that of an equivalent thickness of bulk-core material whenever the coating material is less absorbing than the core. An approximate expression for α, valid for $n_2 \simeq n_1$, is given by

$$\alpha = (\alpha_1 P_1 + \alpha_2 P_2)/P_{\mathrm{T}}, \tag{7.1}$$

where α_1 and α_2 are the bulk absorption coefficients of the core and coating materials, respectively, P_1 and P_2 represent the total time-average flux conducted inside and outside the core, and P_{T} represents the total flux guided by the system. A more general approximation which reduces to Eq. (7.1) as n_2 approaches n_1 is given by Elsasser (*4*).

Exactly how much of the total energy is carried in the coating depends upon the modes which are excited in the fiber. As the angle at which the plane waves constituting a mode (*5*, *6*) approaches the critical angle, a greater fraction of the total energy of the mode is propagated outside the fiber core. When the fiber diameter is large and it is illuminated by focusing a cone of light at its entrance end, most of the modes which are excited correspond to angles considerably greater than the critical angle, and the fraction of the total energy carried in the coating is generally insignificant. This case has been treated extensively in the literature (*7*). As the fiber diameter decreases, the number of modes which it can propagate also decreases. When the parameter R is less than 2.405, then the radiation which may be efficiently propagated by the fiber can travel at a single angle only. When this is the case, the quantitative correlation of theory and experiment becomes entirely unambiguous. With larger fibers ($10 \gtrsim R \gtrsim 2.5$), it is difficult to determine precisely how much of the total energy is being propagated in each of the possible modes and, for this reason, most of the numerical work in this section has been limited to cases where the diameter is such that $R < 2.405$.

A general expression for the flux density inside the core can be easily derived (*8*, *9*). It is given by the time-average axial component of the Poynting vector

$$\bar{S}_z = \tfrac{1}{2}\,\mathrm{Re}(\mathbf{E} \times \mathbf{H}^*)_z .$$

For radiation guided at a characteristic angle allowable for values of $R < 2.405$, the field components may have an infinite variety of forms due to the degeneracy of the waveguide solutions. All forms lead to the same result for the fractional flux inside the core, however, and only the simplest ones will be employed here. These are given in Eq. (7.2), where complex transverse components $E_+ = E_x + iE_y$, $E_- = E_x - iE_y$, etc., are employed [Eq. (7.2) follows directly from the general expressions of Eq. (4.58) specialized to the $HE_{-1,1}$ mode.] The z axis of the coordinate system is assumed to be coincident with the fiber axis, and

$$\begin{aligned}
&E_+ = bf_0; \qquad E_- = -\frac{1-A}{1+A}bf_{-2}; \qquad E_z = \frac{-i\beta}{h}\frac{b}{1+A}f_{-1};\\
&H_+ = \frac{ih}{\omega\mu}\frac{(k_1{}^2/h^2)+A}{1+A}E_+; \qquad H_- = \frac{ih}{\omega\mu}\frac{(k_1{}^2/h^2)-A}{1-A}E_-; \qquad (7.2)\\
&H_z = \frac{-ih}{\omega\mu}AE_z;
\end{aligned}$$

where

$$\begin{aligned}
f_p &= J_p(\beta r)\exp(i\omega t + ip\phi - ihz),\\
\beta^2 + h^2 &= k_1{}^2 = \omega^2\varepsilon_1\mu,\\
A &= -\frac{(1/u^2)+(1/q^2)}{[J_1'(u)/uJ_1(u)] + [K_1'(q)/qK_1(q)]},\\
u &= \beta d/2; \qquad q = \tfrac{1}{2}\beta' d; \qquad h^2 - (\beta')^2 = k_2{}^2 = \omega^2\varepsilon_2\mu.
\end{aligned}$$

J and K are Bessel functions of the first and second (modified) kind, respectively, the subscripts denoting their order. b is an arbitrary constant, possibly complex.

The time-average axial component of the Poynting vector is given by

$$\bar{S}_z = \tfrac{1}{4}\,\mathrm{Re}[i(E_+H_+{}^* - E_-H_-{}^*)]. \tag{7.3}$$

By evaluating the integrals

$$\int_0^{2\pi}\int_0^{d/2} S_z r\,dr\,d\phi = P_1; \qquad \int_0^{2\pi}\int_{d/2}^{\infty} S_z r\,dr\,d\phi = P_2,$$

one can readily derive a general expression giving the ratio P_2/P_1:

$$\frac{P_2}{P_1} = -\frac{u^2}{q^2}\,\frac{(k^2+\xi^2A)\left[\dfrac{K_0{}^2(q)}{K_1{}^2(q)} - 1\right] + \gamma(k^2-\xi^2A)\left[\dfrac{K_0{}^2(q)}{K_1{}^2(q)} - 1 - \dfrac{4}{q^2}\right]}{(1+\xi^2A)\left[\dfrac{J_0{}^2(u)}{J_1{}^2(u)} + 1\right] + \gamma(1-\xi^2A)\left[\dfrac{J_0{}^2(u)}{J_1{}^2(u)} + 1 - \dfrac{4}{q^2}\right]}. \tag{7.4}$$

Here, $k^2 = k_2^2/k_1^2$ and $\xi^2 = h^2/k_1^2$. The constant $\gamma = (1 - A)/(1 + A)$ is negligibly small if $k_2^2/k_1^2 > 0.95$. When this is the case, the second terms of the numerator and denominator may be ignored.

From Eq. (7.4), the ratios P_1/P_T and P_2/P_T required for Eq. (7.1) can be calculated. The quantities q, ξ, and A of Eq. (7.4) all depend on the eigenvalue u of the mode. This eigenvalue also determines the phase velocity ω/h and the characteristic angle θ of the mode as follows:

$$h = [k_1^2 - (2u/d)^2]^{1/2}; \qquad \theta = \cos^{-1}[u/(\pi n_1 d/\lambda)]. \tag{7.5}$$

This eigenvalue u, in turn, is given by the smallest (nonzero) value of u which satisfies the following characteristic equation, which was discussed at length in Chapter 4:

$$\frac{J_0(u)}{J_1(u)} = \frac{1}{u} + \left(1 - \frac{\delta}{2}\right)\left(\frac{uK_0(q)}{qK_1(q)} + \frac{u}{q^2}\right) - \left[\frac{R^2}{q^4}\left(\frac{R^2}{u^2} - \delta\right) + \frac{\delta^2}{4}\left(\frac{uK_0(q)}{qK_1(q)} + \frac{u}{q^2}\right)^2\right]^{1/2}, \tag{7.6}$$

where

$$\delta = 1 - (k_2^2/k_1^2),$$
$$R = (\pi d/\lambda)(n_1^2 - n_2^2)^{1/2},$$
$$q^2 = R^2 - u^2.$$

If either R or δ is varied, the eigenvalue u changes, and the value of P_1/P_2 in Eq. (7.4) changes. By solving for the first root of Eq. (7.6) as a function of R for a number of values of δ, a plot of P_2/P_1 as a function of R for these values of δ can be obtained.

By using the expressions discussed in the previous section, the terms P_1/P_T and α/α_1 were computed as a function of R for the modes $HE_{1,1}$, $HE_{1,2}$, $HE_{1,3}$, $HE_{1,4}$, $HE_{1,5}$, and $TE_{0,1}$. These calculations were made for different values of α_2/α_1 and $\delta = 1 - (n_2^2/n_1^2)$.

Figure 7-1 gives a plot of P_1/P_T as a function of R for $HE_{1,m}$ modes, where $m = 1$, 2, 3, 4, and 5. These curves pertain to the n_2/n_1 value of approximately 0.8. It can be seen that as the value of R goes down, more energy is conducted outside (for $m = 2, 3, 4, \ldots$) until the mode is cut off. For the dominant $HE_{1,1}$ mode, on the other hand, there is no cutoff theoretically. Once P_1/P_T is known, the effective absorption coefficient α can be calculated from Eq. (7.1).

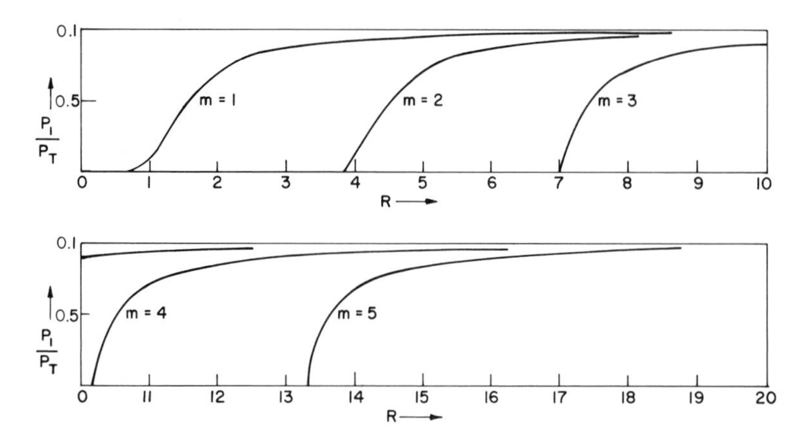

Fig. 7-1. Ratio of the power transmitted inside the fiber core to total power transmitted in fiber of characteristic term R for $HE_{1,m}$ mode, where $m = 1$, 2, 3, 4, and 5.

In Fig. 7-2, the ratio P_1/P_T in the $HE_{1,1}$ mode is plotted as a function of R for $\delta = 0.1$. This curve is also a good approximation to those which would be obtained for smaller values of δ and may, therefore, be used with reasonable accuracy for all values of δ in the range of $0 < \delta \leq 0.1$. For example, the value of P_1/P_T for $R = 1$ and $\delta = 0.01$ is actually 0.17 rather than the value 0.135 obtained from Fig. 7-2. For larger values of R, the discrepancies are smaller. Thus, at $R = 2.4$, $P_1/P_T = 0.827$ for $\delta = 0.01$ as opposed to 0.824 on the graph. The curve inflects at a value of R in the neighborhood of $R = 1.15$. This is to be expected, since the dominant $HE_{1,1}$ mode under consideration is never cut off. The ordinates of the P_1/P_T curve are specified at the left of the plot.

Also plotted in Fig. 7-2 are three linear functions of R, giving $n_1 d/\lambda$ values corresponding to three different values of δ. Each linear function is marked by the pertinent value of δ, which in turn fixes the ratio of n_2 to n_1, as indicated. These curves permit one to read P_1/P_T as a function of $n_1 d/\lambda$, which is the optical diameter of the fiber expressed in terms of the number of wavelengths needed to span the diameter. For example, the curve corresponding to $\delta = 0.1$ can be used to evaluate the characteristics of an infrared-transmitting arsenic trisulfide fiber coated in an arsenic–sulfur glass of lower index ($n_2 = 2.35 = 0.9475 n_1$).

With the aid of the graphs in Fig. 7-2 and Eq. (7.1), the plots of α/α_1 can be generated. In Fig. 7-3, each curved line represents a function of R that specifies the ratio of the absorption coefficients of the coating and core materials. The straight lines again indicate the optical diameter of the fiber as a function of R. For different ratios of the absorption coefficients of the

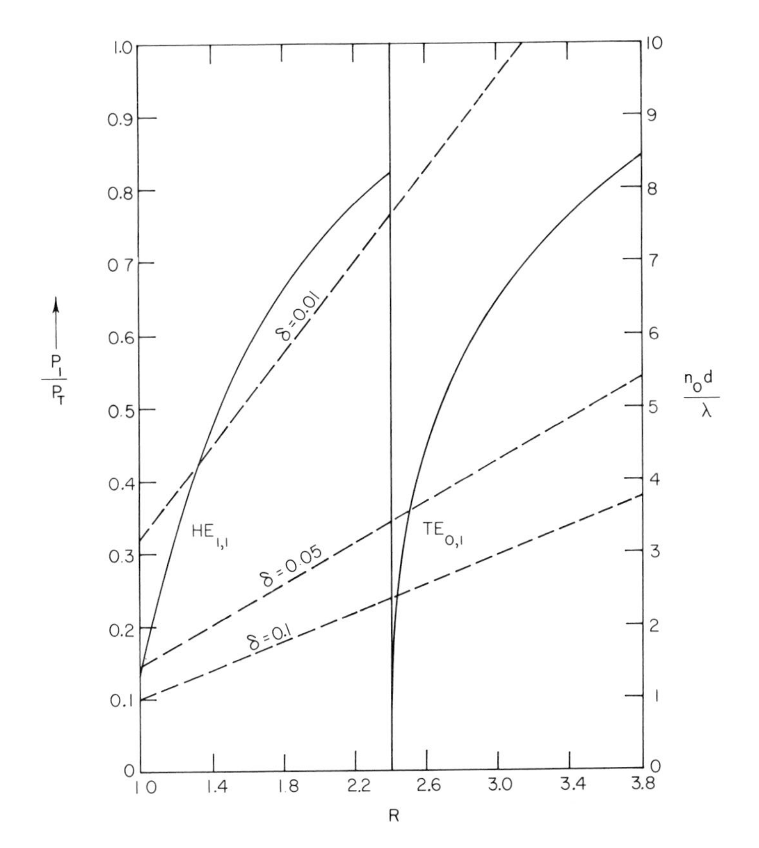

Fig. 7-2. Ratio of power transmitted inside the fiber core to the total power transmitted in $HE_{1,1}$ and $TE_{0,1}$ modes (nonlinear curve plotted to scale at left), and optical diameter of core (linear curves plotted to scale at right), as a function of R.

coating and core materials, seven different curves are plotted. The pertinent curves are appropriately designated. For example, for $\delta = 0.1$, the lowest linear function applies. Let $n_1 d/\lambda = 2$. Then, $R \simeq 2$ and $\alpha/\alpha_1 \simeq 0.84$. This means that if the bulk-core material absorbs 20%/mm, the coated fiber absorbs 16.8%/mm. The large changes in transmission that can be effected in small-diameter fibers are thus obvious.

Figure 7-2 shows another curve of P_1/P_T as a function of R. This curve gives the energy-guiding characteristics of a fiber excited in the transverse electric mode ($TE_{0,1}$). This mode is cut off at $R = 2.405$. In the range of R values $2.405 < R \leq 3.8$, the relative flux inside and outside the core changes very significantly, and corresponding changes in the absorption rates can be readily calculated from Eq. (7.1).

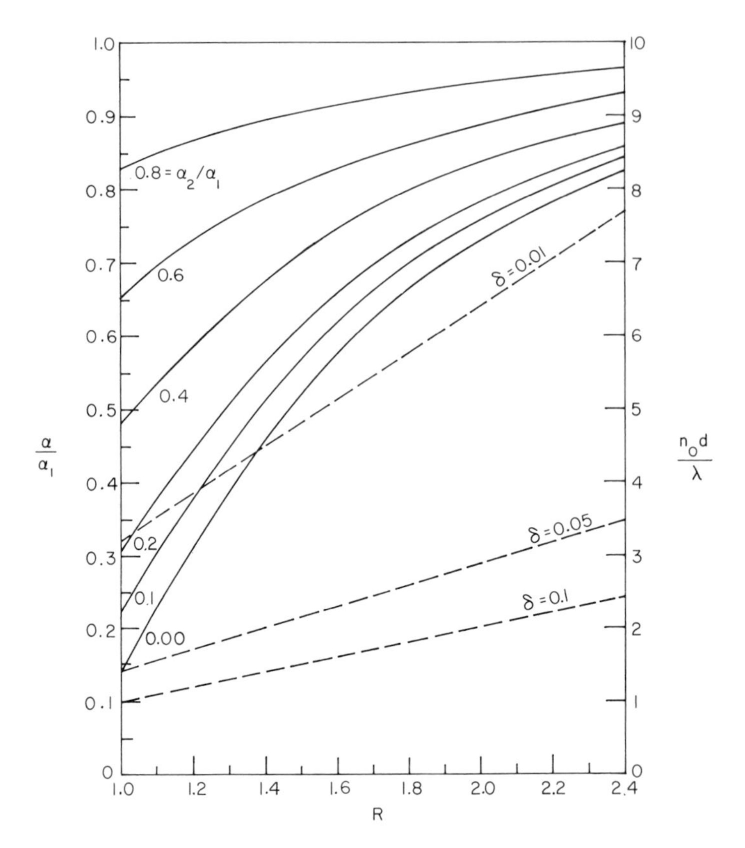

Fig. 7-3. Ratio of effective absorption coefficient of fiber to that of bulk core material for seven values of the ratio of absorption coefficients of coating and core materials (nonlinear curves plotted to scale at left), and optical diameter of core for three values of $\delta = 1 - (n_2^2/n_1^1)$ (linear curves plotted to scale at right), as a function of R.

2. Measurements of Enhanced Transmission

For a quantitative correlation of theory, several fibers (with cores of Corning Signal-Yellow filter glass No. 3-77, coated in soda-lime glass) were tested to determine their practicability. As can be seen from the theoretical curves in Figs. 7-1–7-3, the greatest changes in effective absorption occur over a very small range of d/λ. In order to confirm these predictions experimentally, it is necessary to work with fibers between 1 and 2 μm in diameter and to measure these diameters accurately.

A qualitative confirmation of the theory was accomplished with fibers of core diameters 50, 5, 4, 3, and 2 μm, made of filter glass cores and transparent coatings. Two multiple fibers consisting of five fibers of the absorbing glass were formed. The first multiple fiber was approximately 1 in. long and the second approximately 4.5 in. By illuminating these two multiple fibers with a uniform beam of sodium light large enough to cover all the fibers simultaneously, one can visually compare the transmittance of the five tubes of different diameter in each bundle, and also the transmittance of fibers of the same diameter in the two lengths. In the small-diameter fibers, only simple, lower-order modes are excited. These were studied with a polarizer and analyzer at large magnification. The 5-μm fiber propagated predominantly the $HE_{1,2}$, $EH_{1,1}$, and $HE_{3,1}$ modes. The 4- and 3-μm fibers show predominant excitation in the $TE_{0,1}$, $TM_{0,1}$, and $HE_{2,1}$ modes, while the 2-μm fiber supports only the $HE_{1,1}$ modes.

It is interesting to note that the modes that a given fiber displays in this experiment are the highest-order modes that it will support. Since all fibers are excited by a convergent cone of light wide enough to excite all modes that they are capable of supporting, one might expect that the exit ends of all the fibers would appear more or less uniformly illuminated. That they do not might be interpreted as a qualitative confirmation of the theory. Since only the highest-order modes carry a significant fraction of their total energy outside the fiber, only these modes have an effective attenuation that is considerably less than that derived from the absorption properties of the fiber core alone. Thus, one might assume that all modes that a given fiber could support are excited, but only the highest-order modes are in evidence after the energy has been propagated down the fiber. However, one cannot assume that all modes that a fiber can support are excited with the same intensity. On the contrary, there is considerable evidence to indicate that modes near cutoff are preferentially excited in nonabsorbing fibers, because their substantial surface waves increase the effective diameter of the guide for their excitation. It is, therefore, likely that both these phenomena, i.e., preferential excitation and absorption, were operative in providing the observed patterns at the exit ends of the fibers.

In order to quantitatively determine the absorption coefficient, comparative transmission measurements were made on fibers of 2 and 50 μm diameter. The fiber cores were made of Corning Signal-Yellow glass. Figure 7-4 illustrates the optical arrangement for this experiment. Radiation from an Osram mercury-arc source was collimated, chopped, dispersed by a diffraction grating, and focused on a 50-μm pinhole. The approximately 1-μm image of this pinhole, formed by an 47.5$\times$ apochromatic objective (0.95 N.A.) in the plane of the entrance end of the fibers, was centered on the 2-μm-diameter fiber of absorbing glass ($n_1 = 1.525$) embedded in a 200-μm-

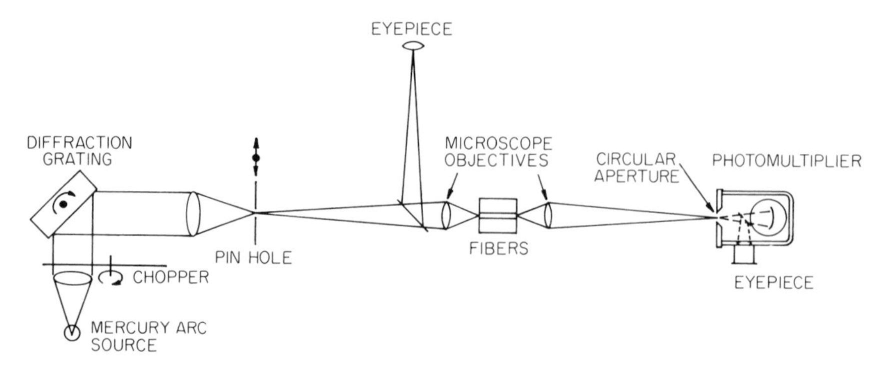

Fig. 7-4. Optical arrangement used for measurement of effective absorption coefficient of fibers.

diameter coating ($n = 1.5188$). The centering of the pinhole image with respect to the fiber was accomplished by moving the 50-μm aperture until the energy transmitted by the fiber gave a maximum signal on a photomultiplier. A 47.5$\times$ apochromatic objective imaged the exit ends of the fiber in the plane of an 0.044-in. aperture behind which the photomultiplier was located. The photomultiplier signal was amplified and rectified and used to drive the pen of a strip-chart recorder. The signal was also maximized by lateral motion of the photomultiplier tube and by rotation of the grating to put maximum flux on the fibers for each of the two mercury lines.

This procedure was followed for each of two wavelengths (5770 and 5461 Å) and two lengths (0.57 and 1 in.) of the 2- and 50-μm-diameter fibers. The relative positions of the source, chopper, grating, condenser, entrance aperture, and illuminating microscope were constant for all runs, except for fine focusing (a few microns) of the microscope objectives and fine adjustments of the 50-μm-diameter aperture for centering the image on the fiber. Because the numerical aperture of the illuminating optics (0.95) was much greater than that of the fiber (0.136), excitation of the coating could not be obviated. The 0.044-in. aperture in front of the phototube eliminated most of the signal from the coating (only 1/10,000 of coating area was imaged at tube).

The measurements gave ratios of effective absorption to bulk absorption of 0.62 ($\alpha_1 = 1.30$) and 0.61 ($\alpha_1 = 2.2$) for wavelengths 5770 and 5461 Å, respectively. Table I shows the detailed results of the measurements. Because of the high sensitivity of the flux measurements to minute motions, the values for the ratios given in columns 6 and 7 may be as much as 20% in error.

It is to be noted that two basic sources of error exist in this experiment. First, the accuracy in the determination of the fiber diameter, as shown in

Table I

THEORETICAL AND EXPERIMENTAL EFFECTIVE ABSORPTION COEFFICIENT FOR TWO FIBERS[a]

λ (microns)	d (microns)	Experimental coefficient: Signals (microvolts) 1.0 in	0.57 in	α (in^{-1})	α/α_1	$\alpha_{2\mu}/\alpha_{50\mu}$	Theoretical coefficient: α/α_1	d
0.577	2±0.2	0.04	0.06	0.81	0.62	0.36	0.61	2.2
	50	0.01	0.03	2.22			0.53	2.0
							0.43	1.8
0.546	2.0	0.04	0.08	1.34	0.61	0.48	0.65	2.2
	50	0.01	0.02	2.86			0.58	2.0
							0.49	1.8

[a] Bulks properties: core: $n = 1.5252$, $\alpha_{(0.577\mu)} = 1.3\ in^{-1}$, $\alpha_{(0.546\mu)} = 2.2\ in^{-1}$; coating: $n = 1.5188$.

the theoretical column, plays a very significant role in the calculation of the term α/α_1. A 10% inaccuracy in the fiber diameter can contribute almost 20% inaccuracy in α/α_1, and, because of the resolution limit of high-power microscope objectives, this magnitude of error is probable. A second source of error lies in the change of the refractive index and perhaps even the absorption coefficient caused when bulk material is heated and drawn into fibers. Previous investigations (*10*) have shown that a measurable refractive-index change occurs in the fiber-drawing process. It is for these reasons that not only is the theoretical value of α/α_1 in variance with the experimental measurements, but also that the α values for a 50-μm-diameter fiber are different from the bulk absorption coefficient. However, within these experimental errors, the results tabulated in Table I serve to verify the theoretical prediction.

B. DISTRIBUTED COUPLING IN FIBER WAVEGUIDES

The coherent exchange of energy between parallel waveguides and transmission lines at microwave frequencies has been studied extensively in the past. Generalized theories of mode coupling have been developed (*11*) and are elucidated in several well-known books (*12–14*). More recently, observations of this phenomenon at optical frequencies in glass-fiber waveguides have also been reported (*15–18*). The general phenomenological theory has been restated in connection with the optical work (*15–17*); however, there has not been a quantitative comparison between theory and optical experiments. The published literature contains two theories (*19*, *20*) that yield specific closed-form expressions for the coupling strength between two parallel, circularly cylindrical dielectric waveguides. Both theories agree well with microwave experiments (*19*, *20*). They have been further compared in a more recent study (*21*), where general relationships between the coupling coefficients and the many parameters that govern them have been investigated. A physical discussion of this phenomenon, applied to planar waveguides, has been given in Chapter 3, Section C.

Some experimental observations of coherent interactions between parallel-fiber-laser resonators have also been described (*22*). We will discuss here several experiments that have been designed with the aid of theoretical predictions of the strengths of coupling between the resonators. The experiments show that if the cavity end mirror is not in close contact with the fibers, the agency of the coupling is not necessarily the evanescent surface waves of the modes of the two resonators. Phase locking also occurs by way of end-mirror reflection from each fiber into the other. Evidence of this type of coupling will be described.

1. Theory of Coupling

A pair of parallel fibers in a glass of lower refractive index constitutes a composite dielectric waveguide with twice as many modes as a single fiber. A launching condition that would provide single, lowest-order mode ($HE_{1,1}$) propagation in an isolated fiber (*18*, *23–25*) will excite two modes in the pair, having slightly different phase velocities. The interference of energy in these two modes provides for a length-dependent modulation of the distribution of the radiant energy between the pair. With the proper choice of length for given refractive indices, wavelength, and guide geometry, all of the energy will appear at the exit end of the unexcited fiber. This occurs when the length of the pair is some odd multiple of one-half of the so-called beat length. At odd multiples of one-quarter of the beat length, the energy is equally distributed between the pair.

In the far field of the coupled pair of fibers, we observe interference effects similar to Young's double-slit fringes, with an envelope that is the radiation pattern of the single fiber. These latter patterns have been studied, both theoretically and experimentally (*18*).

We make use of approximate, analytic expressions for the pair of modes in the two-fiber composite waveguides that correspond to a particular mode in the isolated fiber. As noted by many authors (*11–16*, *19*, *20*), these take the form of symmetric and antisymmetric combinations of the single-fiber modes. They can be described as

$$\begin{aligned} \psi_S &= C_S[\psi_0(1) + \psi_0(2)] \exp[i(h_0 + \Delta h)z - i\omega_0 t] \\ \psi_A &= C_A[\psi_0(1) - \psi_0(2)] \exp[i(h_0 - \Delta h)z - i\omega_0 t]. \end{aligned} \tag{7.7}$$

In these equations, ψ_S and ψ_A describe a component of the electric or magnetic field of one or the other mode. The subscripts S and A designate the symmetric and antisymmetric modes, respectively, either of which can be launched with arbitrary amplitude and phase, as specified by the complex constants C_S and C_A. The term ψ_0 describes the transverse dependence of the corresponding field component for the mode on the isolated fiber that propagates at a phase velocity ω_0/h_0. The arguments 1 and 2 of ψ_0 represent transverse coordinates measured from the centers of fibers 1 and 2, respectively. The symmetric mode is characterized by equal-amplitude and in-phase vibrations of the fields at points that are mirror images in the plane that bisects the fibers. For the antisymmetric mode, the vibrations are π radians out of phase. Any unequal distribution of energy at the input plane and/or any other relative phase condition will result in the excitation of both modes and a z-dependent modulation of the input power between the fibers. In particular, a launching condition that puts all of the power on guide 1 is seen to be equal-amplitude, in-phase excitation of both the symmetric and antisymmetric modes. The power on guide 1 varies as $\cos^2(\Delta hz)$ and on guide 2 as $\sin^2(\Delta hz)$. The beat length λ_B is defined as the length required for all of the input energy to transfer over to the second fiber and back again to the first. Thus, $\lambda_B = \pi/\Delta h$. It can be seen from Eq. (7.7) that when the lauching condition is such that all the energy is initially incident on one fiber, the field vibrations on the second fiber are always $\pi/2$ out of phase with those on the first. At successive odd multiples of $\lambda_B/4$, one fiber first leads, then lags the other by this phase difference, with the energy being equally distributed between them. Therefore, in the far field of the output ends of two coupled fibers of the appropriate length, these phase differences will be manifested by an asymmetry of the radiation pattern, with neither a maximum nor a minimum occurring in the forward direction.

When standing waves in coupled-fiber resonators, rather than traveling waves in coupled waveguides, are under consideration, the expressions for ψ_S and ψ_A in Eq. (7.7) must be modified. The imposition of boundary conditions in the third dimension provides for frequency selection. The symmetric and antisymmetric modes of the coupled resonator are then those with the same longitudinal dependence (e.g., sin $h_0 z$ with h_0 equal to $N\pi z/L$) but with different frequencies (slightly greater or slightly less than the frequency of the corresponding mode of the isolated fiber resonator). If only one pair of such modes were resonating, the energy would beat in time between the two fibers, with the beat frequency being the difference between the frequencies of the symmetric and antisymmetric modes. To calculate this frequency, we note that $\Delta\omega/\omega_0 = \Delta h/h_0 = \lambda_g/2\lambda_B$, where λ_g is the effective guide wavelength for the corresponding waveguide mode at frequency ω_0.

A more complicated behavior is to be expected from coupled laser resonators driven by randomly localized noise sources. In this case, many pairs of symmetric and antisymmetric modes will be running simultaneously, with each pair yielding a different beat frequency because of the dispersion of the fibers. As in other pulsed resonators, the detailed theory of the temporal behavior of single and coupled fiber-laser resonators is a subject of current research.

Investigations of the coherent coupling of optical energy between two parallel glass-fiber waveguides have given rise to several suggested optical switching configurations (*16*, *17*). Quasimonochromatic light injected into one of the fibers can be caused to exit from either fiber. This would be accomplished by an electrooptical or magnetooptical variation in the difference in the refractive index between the fiber core and coating materials. Such a variation modulates the coupling strength between the fibers, thus changing the beat length λ_B of the pair, i.e., the length required to obtain complete transfer of the injected flux from the excited fiber to its neighbor and back again to the excited fiber. Here, we will provide a theoretical basis for evaluating these suggestions.

A numerical study of the predictions of the coupling strength between circular dielectric waveguides by Jones (*20*) and Bracey *et al.* (*19*), has been conducted, and plots have been made of the beat length as a function of the several parameters that affect it. These are shown in the curves of Fig. 7-5. The numerical study is confined to the range of values of the fundamental parameter $R = (\pi d/\lambda_0)(2\bar{n}\,\Delta n)^{1/2}$ over which single-mode operation of the waveguides is assured, since multimode guides are clearly not advisable for switching applications. From the ordinate $C(R)$, the beat length $\lambda_B = C(R)\lambda_0/(\bar{n}\,\Delta n)$ can be calculated for any desired value of the free-space wavelength λ_0, the average index of refraction of the core and coating glasses $\bar{n} = (n_1 + n_2)/2$, and the index difference $\Delta n = n_1 - n_2$. Each curve pertains to a particular value of the ratio t/d of the center-to-center spacing

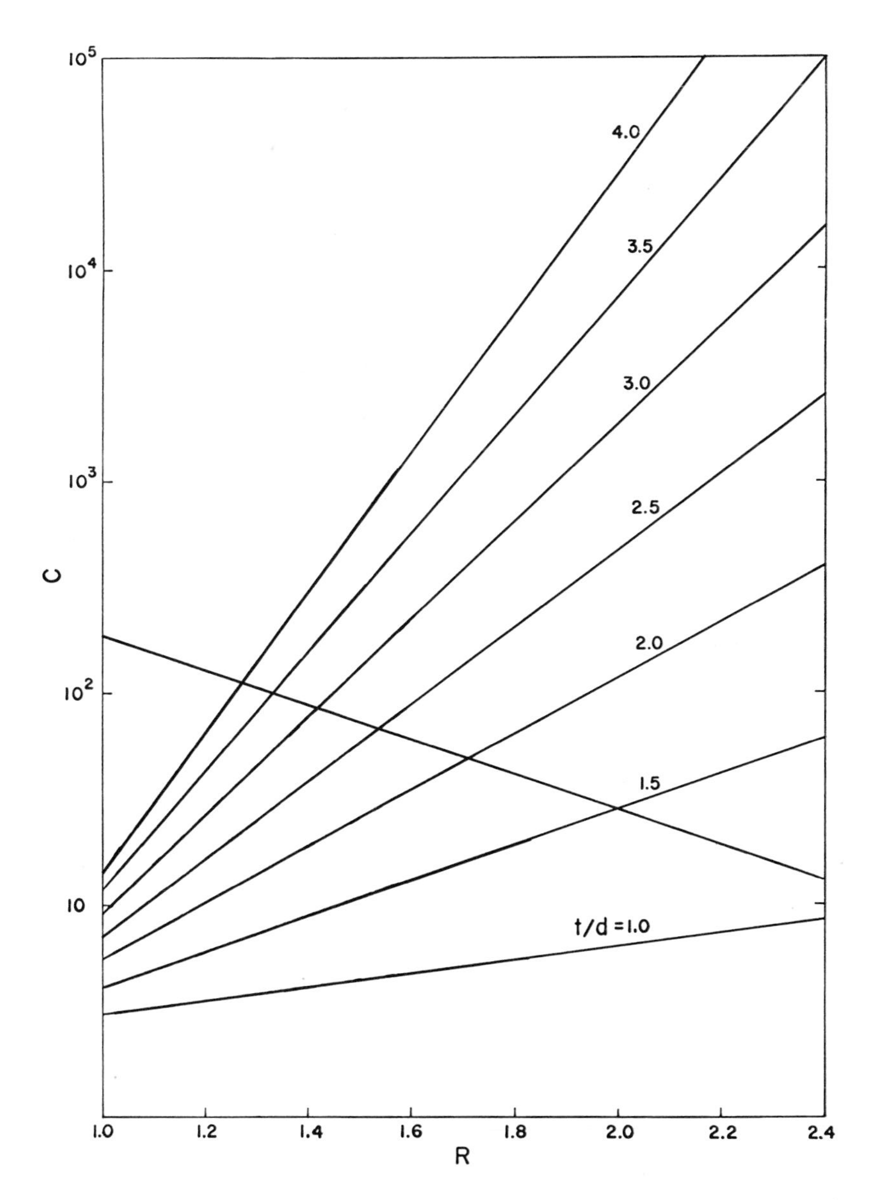

Fig. 7-5. Plots of $C(R)$ versus R for coupled pairs of single-mode glass fibers with the ratio of center-to-center spacing to diameter (t/d) as parameter. The beat length λ_B is calculated from the ordinate according to the equation $\lambda_B = C(R)\lambda_0/(\bar{n}\,\Delta n)$. The fiber parameter enters through the abscissa $R = (\pi d/\lambda_0)(2\bar{n}\,\Delta n)^{1/2}$. The accuracy of the curves is better than 10% above the negatively sloped line.

of the fibers to their diameter. The diameter-to-wavelength ratio enters through the abscissa R. The curves were plotted from numerical predictions based on the theories of Jones and Bracey *et al.*, for the case $n_1 = 1.53$ and $n_2 = 1.52$. Other numerical calculations have shown that these curves are applicable for the range of core indices from 1.53 to 1.75. It is apparent that beat lengths ranging from fractions of millimeters to meters are possible with glass fibers.

The line with a negative slope that crosses the several curves in Fig. 7-5 defines the accruacy of these curves. Above the line, the curves are expected to be accurate to at least 10%. In this region, the curves can clearly be represented by

$$\lambda_{\mathrm{B}} = C_0 e^{\alpha R} \lambda_0 / (\bar{n}\, \Delta n), \tag{7.8}$$

where C_0 is a constant and α is a function of t/d but not of R. In particular, we can write

$$\alpha = 0.74 + 2.3[(t/d) - 1]. \tag{7.9}$$

With the knowledge of the beat length λ_{B} provided by Fig. 7-5, it becomes a simple matter to determine a switching length L_{S}. This is defined as the length of the pair required such that an external field can cause the output light to switch from one fiber to the other. More specifically, the external field must change the index difference Δn by an amount sufficient to make a fiber pair that is m half-beats long before switching and $m + 1$ half-beats long after switching, where m is any positive integer or zero. Since the intensity on a fiber varies as $\cos^2(\pi L_{\mathrm{S}}/\lambda_{\mathrm{B}})$, we therefore have the condition

$$(1/\lambda_{\mathrm{B}_2}) - (1/\lambda_{\mathrm{B}_1}) = 1/2L_{\mathrm{S}}, \tag{7.10}$$

where λ_{B_1} and λ_{B_2} are the beat lengths before and after switching. With the substitution of Eq. (7.8) into Eq. (7.10), we obtain

$$L_{\mathrm{S}} = \frac{\lambda_{B_1}/2}{[1 - (\Delta R/R_1)]^2 e^{\alpha \Delta R} - 1}, \tag{7.11}$$

where R_1 is the value of R before switching and ΔR is the change of R resulting from switching.

The meaning of Eq. (7.11) can be seen from the following considerations. From the definition of R, we have

$$\Delta R/R_1 = \delta(\Delta n)/(2\, \Delta n), \tag{7.12}$$

where $\delta(\Delta n)$ indicates the field-induced variation of the original index difference Δn. For optical fibers fabricated by conventional drawing techniques, the assumption $\Delta n \geq 10^{-3}$ is reasonable. On the other hand, considerations of the magnitude of refractive-index changes that can be effected by external fields suggest that the assumption $\delta(\Delta n) \leq 10^{-4}$ is more than adequate. The maximum value of $\Delta R/R_1$ is accordingly 0.05. For this value of $\Delta R/R_1$, we can calculate, using Eq. (7.11) and the curves of Fig. 7-5, that the shortest lengths over which switching can be obtained correspond to the values of C and R lying along the negatively sloped line in Fig. 7-5. (Below the line, the theory is not sufficiently accurate to give dependable estimates of the effect of small changes of Δn.) For $\bar{n} = 1.5$ and $\lambda_0 = 5000$ Å, the corresponding switching lengths lie between 4 and 5 cm and are thus almost independent of t/d.

For most optical fibers, the assumption $\alpha\,\Delta R \ll 1$ is reasonable. Equation (7.11) then reduces to

$$L_S \simeq C\lambda_0/\{\bar{n}\,\delta(\Delta n)[\alpha R_1 - 2]\}. \tag{7.13}$$

This equation is essentially independent of the beat length. In the region above the 10% accuracy line in Fig. 7-5 [where $(\alpha R_1 - 2) > 1$], we find again that the shortest switching lengths correspond to the values of C and R_1 along the 10% line. Assuming $n = 1.5$ and $\lambda_0 = 5000$ Å, we find that an index variation $\delta(\Delta n) = 10^{-4}$ yields $L_S \simeq 5$ cm, almost independent of t/d. More reasonable values for glass undergoing a Kerr-type, field-induced index variation would provide $\delta(\Delta n) \simeq 10^{-6}$. The corresponding minimum switching lengths would be about 5 m at $\lambda_0 = 5000$ Å.

Equation (7.11) can lead us to conclude that short switching lengths might be achieved by using fibers with a large Δn and a very large t/d. The latter would correspond to a nearly vertical curve in the vicinity of $R = 0.8$ in Fig. 7-5. However, the fabrication of such a coupler is quite impractical, because, in this case, the slightest variation of the dimensions of the guides would completely change the predicted performance, and even the most minor imperfections in surface quality would cause such coupling of the waveguide modes to the continuous-mode spectrum of the radiation field as to negate the entire analysis. We thus conclude that large modulation depths cannot be achieved with Kerr-type modulation of the coupling between short-glass fibers with values of t/d and d/λ_0 corresponding to points above the negatively sloped line in Fig. 7-5. A more accurate theory is required to obtain good estimates below this line.

2. Experiments in Coupling

The method of constructing closely coupled fiber waveguides does not lend itself to the kind of experimental study of this effect that is done at microwave frequencies, where the separation of the guides can be varied continuously. In order to obtain single-mode, glass-coated glass guides, it is necessary to work with diameters of a few microns or less. Furthermore, the theory (*19–21*) indicates that it is necessary that the center-to-center spacing of the guides be of the order of the fiber diameter, to provide significant coupling over reasonable lengths. Such small sizes make it totally impractical to attempt to vary the spacing between the guides by manipulating separate mounts for each guide. The coupled pair must thus be fabricated and mechanically supported as a unit. Several standard fiber optics techniques have been found to yield coupled pairs. The ultimate result is a pair of glass guides of fixed length, diameter, spacing, and refractive indices. The remaining parameter with which the theory can be tested is the wavelength, which is varied by employing a monochromator to yield any desired narrowband excitation of the waveguide pair.

The experimental setup used for these studies (*26*) of the coupling phenomenon is essentially the same as illustrated in Fig. 7-4. A 75-W PEK xenon arc served as the primary light source. A condenser focused the light onto the entrance slit of an Engis monochromator having a dispersion of 33 Å/mm The *f*/4 output beam from the 0.75-mm slit was relayed to a 50-μm aperture by a Kodak Aero-Ektar lens. The flux through the aperture is imaged onto the entrance end of one of the two fibers by a 10$\times$ microscope. The output ends of the fibers are viewed, photographed with the aid of microscopes, or scanned with a photomultiplier system.

a. Spatial Beats

Figure 7-6 gives a plot of the normalized power output of an excited (solid curve) and a parallel unexcited (dashed curve) fiber as a function of wavelength over a range from 550 to 600 nm. The data are normalized so that the total power from the two guides is unity. The fiber cores were 4.38 μm in diameter, with a center-to-center spacing of 8.43 μm. Their length was 9.2 cm. The launching light is unpolarized, which is partially responsible for the modulation being less than full. The modulation can be increased by 5% by using light polarized perpendicular to the midplane between the fibers. For the other linear polarization, the results are very close to those for unpolarized excitation. The precise wavelengths that give

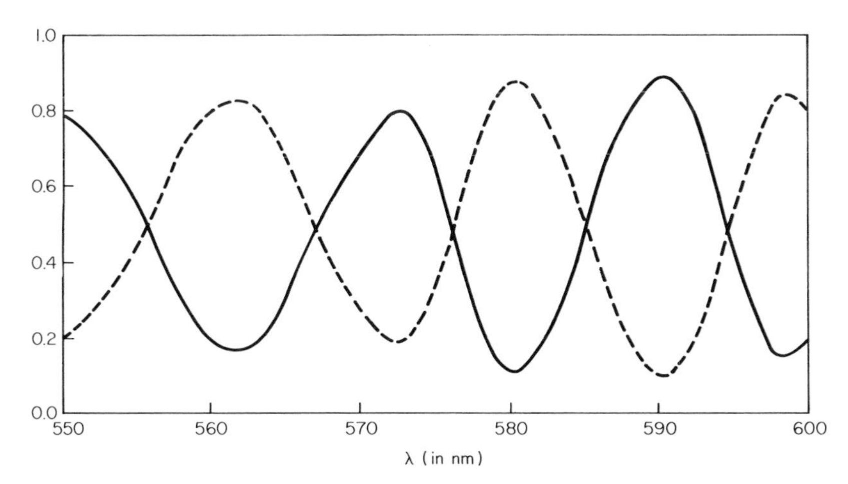

Fig. 7-6. Normalized measured output of excited (——) and unexcited (- - - - -) fibers in coupled pair versus wavelength in nm. Fibers are 9.2 cm long, 4.38-μm in diameter, and 8.38 μm in center-to-center spacing. Fiber numerical aperture$(2\bar{n}\,\Delta n)^{1/2} =$ 0.0822.

maxima and minima in the output of the fibers are dependent on polarization, as predicted by the approximate theories (*19*, *20*). The experimental fibers give significantly larger variations with polarization than are expected by theory, however, and also indicate that none of the two symmetric and two antisymmetric modes is exactly linearly polarized. If the exact polarization of the true modes of the experimental fibers could be imposed on the input face, it would be expected that higher modulation could be achieved. The effective bandwidth of the launching light for this experiment was approximately 20 Å; accordingly, it should have a noticeable effect on the measured modulation in the experiment of Fig. 7-6, where full exchange occurs over 100 Å. From these and other data, it is also apparent that the two guides are not of perfectly resonant dimensions. It is generally the case that the longer the beat length (the higher the frequency), the less the measured modulation. However, the unexcited fiber does not exhibit any zeros of output when the wavelength is varied, as theory (*11*) predicts it should, if dimensional tolerances were the only cause of incomplete modulation. We therefore believe that several of the causes mentioned combine to give the measured modulation. In Fig. 7-6, the distortion and diminution of the maxima at 572 nm was caused by a nonlinearity in the strip-chart recording.

Figure 7-7 shows a trace of the output of two fibers as a function of wavelength for another coupled waveguide assembly 2.1 cm long. The core diameters were 4.73 μm, and the center-to-center spacing was 9.65 μm. It is clear from the curves that these fibers did not have the same cross section over their interaction length or, more simply, were not of resonant dimensions. The term resonant here means having identical wave numbers for the uncoupled normal modes. From theory (*11*), we estimate for this case, where only about 75% of the total energy is exchanged between the parallel fibers, that the measured beat lengths should be about 10% less than would be predicted for guides of perfectly resonant dimensions.

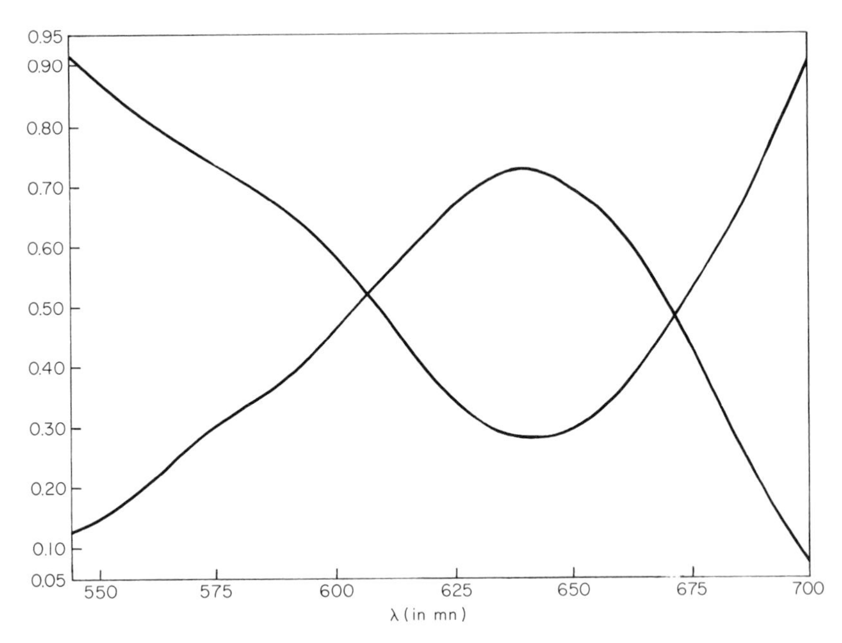

Fig. 7-7. Normalized measured output of excited and unexcited fibers in coupled pair versus wavelength in nm. Fibers are 2.1 cm long, 4.73 μm in diameter, and 9.65 μm in center-to-center spacing. N.A. = 0.0822.

The sine of the angle of maximum acceptance, or numerical aperture [which is given by $(n_1^2 - n_2^2)^{1/2}$, where n_1 and n_2 are the refractive indices of the core and coating material, respectively, and $n_1 > n_2$] of these guides was determined by two independent methods. It was calculated from the cutoff wavelength (510 nm) for the $TE_{0,1}$ mode, using the microscopically measured value of the diameter. It was also determined by measuring the diameter of the incoherent radiation pattern of a large fiber of the same glass constitution when it was illuminated by a cone of light much larger than its numerical aperture. These measurements agree to within 0.2%, yielding a numerical aperture value of 0.0822.

For these values of numerical aperture, length, diameter, and center-to-center spacing, the theoretical curve of Fig. 7-8 was prepared. This shows the number of beat lengths per centimeter as a function of wavelength. It is derived from the theoretical predictions of Jones (*20*) and Bracey *et al.* (*19*) as generalized in Fig. 7-5. The curve indicates that this waveguide pair should be one half-beat length long at 542 nm, two half-beats long at 614 nm, three at 667 nm, and four at 712 nm. The experimental curves of Fig. 7-7, on the other hand, show that the fiber is actually N half-beats long at 540 nm and $N + 2$ half beats at 710 nm, where N is an integer. To determine its value (one, in this case), it is sufficient to examine a longer fiber pair of the same constitution. This was done with fibers 9.2 cm long. In this case, the

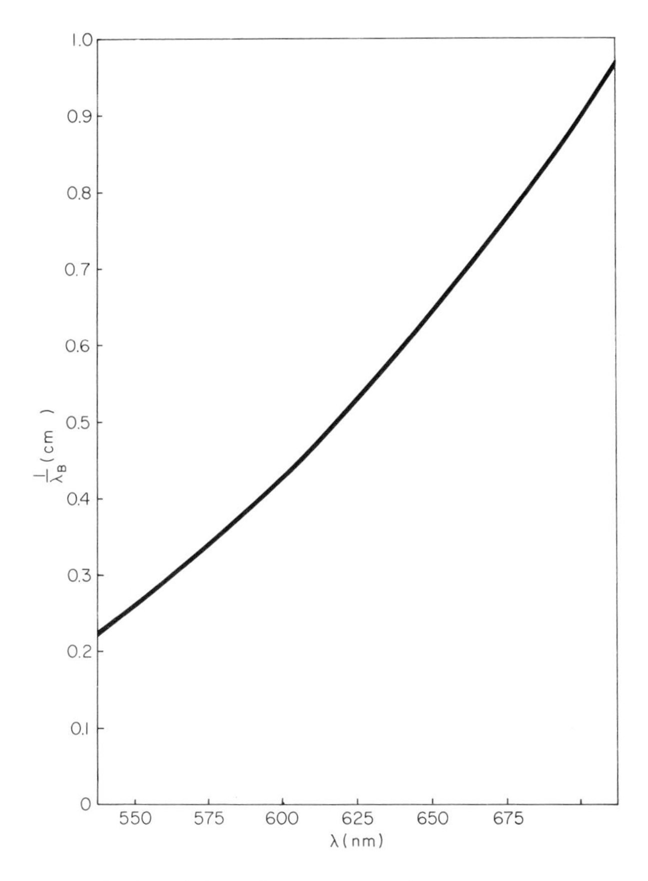

Fig. 7-8. Theoretical prediction of the number of beat lengths per cm versus wavelength for coupled fibers supporting the $HE_{1,1}$ mode. $d = 4.73\ \mu m$, $t = 9.65\ \mu m$, and N.A. $= (2\bar{n}\,\Delta n)^{1/2} = 0.0822$.

excited fiber reaches a maximum of output at 632 nm and its next minimum at 655 nm. The theoretical curve of Fig. 7-8 predicts that these fibers should be 10 half-beats long at 630 nm, 11 at 643 nm, and 12 at 655 nm. Comparing theory with experiment for the two fiber pairs, we conclude that $N = 1$ and that the theoretically predicted coupling strengths are 50–100% greater than the measured values.

This result demonstrates that optical directional couplers and coupled laser resonators can be designed on the basis of theoretical predictions, a fact which was not at all obvious prior to this demonstration, because of the difficulties of fabricating and measuring these microscopic waveguides. It should be recognized that the fibers studied are of the order of 10^5 wavelengths long at optical frequencies. The dimensional tolerances required in such fibers to yield a large modulation depth are thus quite severe. It is significant that the techniques of fiber optics can provide such tolerances. When compared with the predictions of Miller (*11*), the experimental results indicate that the dimensional tolerances of these fiber pairs are much better than 1%.

b. Radiation Field of Coupled Waveguides

The mutual coherence of the radiation from the ends of two fibers propagating the symmetric and antisymmetric forms of coupled $HE_{1,1}$ modes predicted by theory should be unity. Experimentally, we observe a very high degree of coherence, as indicated by the far-field interference patterns of Fig. 7-9. The lower photographs are greatly magnified photomicrographs of the output ends of the two fibers. The corresponding field patterns are shown directly above them. When the power is equally divided between the pair at their output end, the contrast in the interference pattern is nominally unity. The dark lines in the photographs of the radiation field are shadows of a 0.5-mm wire placed just in front of the film plane to locate the forward direction. We note that the patterns for the three conditions of equipartition of power are asymmetric about the forward direction and are inverted about it from one such condition to the next as the wavelength is varied. This agrees with the theoretical prediction inferred from Eq. (7.7) that the field vibrations on the two fibers are $\pi/2$ out of phase.

The far-field photographs were obtained by letting the fibers radiate directly into a camera housing to expose Polaroid film (3000 ASA) located 57 mm from the end plane of the fibers. The fibers used in this experiment were the same as were used to obtain the curves of Fig. 7-6.

Photometric traces of the far field of these fiber pairs were obtained. In this case, an EMI 9558 photomultiplier with an S-20 photocathode, its

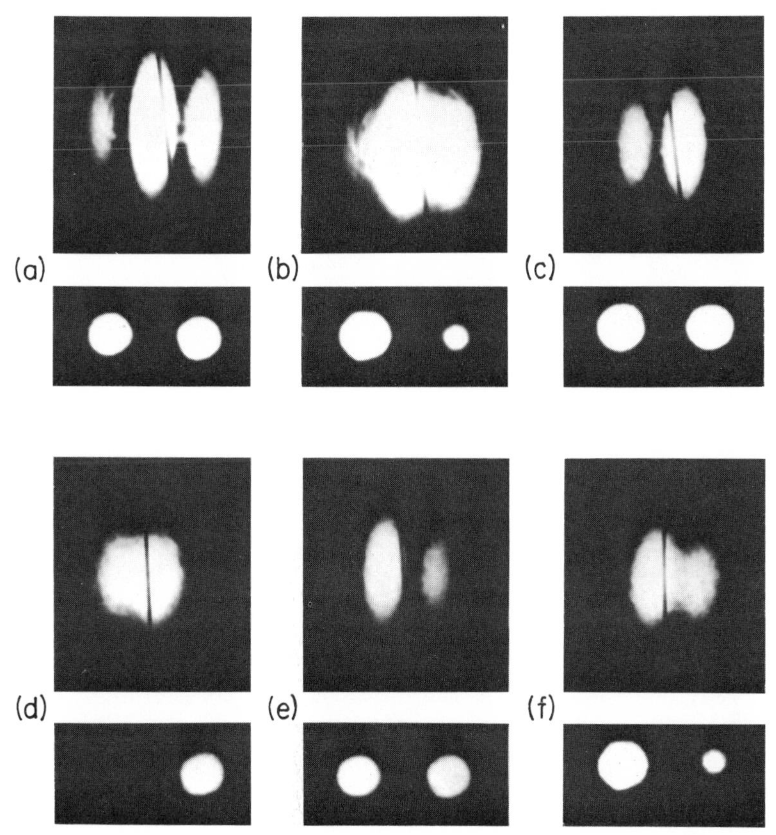

Fig. 7-9. Images of output ends of coupled pair (lower photos) and corresponding far-field radiation patterns for successive wavelengths of equipartition of energy alternating with wavelengths of maximum transfer. Dark line is shadow of wire indicating the forward direction. (a) 572, (b) 577, (c) 583, (d) 586, (e) 591, (f) 597 nm. Fiber parameters as in Fig. 7-8.

entrance face limited with an 0.5-mm aperture, was used to make a linear scan of the field at a distance 57 mm from the end plane of the fibers. The results are shown in Fig. 7-10. Successive conditions of equipartition of the input power are shown at the right; successive conditions of maximum power on either one of the guides are shown at the left. To get an adequate signal-to-noise ratio, the size of the scanning aperture had to be selected too large to resolve the sharp minima; small-angle spatial integrations of the pattern are accordingly apparent. In the vicinity of 582.5 nm, all of the power is radiated from the excited fiber, with the power from its neighbor being negligible. It was found impractical to attempt to set the monochromator at precisely the appropriate wavelength to yield the more nearly sym-

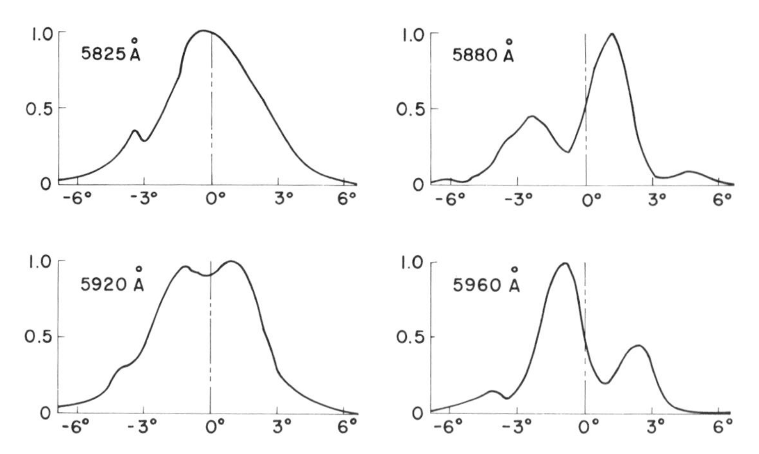

Fig. 7-10. Photometric traces of far field for successive wavelengths of equipartition (right) and maximum transfer (left) showing asymmetries expected for $\pi/2$ phase difference between fibers.

metric $HE_{1,1}$ radiation pattern expected for this case. The asymmetry of the first trace is thus partly due to the improper setting of the wavelength and is presumably also a manifestation that an exact theory would not predict complete power exchange between the guides nor, consequently, complete symmetry in the radiation pattern when the power in the excited fiber is maximum.

c. Coupled Fiber Lasers

The development of neodymium-doped glasses has made possible the study of coupling in fiber-laser arrays. The experiments to be described here are intended to (1) demonstrate qualitatively that a pair of such fibers, when placed in a cavity and optically pumped above threshold, can constitute a phased-locked array which thus radiates coherently; and (2) demonstrate that the locking mechanism can be either external (through the cavity reflectors) or internal (through evanescent-wave coupling). The phase-locking can be demonstrated by observing the simultaneity of the multiple-pulse outputs of the two fibers (*22*) and by recording the interference pattern in the radiation field of the pair. To determine the origin of the coupling or locking mechanism, on the other hand, a study of the effects of variations in fiber size and end-reflector placement is required. Two sets of dual-fiber resonators are sufficient for this study. Core diameters and center-to-center spacings are chosen such that theory would predict a high degree of evanescent-wave coupling for one fiber pair but negligible coupling for the second.

The latter pair, therefore, should not manifest pulse-to-pulse simultaneity or far-field interference when the cavity reflectors are in intimate contact with its ends. Upon removal of a reflector a small distance from the end, however, the output of each fiber is given a path to the adjacent one. Observation of phase-locking would then demonstrate the external coupling mechanism.

The materials selected for use in the fiber-laser resonators were Corning code 0580 neodymium-doped, soda-silicate glass for the active cores and Kimble R6 soda-lime glass for the nonreactive coating. Respective characteristics are:

Refractive index: 1.533 (core) and 1.520 (coating).

Softening point: 708°C (core) and 780°C (coating).

Expansion coefficient: 99×10^{-7}/°C (core) and 93×10^{-7}/°C (coating).

Spectral lifetime of Corning 0580: 320 μsec.

Broadened absorption and doping characteristics of 0580: 324 Å and 4.7% Nd_2O_3 by weight.

Each of the two fiber pairs was cut to 45.7 cm in length, and the ends were polished perpendicular to their axes. The fibers were side-pumped along 30 cm of their length by an EG & G XE1-12 xenon flashlamp. The pumping cavity consisted of the flashlamp and a rigidly mounted support tube, in which the fiber could be placed, with the assembly being wrapped in aluminum foil. The power supply for the flashlamp basically consisted of two 25-μF capacitors connected in parallel, discharging through the flashlamp. The trigger voltage came from a separate dc supply so that its amplitude was independent of the high-voltage setting. The Fabry-Perot cavity was established by using an x, y, z micropositioned front-surface mirror that had 85% reflectance at 1.06 μm at one end of the fibers and Fresnel reflection at the glass–air interface at the other end. By means of a 60× objective, a 15× eyepiece, and a front-surfaced 45° beam splitter, the output face of each fiber was imaged onto a separate DuMont 6911 photomultiplier tube. These outputs were displayed on a Tektronix dual-beam oscilloscope, type 555. The light from each of the cores could thus be individually monitored. Additionally, the total radiation pattern was observed by imaging the near- and far-field radiation of the fibers on an RCA 6914 image converter. The image could then be recorded photographically.

Experimental data obtained with a dual, $3\frac{1}{2}$-μm-diameter set of fibers with a center-to-center spacing of 7 μm is shown in Fig. 7-11. Figure 7-11(a) shows the mode structure of such a fiber pair under 7° end illumination from a broadband tungsten source. Figures 7-11(b–f) represent a series of time traces of the combined output of both fiber lasers as a function of pump power. The photographs in this figure are of particular interest in that they manifest a high-frequency, self-Q-switching effect (*27*). In this instance, the fiber pair was anchored securely at the Fresnel-reflection end but was free to move at the opposite end with respect to an external mirror. Several interesting features are evident: (a) there is generally large spiking in clusters; (b) there are two decay or spurious oscillatory groupings following the large spiking clusters; (c) the large spiking clusters increase in duration sequentially; (d) the oscillation of each cluster shows a tendency to be

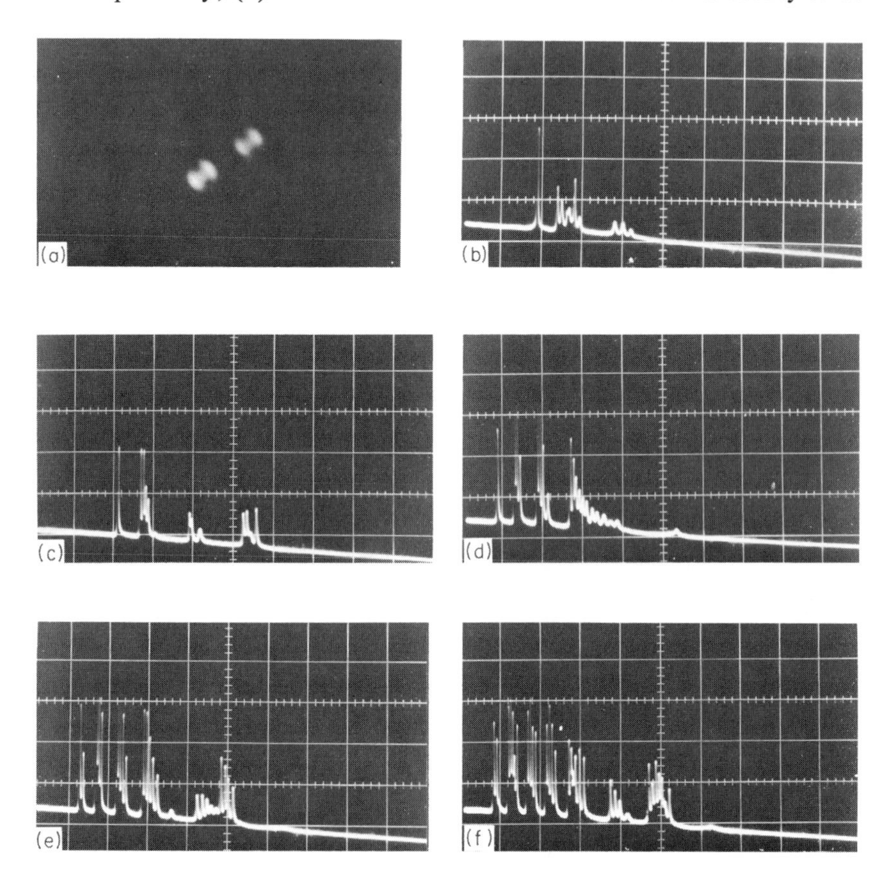

Fig. 7-11. Time scale is 20 μsec/cm. Clustered oscillatory behavior in 3.5-μm, dual-fiber resonators. (a) Mode structure of dual fiber under broadband illumination; (b) 166-J E_{In}, 75-μsec sweep delay; (c) 170-J E_{In}, 75-μsec sweep delay; (d) 182-J E_{In}, 80-μsec sweep delay; (e) 196-J E_{In}, 75-μsec sweep delay; and (f) 202-J E_{In} 75-μsec sweep delay.

damped exponentially; and (e) the numbers of clusters increase as pump power is increased.

An identical experiment performed with a single-cored, $3\frac{1}{2}$-μm fiber laser gave similar results, as shown in Fig. 7-12. We further observed that the pulse clustering could be eliminated by anchoring both ends of the fibers. We therefore conclude that a natural vibratory motion of the fibers during the pumping cycle periodically alters the cavity Q condition. The frequency of oscillation is approximately 5×10^4 Hz.

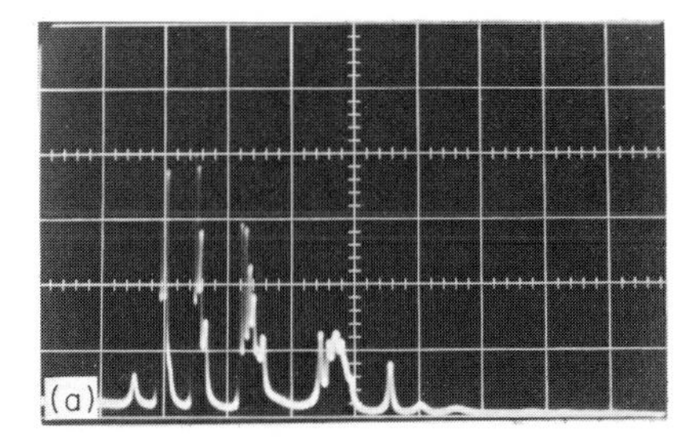

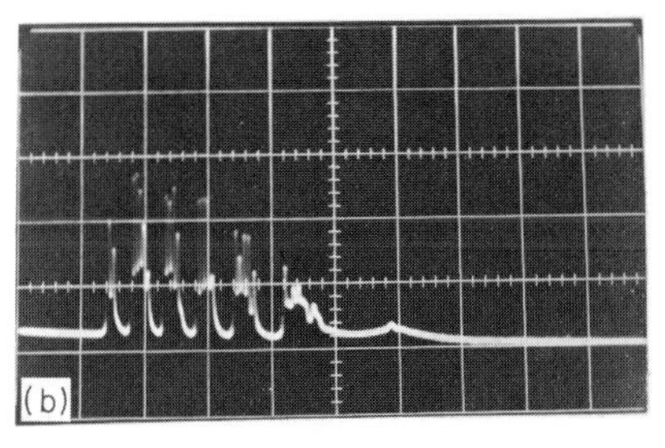

Fig. 7-12. Clustered oscillatory behavior in 3.5-μm single-fiber resonator. (a) 41 J, 20 μsec/cm, 80-μsec sweep delay; and (b) 42 J, 20 μsec/cm, 80-μsec sweep delay.

In order to confirm the evanescent-coupling theory in terms of active paired resonators, the output of each core of the $3\frac{1}{2}$-μm dual fiber was monitored separately and displayed on a dual-beam scope. Additionally, photoelectronic images of the radiation pattern were made at successive intervals into the field. The interference observed is shown in Fig. 7-13. It is remarkable that a high degree of phase-locking exists over the numerous random outputs from each fiber, as evidenced by the high contrast of interference fringes in the far field. Figure 7-14 represents the scope traces of the individual cores during the pumping cycle. The temporal output when the fiber was in optical contact with the end reflector is shown in Figs. 7-14(d, e). Figures 7-14(a, b) give the time traces when the end reflector was removed 0.4 mm from the fiber end. In both cases, the far-field interference patterns shown by Figs. 7-14(c, f) were obtained. Examinations of the time traces show excellent correlation between spiking throughout the pumping cycle. Time base and pump powers are indicated in the figures.

Observations were also made of the radiation field of each core separately in the $3\frac{1}{2}$-μm coupled lasers. This was accomplished by blocking off the radiation of one of the fiber lasers at its distal end. The resulting pattern, shown in Fig. 7-15(b), has the appearance of the $HE_{1,1}$ mode, the only mode that a fiber of this dimension and numerical aperture is capable of

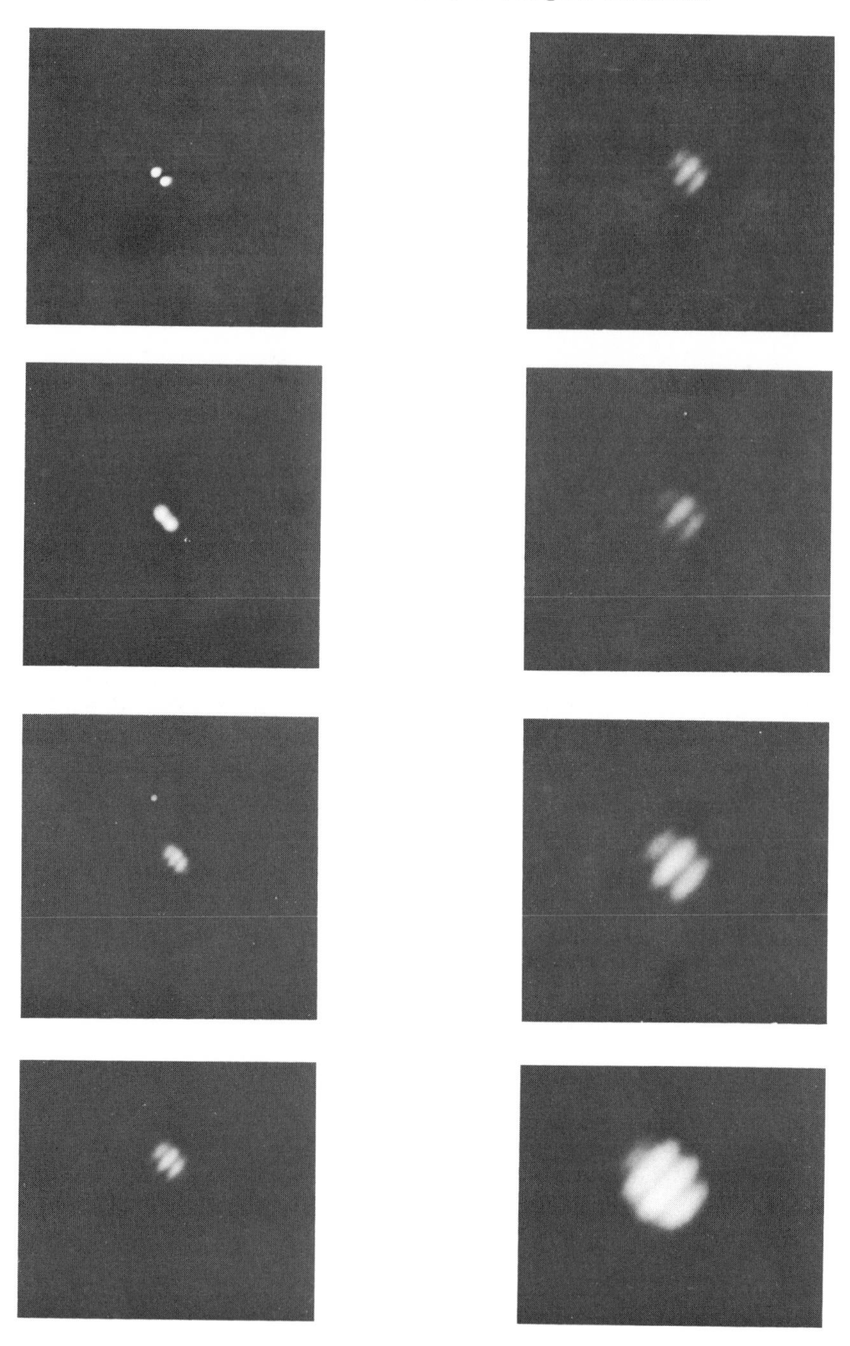

Fig. 7-13. Interference pattern of 3.5-μm, dual-fiber resonators from near- to far-field positions (top to bottom, left to right): focus, 0.1, 0.15, 0.2, 0.3, 0.35, 0.4, and 0.5 mm.

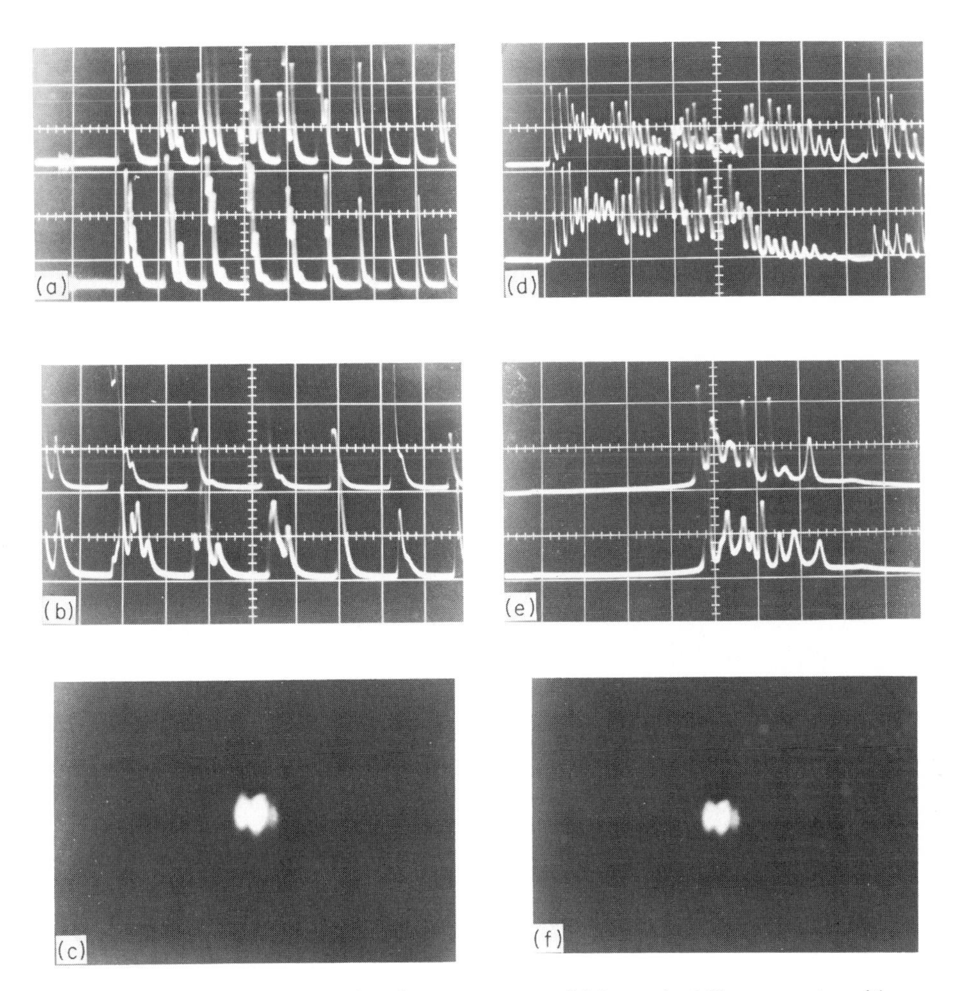

Fig. 7-14. Time trace and interference patterns of 3.5-μm, dual-fiber resonator with: (a)–(c) fiber allowed to oscillate; reflector not in contact with fiber; and (d)–(f) fiber not physically oscillating; reflector in contact with fiber. Time base: (a) 10 μsec/cm, 145 J; (b) 5 μsec/cm, 132 J; (d) 10 μsec/cm, 145 J; and (e) 5 μsec/cm, 127 J.

supporting. The fine structure is explained by slight diffraction around the blocking edge and small disparities in the intervening optics. The interference patterns of Figs. 7-15(a, c) are those before and after the wedge was removed and confirm the coherence properties of the fibers during the cycle; pump power and reflector placement were constant throughout this series.

The trace and photograph of Fig. 7-16 show the output and time trace of a pair of 7-μm parallel fiber lasers whose center-to-center spacing is 14 μm.

Theory predicts that low-order-mode coupling should be very weak in fibers of these dimensions, the mode energy being confined almost completely to the cores of the fiber. An experiment was performed in which such a fiber pair was excited above threshold with the high-reflecting cavity mirror placed in contact with the end of the fibers. The resonant cavity was completed, as before, by using the Fresnel reflection of the glass–air interface

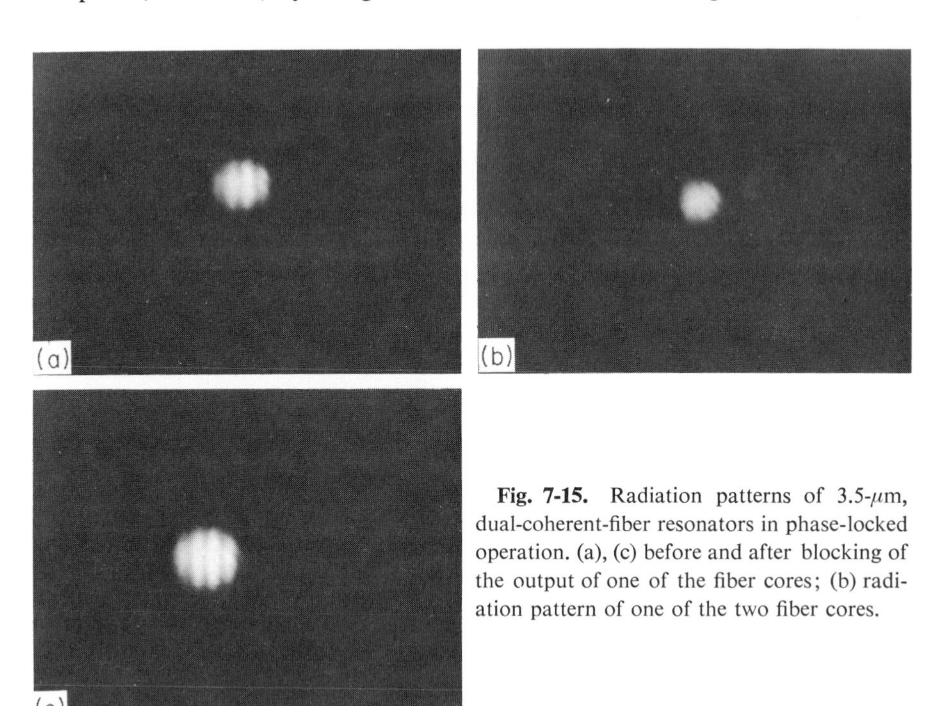

Fig. 7-15. Radiation patterns of 3.5-μm, dual-coherent-fiber resonators in phase-locked operation. (a), (c) before and after blocking of the output of one of the fiber cores; (b) radiation pattern of one of the two fiber cores.

at the other end of the fibers. Figure 7-16(a), which is a photograph of the far-field radiation pattern, gives no evidence of an interference effect that would have been produced if the two cores were phase-locked and radiated coherently. The attendant time trace of Fig. 7-16(b) further demonstrates that each core oscillated independently. The initial variations in the upper time trace are due to some scattering of energy from one core into the phototube that monitored the output of its neighbor. It is also interesting to note the delay between the initiation of independent oscillation in each of the cores.

On the other hand, the photographs of Fig. 7-17 indicate, both by the time trace of the spiking characteristics and by the interference phenomenon in the far field, that coupling can be achieved in fiber lasers of larger

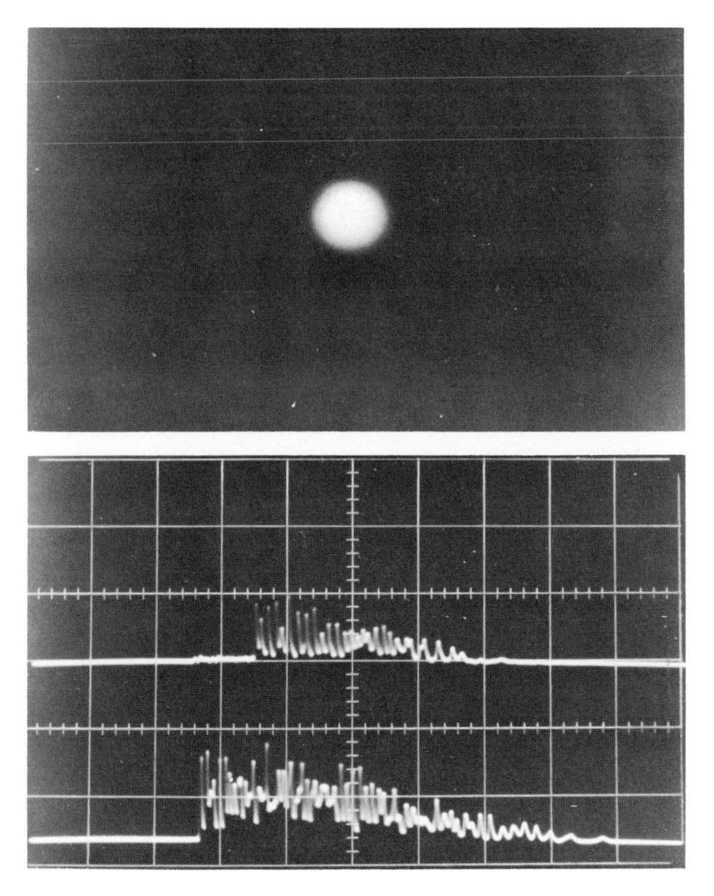

Fig. 7-16. Radiation output and time trace of 7-μm, dual resonator with resonator reflector in contact with fiber.

dimensions, although in a different manner. In this experiment with 7-μm fibers, the resonator reflector was moved by measured intervals from the end of the fiber pair. The interference patterns indicate that the fiber lasers radiated coherently throughout the laser cycle and that the coupling phenomenon was external through the cavity mirror. The time trace shows a close correspondence of pulse simultaneity, which further verifies the phase-locking mechanism, and also indicates the periodic clustered oscillation which occurs when the end of the fiber, as seen by the external cavity reflector, is allowed to vibrate. Similar pulse-to-pulse correlations were seen to occur when the large fiber pair was held immobile, though the cluster phenomenon was absent, as expected.

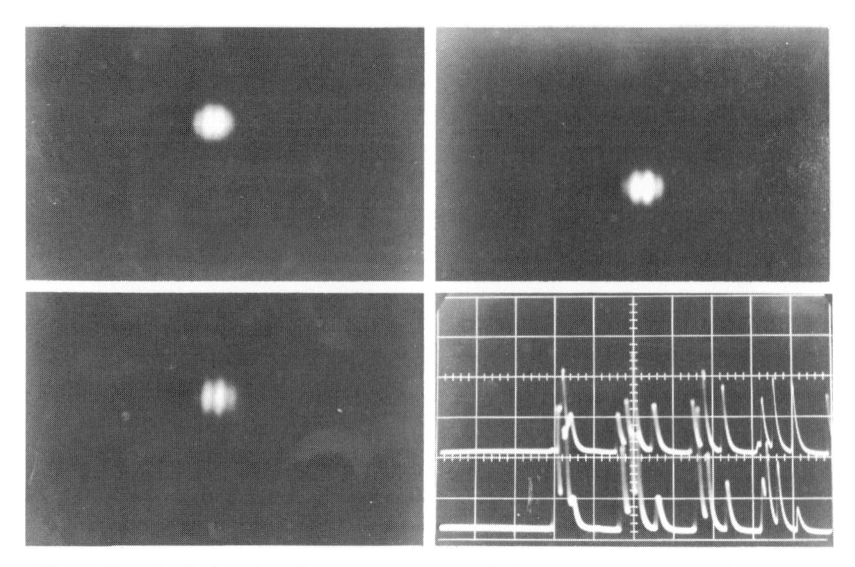

Fig. 7-17. Radiation interference patterns and time trace, demonstrating coherent phase-locked output of 7-μm, dual resonator; cavity reflector not in contact with fiber.

d. Thermally Induced Beating

It has been pointed out that the coupling and, therefore, the beat length are a function of the fiber diameter a, the fiber spacing d, the length L, the refractive index of the fiber cores n_1 and the coating n_2, the wavelength of the exciting radiation, and the particular mode excited in the waveguide. The optical beat phenomenon in two closely spaced glass fibers can also be demonstrated (*28*) by changing only the temperature of the materials, which results in changes of the physical and optical characteristics of the system. A simple experimental system (shown in Fig. 7-18) was set up in which light (6328 Å) from an He–Ne gas laser was directly incident on the entrance end of one of a pair of fibers without any auxiliary optics. To illuminate only one of the fibers by laser light, it was necessary to cover the rest of the entrance end of the two-fiber assembly with an opaque material. This was done by placing a sharp razor blade over one fiber core, while observing them under a microscope. After each cycle of the experiment, the position of the razor blade was similarly checked, to confirm that no displacement had occurred. The usual method of exciting one fiber, by imaging a pinhole with a reversed-microscope system, was not possible in this experiment because the position of the fiber shifted slightly in its support, owing to thermal changes. The output end of the fiber array was viewed with a 44×

microscope objective to observe the beating effect. A heater was set up to vary the temperature, and a thermocouple was attached to the fiber assembly to measure the temperature. The fiber pair was 5 cm long, and the fiber cores, drawn from UK-50 glass ($n_1 = 1.523$), were approximately 5 μm in diameter and had a center-to-center spacing of approximately 15 μm. The embedding glass was soda lime ($n_2 = 1.520$), thereby giving a numerical aperture of 0.1. The fibers were slightly elliptical in cross section, the major axis being approximately 25% longer than the minor axis.

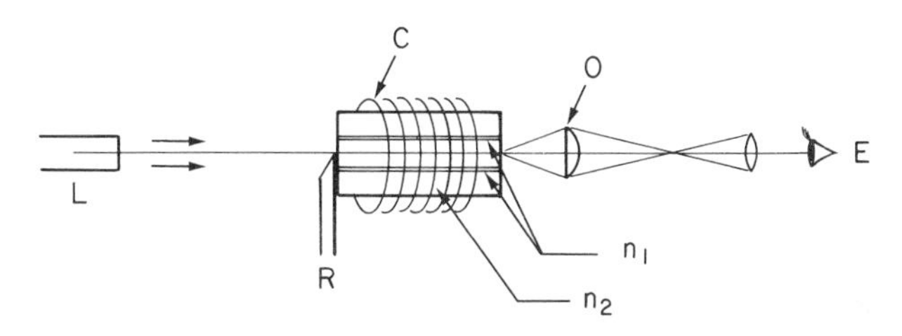

Fig. 7-18. Optical setup for observation of thermal variation of the beat length in coupled fibers.

For the two-fiber assembly used, the excitation conditions were such as to yield almost equal radiant flux from both fibers at room temperature. At an increased temperature (240°C), the light appeared only at the output of the unexcited fiber, and with a further increase (to 310°C), there was approximately equal flux in both fibers. Finally, at approximately 350°C, most of the energy appeared at the output of the excited fiber. Photographs of the output for these conditions are shown in Fig. 7-19. We also note that there was a change of mode, from a two-lobe pattern ($TM_{0,1}$ or $TE_{0,1} + HE_{1,2}$) to a single spot ($HE_{1,1}$) between Figs. 7-19a and 7-19c.

It should be noted that the output-mode variations are caused by the difference in thermal environment rather than by any angular shift of the fiber with respect to the incident laser beam during the experiment. This was confirmed by making an angular variation, in an attempt to generate the same mode prior to changing the thermal surrounds. This gave no evidence of the appearance of the single-spot mode pattern that was generated during the thermal-switching experiments. Also, the fiber was finally set up with its face perpendicular to the incident beam. Any angular shift of the fiber during the experiment would tend to excite higher-order rather than lower-order modes.

The temperature variation causes two changes: one is the variation of the dimensions (a, d, and L) of the test piece in accordance with the thermal

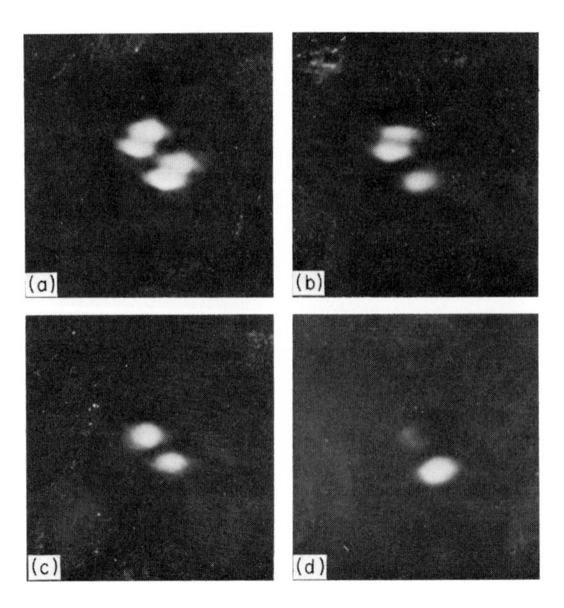

Fig. 7-19. Thermal variation of beat length in two closely spaced glass fibers: (a) at room temperature; (b) $T = 240°C$; (c) $T = 310°C$; and (d) $T = 350°C$.

expansion coefficient of the materials, and the other is the thermal variation of the refractive indices (n_1 and n_2) of the core and coating glasses.

The expected dimensional changes are of the order of magnitude of 10^{-2}–10^{-1} in terms of the aspect ratio L/a (the expansion coefficients of the cores and coating are 7.6×10^{-6} and 9.3×10^{-6} cm/°C, respectively, between 0° and 300°). As shown in the data obtained by Kapany *et al.* (26), a geometric change (wavelength variation) of the order of more than 10^{-2} is required to produce beating. It is therefore obvious that the geometric deformation may be sufficient to produce the observed beat phenomenon.

Furthermore, as the temperature increases, the refractive index of glass decreases greatly in value. The refractive-index variations of the fiber materials are estimated to be approximately 5×10^{-3} (at approximately 300°C). As pointed out by several investigators, this amount of index variation is also sufficient to explain the observed beat phenomenon.

REFERENCES

1. D. G. Kiely, "Dielectric Aerials." Methuen, London, 1953.
2. C. M. McKinney, Thesis, Univ. of Texas (1950).
3. H. J. Wegener, Thesis, Technische Hochschule, Berlin (1944).
4. W. M. Elsasser, *J. Appl. Phys.* **20**, 1188 (1948).

5. N. S. Kapany and J. J. Burke, *J. Opt. Soc. Amer.* **51**, 1067 (1961).
6. N. S. Kapany and J. J. Burke, *Solid State Design* **3**, 35 (1962).
7. N. S. Kapany, *J. Opt. Soc. Amer.* **49**, 770 (1959).
8. S. A. Schelkunoff, "Electromagnetic Waves," pp. 77–81. Van Nostrand–Reinhold, Princeton, New Jersey, 1943.
9. J. A. Stratton, "Electromagnetic Theory," pp. 131–137. McGraw-Hill, New York, 1941.
10. N. S. Kapany, *J. Opt. Soc. Amer.* **47**, 413 (1957).
11. S. E. Miller, *Bell Syst. Tech. J.* **33**, 661 (1954).
12. W. H. Louisell, "Coupled Mode and Parametric Electronics." Wiley, New York, 1960.
13. S. A. Schelkunoff, "Electromagnetic Fields." Blaisdell, New York, 1963.
14. C. C. Johnson, "Field and Wave Electrodynamics." McGraw-Hill, New York, 1965.
15. E. Snitzer, *in* "Advances in Quantum Electronics" (J. R. Singer, ed.), p. 348. Columbia Univ. Press, New York, 1961.
16. E. Snitzer, *in* "Optical Processing of Information" (D. K. Pollock, C. J. Koester, and J. T. Tippett, eds.), p. 61. Spartan Books, Baltimore, Maryland, 1963.
17. N. S. Kapany, G. M. Burgwald, and J. J. Burke, *in* "Optical and Electro-Optical Information Processing" (J. T. Tippett, D. A. Berkowitz, L. C. Clapp, C. J. Koester, and A. Vanderburgh, Jr., eds.), p. 305. MIT Press, Cambridge, Massachusetts, 1965.
18. N. S. Kapany, J. J. Burke, and K. L. Frame, *Appl. Opt.* **4**, 1534 (1965).
19. M. F. Bracey, A. L. Cullen, E. F. F. Gillespie, and J. A. Staniforth, *IRE Trans.* **AP-7**, S219 (1959).
20. A. L. Jones, *J. Opt. Soc. Amer.* **55**, 261 (1965).
21. J. J. Burke, *J. Opt. Soc. Amer.* **57**, 1056 (1967).
22. C. J. Koester and C. H. Swope, in Kapany *et al.* (17), p. 253.
23. E. Snitzer, *J. Opt. Soc. Amer.* **51**, 491 (1961).
24. E. Snitzer and H. Osterberg, *J. Opt. Soc. Amer.* **51**, 499 (1961).
25. E. Snitzer, *J. Appl. Phys.* **32**, 36 (1961).
26. N. S. Kapany, J. J. Burke, K. L. Frame, and R. E. Wilcox, *J. Opt. Soc. Amer.* **58**, 1176 (1968).
27. H. C. Nedderman, Y. C. Kiang, and F. C. Unterleitner, *Proc. IRE* **50**, 7 (1962).
28. N. S. Kapany and T. Sawatari, *J. Opt. Soc. Amer.* **60**, 135 (1970).

APPENDIX A

Noncircular Dielectric Waveguides

C. YEH*

A. INTRODUCTION

The guiding properties of a circular fiber have been considered in detail in the text of this book. It is of interest, however, to understand how the propagation characteristics of the guided wave are affected when the circular fiber is deformed and when the circular symmetry no longer exists. Extreme mathematical difficulties are encountered when one tries to solve the problem of wave propagation along a dielectric fiber of arbitrary cross-sectional shape. Fortunately, a deformed circular fiber can, in general, be approximated by an elliptical or a rectangular fiber. Depending upon the eccentricity of the elliptical fiber, it can take the form of a circular fiber or the form of a flat tape fiber. Exact solution, although very involved, does exist for the elliptical dielectric waveguide problem. Since this solution is the only available exact analytic solution for a noncircular dielectric waveguide and since an elliptical fiber can approximate a great number of different shapes, a great deal of emphases will be given to the analytic results for the elliptical dielectric waveguide problem.

Detailed theoretical as well as experimental results for the dominant modes and analytic results for higher-order modes on an elliptical dielectric waveguide were given by Yeh (*1*, *2*). Lynbimov *et al.* (*3*) and Piefke (*4*) also derived the characteristic equations for various modes on an elliptical dielectric guide, but no experimental or numerical results were given. Most

* Department of Electrical Sciences and Engineering, University of California, Los Angeles, California.

recently, numerical solution for the rectangular dielectric waveguide problem has been obtained by Goell (*5*), who expanded the fields in terms of a series of circular harmonics and then matched the fields on a number of points across the rectangular boundary. He not only presented the numerical results for the dominant modes, but also those for higher-order modes on a rectangular dielectric rod. Marcatili (*6*), using approximation based on the assumption that most of the power flow is confined to the waveguide core, has derived in closed form the properties of a rectangular dielectric waveguide.

In this appendix, we shall first consider the exact analytic solution to the elliptical dielectric waveguide problem. Numerical results and detailed discussion for the dominant modes on the elliptical dielectric fiber will then be given. In Section C, the propagation characteristics of various modes on a rectangular dielectric fiber are presented. Comparison between theoretical and experimental results for the dominant modes on rectangular fiber will also be made. Finally, a discussion summarizing the results is given.

B. ELLIPTICAL DIELECTRIC WAVEGUIDES

The elliptical dielectric waveguide consists of a core of elliptical cross section with high dielectric constant ε_1 surrounded by a cladding of lower dielectric constant ε_0. Both regions are assumed to be perfect insulators with the free-space magnetic permeability μ. It is further assumed that the exciting source is so far away that in the region of interest, the surface wave dominates the radiated wave from the source. For given values of ε_1, ε_0, and the dimensions of the elliptical cross section, a finite number of surface wave modes may be guided along this structure. In the following, we shall study the propagation characteristics of these modes.

1. Formulation of the Problem

To analyze this problem, the elliptical cylinder coordinates (ξ, η, z), as shown in Fig. A-1, are introduced. In terms of the rectangular coordinates (x', y', z'), the elliptical cylinder coordinates are defined by the following relations:

$$x' = q(\cosh \xi) \cos \eta, \qquad y' = q(\sinh \xi) \sin \eta, \qquad z' = z, \tag{A.1}$$
$$(0 \leq \xi \leq \infty; \quad 0 \leq \eta \leq 2\pi),$$

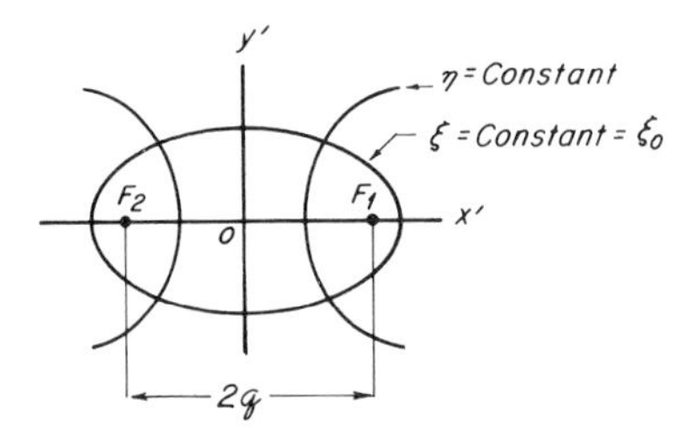

Fig. A-1. Elliptical coordinate system. F_1 and F_2 are the focuses of the ellipse. The distance between focuses is the focal distance $2q$. Ratio (major axis/minor axis) $= \cosh \xi_0$. Area of ellipse $= \pi q^2 \cosh \xi_0 \sinh \xi_0$.

where q is the semifocal length of the ellipse. One of the confocal elliptical cylinders with $\xi = \xi_0$ is assumed to coincide with the boundary of the cylindrical core of high dielectric constant, and the z axis coincides with its longitudinal axis.

The harmonic time dependence of $e^{-i\omega t}$ for all field components is assumed. We shall confine our treatment to waves propagating along the positive-z axis. In complex representation, these assumptions result in a multiplication of all wave functions by $e^{-i\omega t+i\beta z}$. The propagation constant β is to be determined from the boundary conditions.

Because of the cylindrical symmetry, all other components of the field can be expressed in terms of E_z and H_z. The z (axial) components of the field satisfy the wave equation in elliptical coordinates. The appropriate solutions of the wave equation for this problem are as follows.* For region 1 $(0 \le \xi \le \xi_0)$,

$$\{E_{z1} \quad \text{or} \quad H_{z1}\} = \begin{Bmatrix} \mathrm{Ce}_n(\xi, \gamma_1^2)\, \mathrm{ce}_n(\eta, \gamma_1^2) & \text{(even)} \\ \mathrm{Se}_n(\xi, \gamma_1^2)\, \mathrm{se}_n(\eta, \gamma_1^2) & \text{(odd)} \end{Bmatrix}, \tag{A.2}$$

and for region 0 $(\xi_0 \le \xi < \infty)$,

$$\{E_{z0} \quad \text{or} \quad H_{z0}\} = \begin{Bmatrix} \mathrm{Fek}_n(\xi, -\gamma_0^2)\, \mathrm{ce}_n(\eta, -\gamma_0^2) & \text{(even)} \\ \mathrm{Gek}_n(\xi, -\gamma_0^2)\, \mathrm{se}_n(\eta, -\gamma_0^2) & \text{(odd)} \end{Bmatrix}, \tag{A.3}$$

where

$$\begin{aligned} \gamma_1^2 &= \tfrac{1}{4}(k_1^2 - \beta^2)q^2, \qquad & \gamma_0^2 &= \tfrac{1}{4}(\beta^2 - k_0^2)q^2, \\ k_1^2 &= \omega^2 \mu \varepsilon_1, & k_0^2 &= \omega^2 \mu \varepsilon_0. \end{aligned} \tag{A.4}$$

2. Mode Classification

In order to satisfy the boundary conditions at the boundary surface $\xi = \xi_0$, E_z and H_z must both be present. Hence, all modes on an elliptical dielectric guide are hybrid. Physically speaking, the presence of E_z in a

* We shall follow the notation for Mathieu functions adopted by McLachlan (7).

predominantly H wave or vice versa assumes the return path for the electric or magnetic lines of force. (The circularly symmetric TE or TM waves on a circular dielectric rod are exceptions since the electric and magnetic lines of force of these circularly symmetric waves have already formed closed loops.) Due to the asymmetry of the elliptical cylinder, it is possible to have two orientations for the field configurations. Thus, a hybrid wave on an elliptical dielectric rod will be designated by a prescript e or o, indicating an even wave or an odd wave. The axial magnetic and electric field of an even wave are represented by even and odd Mathieu functions, respectively, and those of an odd wave by odd and even Mathieu functions, respectively. The hybrid waves are designated by ${}_{\mathrm{e,o}}\mathrm{HE}_{m,n}$ if the cross-sectional field pattern resembles that of an H wave and by ${}_{\mathrm{e,o}}\mathrm{EH}_{m,n}$ if the cross-sectional field pattern resembles that of an E wave. [A more exact definition based on the relative strength of H_z and E_z components has been given by Snitzer (*8*).] The subscripts m and n denote, respectively, the number of cyclic variation with η and the nth root of the characteristic equation. It is noted that ${}_{\mathrm{e,o}}\mathrm{HE}_{m,n}$ or ${}_{\mathrm{e,o}}\mathrm{EH}_{m,n}$ modes on an elliptical rod degenerate, respectively, to $\mathrm{HE}_{m,n}$ or $\mathrm{EH}_{m,n}$ modes on a circular rod as the eccentricity of the ellipse becomes zero.

3. The Determinantal Equations

a. ${}_{\mathrm{e}}\mathrm{HE}_{m,n}$ Wave

According to the definition given in the previous section, the most general expressions for the axial magnetic and electric fields of an ${}_{\mathrm{e}}\mathrm{HE}_{m,n}$ wave are: For region 1 $(0 \leq \xi < \xi_0)$,

$$H_{z1} = \sum_{m=0}^{\infty} A_m \,\mathrm{Ce}_m(\xi)\,\mathrm{ce}_m(\eta)$$

$$E_{z1} = \sum_{m=1}^{\infty} B_m \,\mathrm{Se}_m(\xi)\,\mathrm{se}_m(\eta) \tag{A.5}$$

and for region 0 $(\xi_0 \leq \xi < \infty)$,

$$H_{z0} = \sum_{r=1}^{\infty} L_r \,\mathrm{Fek}_r(\xi)\,\mathrm{ce}_r^*(\eta)$$

$$E_{z0} = \sum_{r=1}^{\infty} P_r \,\mathrm{Gek}_r(\xi)\,\mathrm{se}_r^*(\eta) \tag{A.6}$$

where A_m, B_m, L_r, and P_r are the arbitrary constants and the abbreviations

$$\begin{aligned} \mathrm{Ce}_n(\xi) &= \mathrm{Ce}_n(\xi, \gamma_1^2), & \mathrm{ce}_n(\eta) &= \mathrm{ce}_n(\eta, \gamma_1^2) \\ \mathrm{Se}_n(\xi) &= \mathrm{Se}(\xi, \gamma_1^2), & \mathrm{se}_n(\eta) &= \mathrm{se}_n(\eta, \gamma_1^2) \\ \mathrm{Fe}k_n(\xi) &= \mathrm{Fe}k_n(\xi, -\gamma_0^2), & \mathrm{ce}_n{}^*(\eta) &= \mathrm{ce}_n(\eta, -\gamma_0^2) \\ \mathrm{Ge}k_n(\xi) &= \mathrm{Ge}k_n(\xi, -\gamma_0^2), & \mathrm{se}_n{}^*(\eta) &= \mathrm{se}_n(\eta, -\gamma_0^2) \end{aligned} \tag{A.7}$$

have been used. The transverse components of the field can be derived from Maxwell's equations.

Equating the tangential electric and magnetic fields at the boundary surface, $\xi = \xi_0$, gives

$$\sum_{m=0}^{\infty} A_m \,\mathrm{Ce}_m(\xi_0)\,\mathrm{ce}_m(\eta) = \sum_{r=0}^{\infty} L_r \,\mathrm{Fe}k_r(\xi_0)\,\mathrm{ce}_r{}^*(\eta), \tag{A.8a}$$

$$\sum_{m=1}^{\infty} B_m \,\mathrm{Se}_m(\xi_0)\,\mathrm{se}_m(\eta) = \sum_{r=1}^{\infty} P_r \,\mathrm{Ge}k_r(\xi_0)\,\mathrm{se}_r{}^*(\eta), \tag{A.8b}$$

$$\begin{aligned} &\sum_{m=1}^{\infty} \{A_m[1 + (\gamma_1^2/\gamma_0^2)]\,\mathrm{Ce}_m(\xi_0)\,\mathrm{ce}_m{}'(\eta) + (\omega\varepsilon_1/\beta)B_m\,\mathrm{Se}_m{}'(\xi_0)\,\mathrm{se}_m(\eta)\} \\ &\qquad = -(\gamma_1^2/\gamma_0^2) \sum_{r=1}^{\infty} (\omega\varepsilon_0/\beta)P_r\,\mathrm{Ge}k_r{}'(\xi_0)\,\mathrm{se}_r{}^*(\eta), \end{aligned} \tag{A.8c}$$

$$\begin{aligned} &\sum_{m=1}^{\infty} \{A_m(\omega\mu/\beta)\,\mathrm{Ce}_m{}'(\xi_0)\,\mathrm{ce}_m(\eta) - B_m[1 + (\gamma_1^2/\gamma_0^2)]\,\mathrm{Se}_m(\xi_0)\,\mathrm{se}_m{}'(\eta)\} \\ &\qquad = -(\gamma_1^2/\gamma_0^2) \sum_{r=0}^{\infty} (\omega\mu/\beta)L_r\,\mathrm{Fe}k_r{}'(\xi_0)\,\mathrm{ce}_r{}^*(\eta). \end{aligned} \tag{A.8d}$$

The prime denotes the derivative with respect to ξ_0 or η, as the case may be. It is noted that in contrast with the case of the circular dielectric cylinder, the angular functions in the elliptical case are functions not only of η, the angular coordinates, but also of the electrical properties of the medium in which they apply. Consequently, the summation signs and the angular Mathieu functions in these equations may not be omitted. To eliminate the η dependence, we shall make use of the orthogonality properties of angular Mathieu functions. Multiplying both sides of Eqs. (A.8a) and (A.8d) by $\mathrm{ce}_n(\eta)$ and both sides of Eqs. (A.8b) and (A.8c) by $\mathrm{se}_n(\eta)$, integrating with respect to η from 0 to 2π, and applying the orthogonality relations of the angular Mathieu functions leads to

$$A_n a_n = \sum_{r=0}^{\infty}{}' L_r l_r \alpha_{r,n}, \tag{A.9a}$$

$$B_n b_n = \sum_{r=1}^{\infty}{}' P_r p_r \beta_{r,n}, \tag{A.9b}$$

$$\frac{\omega\varepsilon_1}{\beta} B_n b_n' + \left(1 + \frac{\gamma_1^2}{\gamma_0^2}\right) \sum_{r=1}^{\infty}{}' A_r a_r \chi_{r,n} = \left(-\frac{\gamma_1^2}{\gamma_0^2}\right) \frac{\omega\varepsilon_0}{\beta} \sum_{r=1}^{\infty}{}' P_r p_r \beta_{r,n}, \tag{A.9c}$$

$$\frac{\omega\mu}{\beta} A_n a_n' - \left(1 + \frac{\gamma_1^2}{\gamma_0^2}\right) \sum_{r=0}^{\infty}{}' B_r b_r \nu_{r,n} = \left(-\frac{\gamma_1^2}{\gamma_0^2}\right) \frac{\omega\mu}{\beta} \sum_{r=0}^{\infty}{}' L_r l_r' \alpha_{r,n}, \tag{A.9d}$$

$$(n = 0, 2, 4, \ldots \quad \text{or} \quad n = 1, 3, 5, \ldots)$$

where the abbreviations

$$\begin{aligned} a_n &= \mathrm{Ce}_n(\xi_0), & b_n &= \mathrm{Se}_n(\xi_0) \\ l_r &= \mathrm{Fe}k_r(\xi_0), & p_r &= \mathrm{Ge}k_r(\xi_0) \end{aligned} \tag{A.10}$$

have been used. The prime over a_n, b_n, l_r, or p_r denotes the derivative with respect to ξ_0, while the prime over the summation sign indicates that odd or even integer values of r are to be taken according as n is odd or even. $\alpha_{r,n}$, $\beta_{r,n}$, $\chi_{r,n}$, and $\nu_{r,n}$ are given by the following:

$$\begin{aligned} \alpha_{r,n} &= \left[\int_0^{2\pi} \mathrm{ce}_r^*(\eta)\, \mathrm{ce}_n(\eta)\, d\eta\right] \Big/ \int_0^{2\pi} \mathrm{ce}_n^2(\eta)\, d\eta, \\ \beta_{r,n} &= \left[\int_0^{2\pi} \mathrm{se}_r^*(\eta)\, \mathrm{se}_n(\eta)\, d\eta\right] \Big/ \int_0^{2\pi} \mathrm{se}_n^2(\eta)\, d\eta, \\ \chi_{r,n} &= \left[\int_0^{2\pi} \mathrm{ce}_r'(\eta)\, \mathrm{se}_n(\eta)\, d\eta\right] \Big/ \int_0^{2\pi} \mathrm{se}_n^2(\eta)\, d\eta, \\ \nu_{r,n} &= \left[\int_0^{2\pi} \mathrm{se}_r'(\eta)\, \mathrm{ce}_n(\eta)\, d\eta\right] \Big/ \int_0^{2\pi} \mathrm{ce}_n^2(\eta)\, d\eta. \end{aligned} \tag{A.11}$$

Simplifying Eqs. (A.9) and making the identifications

$$\begin{aligned} g_{m,n} &= [1 + (\gamma_1^2/\gamma_0^2)] l_m \sum_{r=1}^{\infty}{}' \chi_{r,n} \alpha_{m,r} \\ s_{m,n} &= -[1 + (\gamma_1^2/\gamma_0^2)] p_m \sum_{r=1}^{\infty}{}' \nu_{r,n} \beta_{m,r} \\ h_{m,n} &= (\omega\varepsilon_1/\beta)(b_n'/b_n) p_m \beta_{m,n} + (\gamma_1^2/\gamma_0^2)(\omega\varepsilon_0/\beta) p_m' \beta_{m,n} \\ t_{m,n} &= (\omega\mu/\beta)(a_n'/a_n) l_m \alpha_{m,n} + (\gamma_1^2/\gamma_0^2)(\omega\mu/\beta) l_m' \alpha_{m,n}, \end{aligned} \tag{A.12}$$

one obtains

$$\sum_{m=0}^{\infty}{}' \, [L_m g_{m,n} + P_m h_{m,n}] = 0 \tag{A.13a}$$

$$\sum_{m=0}^{\infty}{}' \, [L_m t_{m,n} + P_m s_{m,n}] = 0 \tag{A.13b}$$

$$(n = 0, 2, 4, \ldots \quad \text{or} \quad n = 1, 3, 5, \ldots).$$

Equations (A.13) are two sets of infinite, homogeneous, linear algebraic equations in L_m and P_m. For a nontrivial solution, the determinant of these equations must vanish. The roots of this infinite determinant provide the value from which the propagation constant β can be determined. For example, the infinite determinant for the $m = 1, 3, 5, \ldots$ modes is

$$\begin{vmatrix} g_{1,1} & h_{1,1} & g_{3,1} & h_{3,1} & g_{5,1} & h_{5,1} & \cdots \\ t_{1,1} & s_{1,1} & t_{3,1} & s_{3,1} & t_{5,1} & s_{5,1} & \cdots \\ g_{1,3} & h_{1,3} & g_{3,3} & h_{3,3} & g_{5,3} & h_{5,3} & \cdots \\ t_{1,3} & s_{1,3} & t_{3,3} & s_{3,3} & t_{5,3} & s_{5,3} & \cdots \\ g_{1,5} & h_{1,5} & g_{3,5} & h_{3,5} & g_{5,5} & h_{5,5} & \cdots \\ t_{1,5} & s_{1,5} & t_{3,5} & s_{3,5} & t_{5,5} & s_{5,5} & \cdots \\ \vdots & \vdots & \vdots & \vdots & \vdots & \vdots & \vdots \end{vmatrix} = 0. \tag{A.14}$$

A similar infinite determinant with even indices for the $m = 0, 2, 4, 6, \ldots$ modes can also be obtained.

b. $_{o}\mathrm{HE}_{m,n}$ Wave

The general expressions for the axial magnetic and electric fields of an $_{o}\mathrm{HE}_{m,n}$ wave are: For region 1 $(0 \leq \xi \leq \xi_0)$,

$$\begin{aligned} H_{z1} &= \sum_{m=1}^{\infty} C_m \, \mathrm{Se}_m(\xi) \, \mathrm{se}_m(\eta) \\ E_{z1} &= \sum_{m=0}^{\infty} D_m \, \mathrm{Ce}_m(\xi) \, \mathrm{ce}_m(\eta), \end{aligned} \tag{A.15}$$

and for region 0 ($\xi_0 \leq \xi < \infty$),

$$H_{z0} = \sum_{r=1}^{\infty} G_r \, \mathrm{Gek}_r(\xi) \, \mathrm{se}_r{}^*(\eta)$$
$$E_{z0} = \sum_{r=0}^{\infty} F_r \, \mathrm{Fek}_r(\xi) \, \mathrm{ce}_r{}^*(\eta), \tag{A.16}$$

where C_m, D_m, G_r, and F_r are the arbitrary constants. Upon matching the boundary conditions at $\xi = \xi_0$ and applying the similar mathematical operations as for the ${}_{\mathrm{e}}\mathrm{HE}_{m,n}$ mode, one can easily obtain the characteristic equation for the ${}_{\mathrm{o}}\mathrm{HE}_{m,n}$ wave. For example, the determinantal equation for the ${}_{\mathrm{o}}\mathrm{HE}_{m,n}$ modes with $m = 1, 3, 5, \ldots$ is

$$\begin{vmatrix} g_{1,1}^* & h_{1,1}^* & g_{3,1}^* & h_{3,1}^* & g_{5,1}^* & h_{5,1}^* & \cdots \\ t_{1,1}^* & s_{1,1}^* & t_{3,1}^* & s_{3,1}^* & t_{5,1}^* & s_{5,1}^* & \cdots \\ g_{1,3}^* & h_{1,3}^* & g_{3,3}^* & h_{3,3}^* & g_{5,3}^* & h_{5,3}^* & \cdots \\ t_{1,3}^* & s_{1,3}^* & t_{3,3}^* & s_{3,3}^* & t_{5,3}^* & s_{5,3}^* & \cdots \\ g_{1,5}^* & h_{1,5}^* & g_{3,5}^* & h_{3,5}^* & g_{5,5}^* & h_{5,5}^* & \cdots \\ t_{1,5}^* & s_{1,5}^* & t_{3,5}^* & s_{3,5}^* & t_{5,5}^* & s_{5,5}^* & \cdots \\ \vdots & \vdots & \vdots & \vdots & \vdots & \vdots & \vdots \end{vmatrix} = 0, \tag{A.17}$$

where

$$g_{m,n}^* = [1 + (\gamma_1{}^2/\gamma_0{}^2)] p_m \sum_{r=1}^{\infty} \nu_{r,n} \beta_{m,r}$$
$$s_{m,n}^* = -[1 + (\gamma_1{}^2/\gamma_0{}^2)] l_m \sum_{r=1}^{\infty} \chi_{r,n} \alpha_{m,r}$$
$$h_{m,n}^* = (\omega\varepsilon_1/\beta)(a_n{}'/a_n) l_m \alpha_{m,n} + (\gamma_1{}^2/\gamma_0{}^2)(\omega\varepsilon_0/\beta) l_m{}' \alpha_{m,n}$$
$$t_{m,n}^* = (\omega\mu/\beta)(b_n{}'/b_n) p_m \beta_{m,n} + (\gamma_1{}^2/\gamma_0{}^2)(\omega\mu/\beta) p_m{}' \beta_{m,n}. \tag{A.18}$$

A similar infinite determinant with even indices for the $m = 0, 2, 4, 6, \ldots$ modes can also be obtained.

It can easily be shown that the characteristic equations for the ${}_{e,o}HE_{m,n}$ waves degenerate to the well-known characteristic equation for the $HE_{m,n}$ wave as the elliptical cross section degenerates to a circular one (i.e., as $q \to 0$ and $\xi_0 \to \infty$ such that $qe^{\xi_0}/2 \to r_0$, where r_0 is the radius of the degenerated circle):

$$\alpha_{r,n} \approx \beta_{r,n} = \begin{cases} 1 & \text{when} \quad r = n \\ 0 & \text{when} \quad r \neq n \end{cases}$$

$$\nu_{m,n} \approx \chi_{m,n} = \begin{cases} 1 & \text{when} \quad m = n \\ 0 & \text{when} \quad m \neq n, \end{cases}$$

so, the degenerated infinite determinant becomes

$$\prod_m (g_{m,m}s_{m,m} - h_{m,m}t_{m,m}) = 0$$

or

$$g_{m,m}s_{m,m} - h_{m,m}t_{m,m} = 0$$

for $m = 0, 1, 2, \ldots$.

Decoupling of the infinite determinants also occurs when the dielectric constant of the surrounding sheath is only slightly lower than the dielectric constant of the core. (This applies usually to the optical fiber situation). In this case, the infinite determinants also degenerate to

$$g_{m,m}s_{m,m} - h_{m,m}t_{m,m} = 0 \tag{A.19}$$

for ${}_{e}HE_{m,n}$ modes and

$$g^*_{m,m}s^*_{m,m} - h^*_{m,m}t^*_{m,m} = 0 \tag{A.20}$$

for ${}_{o}HE_{m,n}$ modes, with the radial Mathieu functions in Eq. (A.12) or Eq. (A.18) remaining unchanged.

4. Cutoff Conditions

Surface wave modes may exist only if β^2, γ_0^2, and γ_1^2 are all real and positive. This fact offers a way to determine the upper and lower bounds of the propagation constant β. According to Eqs. (A.4), $\beta^2 \leq k_1^2$ and $\beta^2 \geq k_0^2$; hence $k_0 \leq \beta \leq k_1$. The cutoffs for the various modes are found by solving the determinantal equations in the limit of $\gamma_0^2 \to 0$. The frequency corresponding to $\gamma_0^2 = 0$, called the cutoff frequency of the mode, is

$$\omega_{\text{cutoff}} = 2\gamma_1/\{q[\mu\varepsilon_0(\varepsilon_1/\varepsilon_0 - 1)]^{1/2}\}, \tag{A.21}$$

where γ_1 corresponds to the root of the determinantal equation with $\gamma_0 = 0$. Physically, it means that at or below this cutoff frequency, the structure can no longer support such a mode and thereby ceases to be a binding medium for this mode.

Unlike the case for the circular dielectric rod, no simple cutoff equations can be found for the surface wave modes along the elliptical dielectric rod. Cutoff conditions must still be computed from the very complicated determinantal equations. However, if we impose the condition of small γ_1 as well as small γ_0, an expression for γ_1 as a function of γ_0, ξ_0, m, and $\varepsilon_1/\varepsilon_0$ can be derived from the determinantal equations. The tedious but straightforward procedure as well as the functional expression for γ_1 will not be given here. Only the results will be discussed in the following. It was shown that the imposed small- γ_1 condition for $\gamma_0 \to 0$ is only valid if $m = 1$ for both ${}_{\mathrm{e}}\mathrm{HE}_{m,n}$ and ${}_{\mathrm{o}}\mathrm{HE}_{m,n}$ modes. Furthermore, as $\gamma_0 \to 0$, γ_1 also approaches zero for $m = 1$ modes, although γ_1 approaches zero slower for flatter rods for the ${}_{\mathrm{e}}\mathrm{HE}_{1,1}$ mode while γ_1 approaches zero faster for flatter rods for the ${}_{\mathrm{o}}\mathrm{HE}_{1,1}$ mode. Hence, there are only two nondegenerate modes on an elliptical dielectric rod, namely the ${}_{\mathrm{e}}\mathrm{HE}_{1,1}$ mode and the ${}_{\mathrm{o}}\mathrm{HE}_{1,1}$ mode, that possess zero cutoff frequency. Additional numerical computations also showed that as the elliptical dielectric cylinder becomes flatter, the cutoff frequencies for higher-order even modes (${}_{\mathrm{e}}\mathrm{HE}_{m,n}$) are higher and those for higher-order odd modes (${}_{\mathrm{o}}\mathrm{HE}_{m,n}$) are lower.

5. Propagation Characteristics of the Dominant Modes

Due to the complexity of the dispersion relations [Eqs. (A.14) and (A.17)], computation has only been carried out for the dominant modes. It was found (numerically) that the infinite determinants converged quite rapidly, such that they could be approximated very well by determinants of finite order N; e.g., an 8×8 determinant yielded three-place accuracy for all β's of ${}_{\mathrm{e}}\mathrm{HE}_{1,1}$ or ${}_{\mathrm{o}}\mathrm{HE}_{1,1}$ mode. In general, it was found that the β's for the mth mode are governed principally by the expression

$$\begin{vmatrix} g_{m,m} & h_{m,m} \\ t_{m,m} & s_{m,m} \end{vmatrix} = 0 \tag{A.22}$$

for the ${}_{\mathrm{e}}\mathrm{HE}_{m,m}$ mode and by the expression

$$\begin{vmatrix} g^*_{m,m} & h^*_{m,m} \\ t^*_{m,m} & s^*_{m,m} \end{vmatrix} = 0 \tag{A.23}$$

for the ${}_{\mathrm{o}}\mathrm{HE}_{m,m}$ mode.

Since the propagation characteristics of various modes on a circular dielectric rod are quite well known, we shall emphasize the effects on the propagation characteristics of the dominant modes caused by the deformation of a circular rod into an elliptical rod. In Fig. A-2, the normalized guide wavelength λ/λ_0 for the ${}_{\mathrm{e}}\mathrm{HE}_{1,1}$ mode is plotted as a function of the normalized cross-sectional area (NCSA) for various values of ξ_0 and for $\varepsilon_1/\varepsilon_0 = 2.5$. It is noted that $\lambda = 2\pi/\beta$, $\lambda_0 = 2\pi/[\omega(\mu\varepsilon_0)^{1/2}]$, NCSA $= [2q(\cosh \xi_0)/\lambda_0]^2 \tanh \xi_0 =$ (cross-sectional area)/$(\lambda_0{}^2\pi/4)$, and $\coth \xi_0 =$ (major axis)/(minor axis). As expected, no cutoff frequency exists for this dominant ${}_{\mathrm{e}}\mathrm{HE}_{1,1}$ mode. It can be seen that for very small values of NCSA, $\lambda/\lambda_0 \approx 1$ for all values of ξ_0. This is because when the wavelength λ_0 is much larger than the physical dimensions of the dielectric and most of the energy being

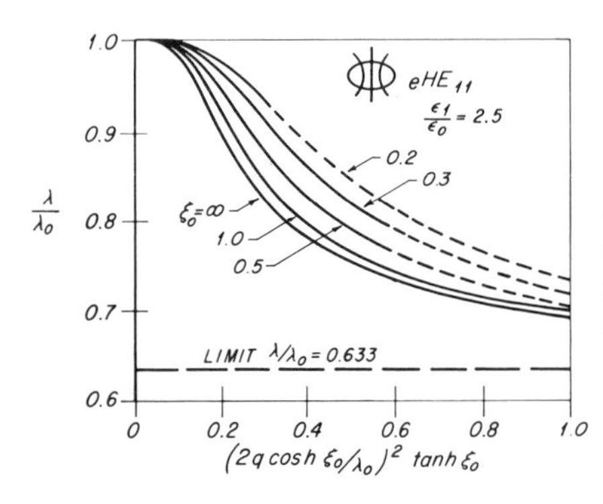

Fig. A-2. Normalized guide wavelength λ/λ_0 as a function of the normalized cross-sectional area $A/(\lambda_0{}^2\pi/4)$ for the ${}_{\mathrm{e}}\mathrm{HE}_{1\,1}$ mode. A is the cross-sectional area of the elliptical guide. [From Yeh (1)].

transported is outside the dielectric rod, the geometry of the cross section is no longer important. As NCSA becomes larger, the effect of varying ξ_0 is more pronounced. For a fixed value of NCSA, λ/λ_0 is smaller for smaller ξ_0 (i.e., for flatter dielectric rod). This behavior suggests that the field intensity is more concentrated in a circular rod, and that more energy of the ${}_{\mathrm{e}}\mathrm{HE}_{1,1}$ mode is carried within the circular rod. Therefore, the circular dielectric rod is a better binding geometry for this mode. As the NCSA becomes very large, the effect of varying ξ_0 on λ/λ_0 again becomes quite small. This is because most of the energy is being carried inside the dielectric rod and the shape of the rod boundary is no longer of importance. As NCSA $\to \infty$ and $\xi_0 \to 0$, the dielectric rod degenerates to a thin sheet of dielectric slab, the ${}_{\mathrm{e}}\mathrm{HE}_{1,1}$ mode degenerates to the TM mode on a plane dielectric slab. Another view of the normalized guide wavelength variation is given in Fig. A-3, in which λ/λ_0 is plotted against the normalized major axis (NMA) for various values of ξ_0 with $\varepsilon_1/\varepsilon_0 = 2.5$. In this figure, the curves are spread out more than those in Fig. A-2. This is because for a

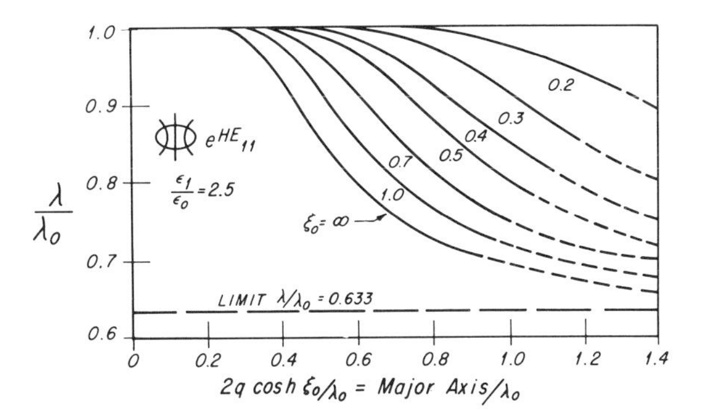

Fig. A-3. Normalized guide wavelength as a function of normalized major axis for the $_{e}HE_{1,1}$ mode. [From Yeh (1).]

fixed value of NMA, there is more binding dielectric material in a circular rod than in a flatter elliptical rod. Since it was found that the effect of the variation of relative dielectric constant $\varepsilon_1/\varepsilon_0$ on the guide wavelength for an elliptical rod is very similar to that for a circular rod, no additional results will be given here.

The normalized guide wavelength for the $_{o}HE_{1,1}$ mode is given in Fig. A-4 as a function of the normalized cross-sectional area for various values of ξ_0 and for $\varepsilon_1/\varepsilon_0 = 2.5$. Again, as expected, no cutoff frequency exists for the $_{o}HE_{1,1}$ mode. Unlike the case for the $_{e}HE_{1,1}$ mode, it appears that the elliptical rod is a better binding geometry for the $_{o}HE_{1,1}$ mode than a

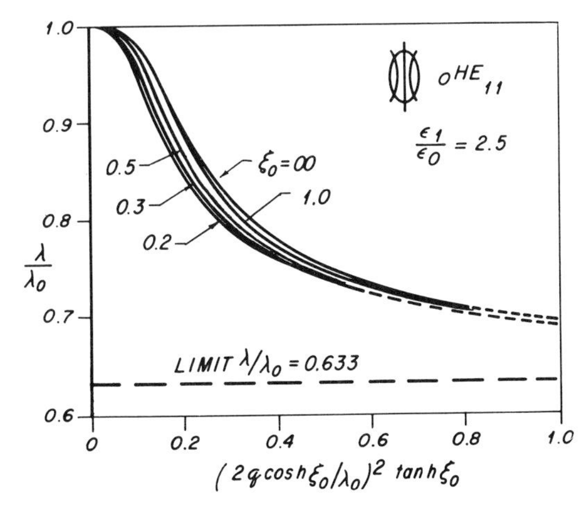

Fig. A-4. Normalized guide wavelength as a function of the normalized cross-sectional area for the $_{o}HE_{1,1}$ mode. [From Yeh (1).]

circular rod, since for a fixed value of NCSA, λ/λ_0 is smaller for a flatter rod. The curves for various values of ξ_0 are quite close to each other, which implies that the field lines are quite uniform for this ${}_0HE_{1,1}$ mode. The small differences between these curves are due to the fact that as a circular rod deforms into an elliptical rod, the electric lines of force are being squeezed together so that the field density is now more concentrated inside the dielectric rod. For a very flat elliptical rod, the electric lines of force are almost uniform within the rod (i.e., the field density distribution is almost uniform); therefore, any further flattening of the rod would not cause further changes in the field density distribution. It should be pointed out that as NCSA $\to \infty$ and $\xi_0 \to 0$, the ${}_0HE_{1,1}$ mode degenerates to a TE mode propagating along a thin dielectric plane slab. Another view of the variation of normalized guide wavelength is given in Fig. A-5, in which λ/λ_0 is plotted against the normalized major axis (NMA) for various values of ξ_0 with $\varepsilon_1/\varepsilon_0 = 2.5$. As in the case for the ${}_eHE_{1,1}$ mode and for the same reason, the curves for the ${}_0HE_{1,1}$ mode in Fig. A-5 are also spread out further than those in Fig. A-4.

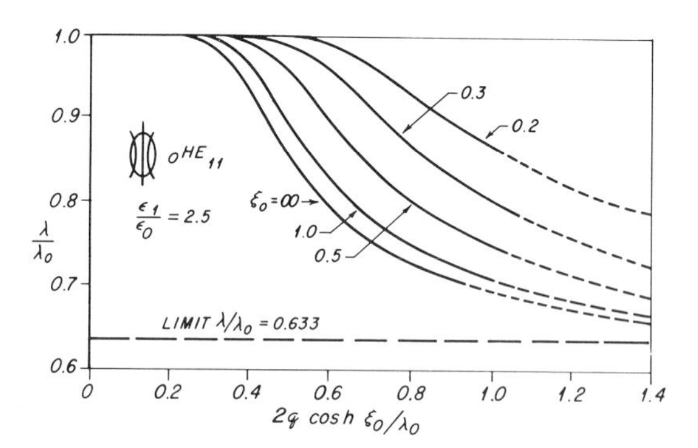

Fig. A-5. Normalized guide wavelength as a function of the normalized major axis for the ${}_0HE_{1,1}$ mode. [From Yeh (1).]

In summary, one notes that the guide wavelength for the dominant ${}_eHE_{1,1}$ mode becomes longer as the rod cross section becomes flatter and that the opposite is true for the dominant ${}_0HE_{1,1}$ mode, although the effect is not as pronounced. Although this conclusion cannot be generalized to apply to other higher-order modes, since the field patterns for these higher-order modes are different than those for the dominant modes, one may, nevertheless, conjecture here that if the field density distribution within the dielectric rod is spread out further as a result of flattening the cross section, then the guide wavelength will generally be longer than that for a circular rod with identical cross-sectional area and that the reverse is true if the field density distribution within the rod is squeezed closer together as a result of flattening the cross section.

6. Field Configurations

The field configuration for a specific mode should be calculated from the mode function for that mode. The field distribution are given by field lines in which the direction of the line at a point gives the direction of the field and the density of lines its magnitude. Unlike the case for the circular dielectric rod, the elliptical dielectric rod cannot support the circularly symmetric modes. On the other hand, the dominant ${}_{e}HE_{1,1}$ and ${}_{o}HE_{1,1}$ modes on the elliptical rod bear very close resemblances to the $HE_{1,1}$ mode on the circular rod. Perspective views of the electric lines of force for these dominant modes are given in Fig. A-6. It is also noted that the field distributions in an elliptical dielectric rod are very similar to those in an elliptical metallic waveguide. The significant difference is that in the former case, the electric and magnetic lines of force must form closed loops outside the dielectric rod, and in the latter case, the electric field is normal to the metallic boundary surrounding the core and the magnetic field is parallel to it. Therefore, the knowledge of field distributions in an elliptical metallic waveguide provides a simple means of sketching the field patterns in an elliptical dielectric waveguide.

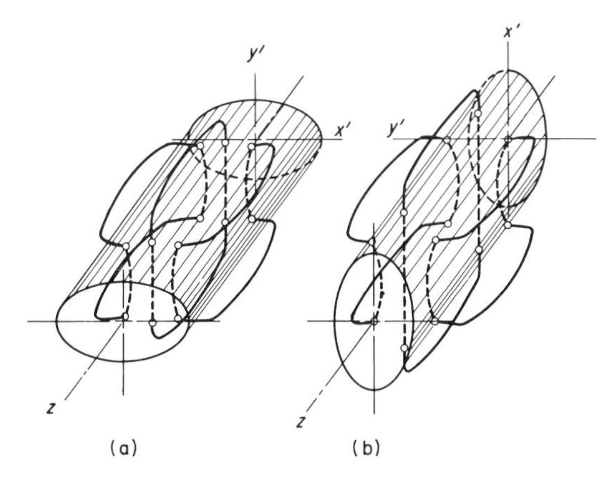

Fig. A-6. (a) A sketch of the electric lines of force for the ${}_{e}HE_{1\ 1}$ mode; (b) a sketch of the electric lines of force for the ${}_{o}HE_{1,1}$ mode.

Since a dielectric rod is an open structure, the field are not confined within the rod. Hence, the knowledge of the external field extent is very important. It is of interest to study the effect of changing the shape of a circular dielectric rod upon the field extent outside the rod. A study has been made on the variation of external field extent for the dominant modes as a circular rod is flattened into an elliptical rod. Results indicate, in general, that the field extent is smaller for $_{\mathrm{e}}\mathrm{HE}_{m,n}$ modes and larger for $_{\mathrm{o}}\mathrm{HE}_{m,n}$ modes as the rod is flattened further. Hence, the cladding on an equivalent elliptical dielectric rod (i.e., equivalent in area with a circular rod) may be decreased or increased according to whether even or odd modes are being transported.

7. Attenuation of the Dominant Modes

Due to imperfection of the dielectric material, power loss for the guided waves along a clad dielectric rod is unavoidable. Fortunately, the power loss per guide wavelength is quite small compared to the power flowing along the rod, and the attenuation constant can be calculated by a perturbation technique. In this technique, it is assumed that the propagation constant β remains unchanged by the presence of small dielectric loss while the mode functions in the case of small dielectric loss differ from those of the lossless case only by a multiplicative attenuation factor $e^{-\alpha z}$, where the attenuation constant α is given by the following formula:

$$\alpha = 4.343\sigma_{\mathrm{d}}(\mu/\varepsilon_0)^{1/2}R \quad (\mathrm{dB/m}) \tag{A.24}$$

with

$$R = \left|\left(\int_{A_i} \mathbf{E}\cdot\mathbf{E}^*\, dA\right)\Big/\left(\frac{\mu}{\varepsilon_0}\right)^{1/2} \int_A \mathrm{e}_z\cdot(\mathbf{E}_{\mathrm{t}} \times \mathbf{H}_{\mathrm{t}}^*)\, dA\right|. \tag{A.25}$$

σ_{d} is the conductivity of the rod (the outside cladding is assumed to be lossless), e_z is the unit vector in the direction of propagation, A_i is the cross-sectional area of the dielectric rod and A is the total cross-sectional area of the guide (including the cladding area), and $\mathbf{E}_{\mathrm{t}}$ and $\mathbf{H}_{\mathrm{t}}$ are the transverse components of the electric and magnetic field of the mode under consideration. The asterisk signifies the complex conjugate. Again we are interested in any variation of α as the shape of the dielectric rod is changed. As an example, we shall provide the results for the dominant modes on an elliptical dielectric rod. It should be pointed out that these results are not easily obtained. Due to the coupling of all coefficients in the expressions for the field components, the equation for R is extremely complicated. In Fig. A-7, R for the ${}_{e}HE_{1,1}$ mode is plotted as a function of the normalized cross-sectional area (NCSA) for various values of ξ_0 with $\varepsilon_1/\varepsilon_0 = 2.5$. For sufficiently large values of NCSA, R tends toward the value $(\varepsilon_0/\varepsilon_1)^{1/2}$, which is the attenuation factor for a plane wave in an infinite medium, for all values of ξ_0; for small enough values of NCSA, R can be made as small as desired. This behavior is attributed to the fact that when the NCSA is sufficiently large, almost all of the energy of the wave is transmitted inside the rod, and for small values of NCSA, most energy is transported outside the rod (i.e., in the lossless cladding of the rod). It is also noted that R tends to the limit $(\varepsilon_0/\varepsilon_1)^{1/2}$ much more slowly as ξ_0 gets smaller and that

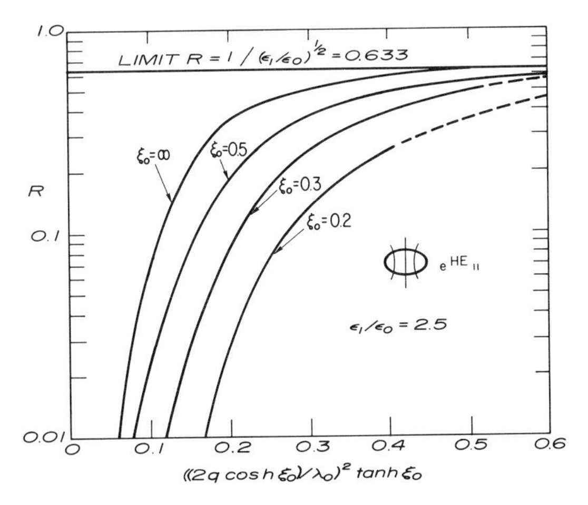

Fig. A-7. Attenuation factor R for the ${}_{e}HE_{1,1}$ mode as a function of normalized cross-sectional area. [From Yeh (2).]

the variation of slopes with respect to NCSA is smaller for flatter rods in the low-loss region. This means that small imperfections in the dimensions of a flatter rod would induce a smaller change in the attenuation. This observation also implies that as the rod becomes flatter, the field for the ${}_{e}HE_{1,1}$ mode is spread out more and the field inside the rod becomes less dense.

Figure A-8 gives the variation of the attenuation factor R for the ${}_{0}HE_{1,1}$ mode as a function of the NCSA for various ξ_0 with $\varepsilon_1/\varepsilon_0 = 2.5$. Unlike the case for the even dominant wave (the ${}_{e}HE_{1,1}$ wave), it appears that the flatter elliptical rod supporting the odd dominant wave does not provide a more favorable attenuation factor. The fact that the curves for various

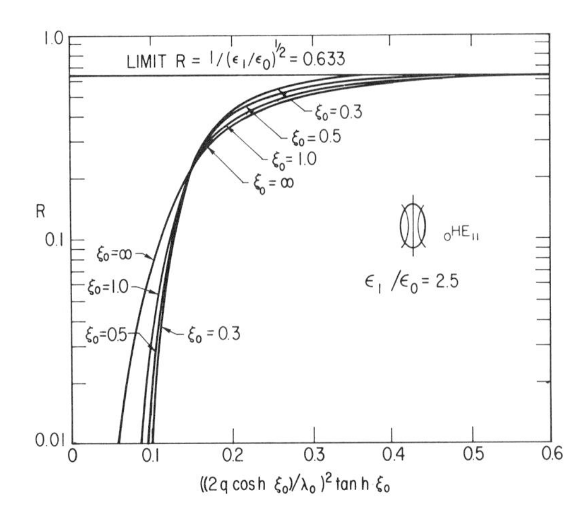

Fig. A-8. Attenuation factor R for the $_0HE_{1,1}$ mode as a function of normalized cross-sectional area. [From Yeh (2).]

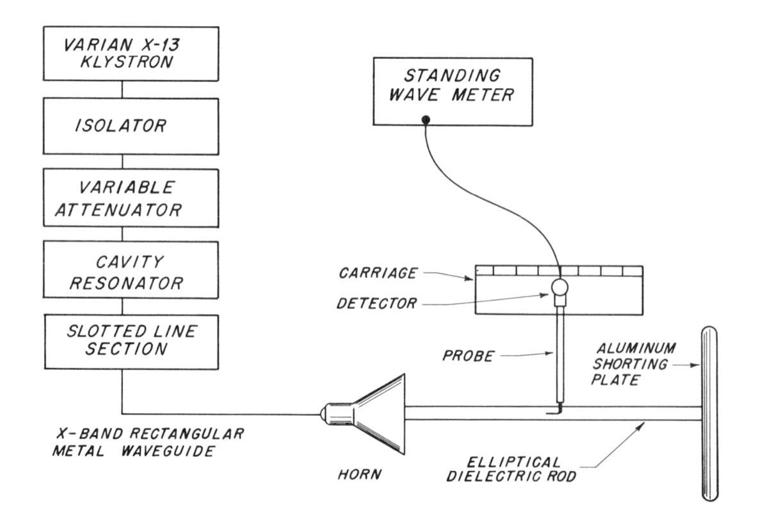

Fig. A-9. Block diagram showing arrangement of components for determining guide wavelength and attenuation of an elliptical dielectric waveguide.

values of ξ_0 in Fig. A-8 are quite close to each other again suggests that the field intensity is quite uniform for the odd dominant wave within the dielectric rod.

The distribution of the transmitted power can also be computed. There is a very close correlation between the percentage of power being carried inside the rod and the value of the loss factor. Numerical computations have been carried out. In general, the results show that more power is carried inside the circular rod than in an elliptical rod of identical cross section for the even dominant mode.

Detailed interpretation of the propagation characteristics for other higher-order modes must be obtained from additional computed results. However, one may conjecture here that the even modes in general will have less loss than the odd modes and that the difference becomes more pronounced for flatter rods.

8. Experimental Verifications

Experiments were performed in the microwave frequency range to verify the above theoretical results on the dominant modes. The apparatus is shown schematically in Fig. A-9. Since a rectangular metal guide operating in the dominant $TE_{1,0}$ mode has an electric field whose configuration is roughly similar to the transverse component of the electric field of the dominant modes on the dielectric guide, the transfer of microwave energy could be made by simply inserting the dielectric rod longitudinally into the metal guide for a short distance. The orientation of the cross section depends upon whether the ${}_{e}HE_{1,1}$ wave or ${}_{o}HE_{1,1}$ wave is desired. To improve matching and to minimize reflection, a flare pyramidal horn whose flare angle was adjusted for best energy transfer was connected to the metal guide. The dielectric rod was tapered to whatever size was required for a given test. The other end of the dielectric rod was machined very flat to accommodate a good contact with a flat aluminum plate which was used as a good shorting device. The aluminum plate was made large enough to intercept practically all of the energy outside the dielectric rod. Elliptical dielectric rods were machined from available rectangular Lucite strips whose dielectric constant was $\varepsilon_1/\varepsilon_0 = 2.5$. The wavelengths and the attenuation factor were measured using the standing wave technique. Results are shown in Figs. A-10 and A-11. It can be seen that very good agreement was found between analytic and experimental results.

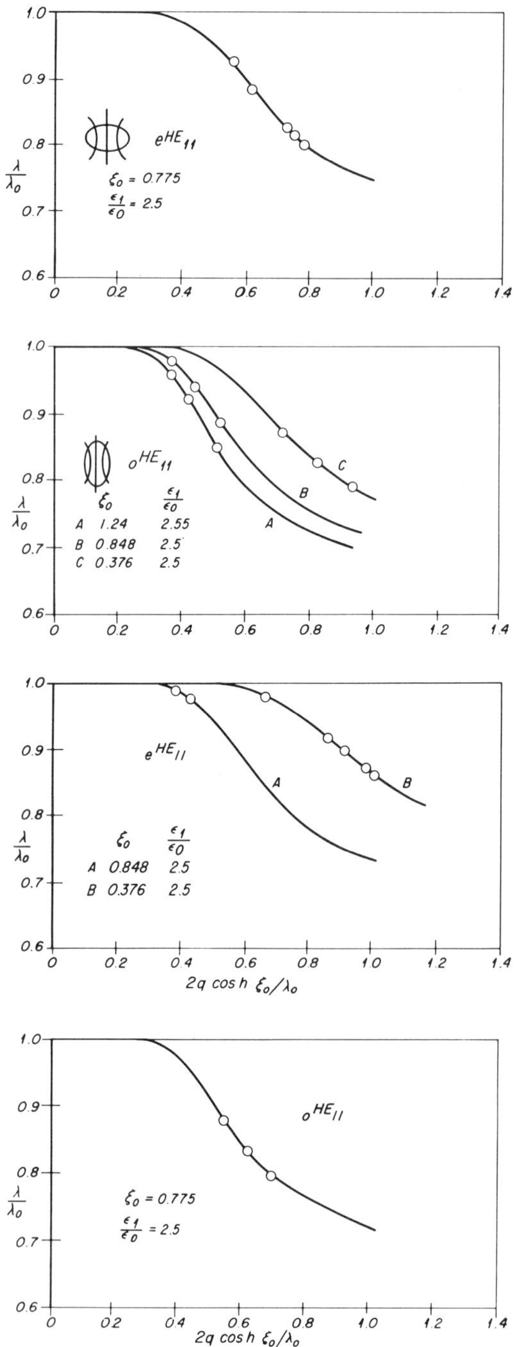

Fig. A-10. Experimental verification of guide wavelength for $_{e,o}HE_{1,1}$ modes. Circles are experimental points. [From Yeh (1).]

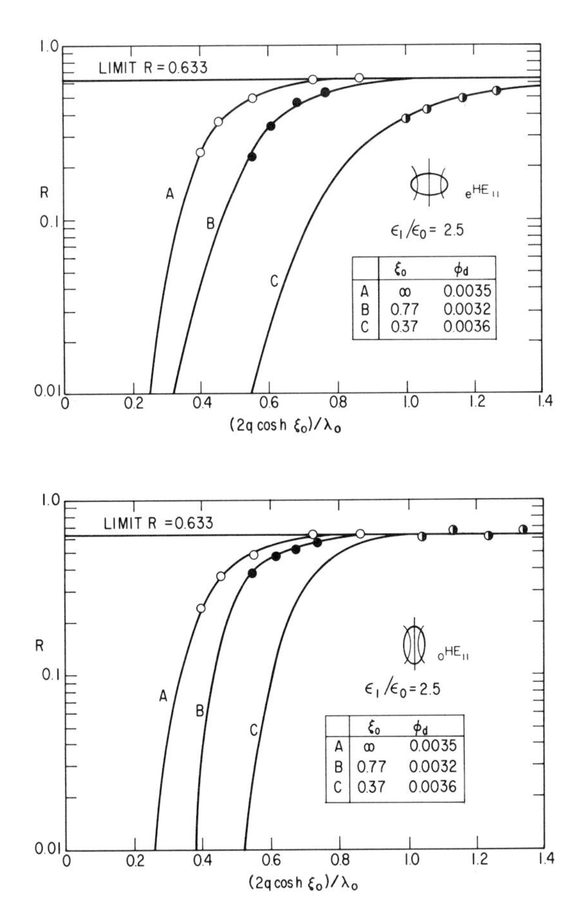

Fig. A-11. Experimental verification of attenuation constant for $_{e,o}HE_{1,1}$ modes. Circles are experimental points. [From Yeh (2).]

C. RECTANGULAR DIELECTRIC WAVEGUIDES

Unlike the case for wave propagation on circular or elliptical dielectric rods, exact analytic results are not available for the rectangular dielectric waveguides. However, a computer analysis of the propagating modes of a reactangular dielectric waveguide has recently been carried out by Goell (*5*). We shall describe in the following some of his results as well as an approximate method for analyzing this problem given by Marcatili (*6*). We shall also correlate some of the earlier experimental data with available theoretical results.

1. Computer Analysis

Goell's computer analysis is based on an expansion of the electromagnetic field in terms of a series of circular harmonics. The electric and magnetic fields inside the dielectric rod are matched to those outside the core at appropriate points on the boundary to yield equations which are then solved on a computer for the propagation constants and field configurations of various modes. The geometry of the problem is shown in Fig. A-12. Expressing the longitudinal electric and magnetic fields in terms of circular harmonics, we have (*9*)

$$E_{z1} = \sum_{n=0}^{\infty} a_n J_n(hr)[\sin(n\theta + \phi_n)]e^{i(\beta z - \omega t)} \tag{A.26}$$

$$H_{z1} = \sum_{n=0}^{\infty} b_n J_n(hr)[\sin(n\theta + \psi_n)]e^{i(\beta z - \omega t)} \tag{A.27}$$

inside the dielectric core, and

$$E_{z0} = \sum_{n=0}^{\infty} c_n K_n(pr)[\sin(n\theta + \phi_n)]e^{i(\beta z - \omega t)} \tag{A.28}$$

$$H_{z0} = \sum_{n=0}^{\infty} d_n K_n(pr)[\sin(n\theta + \psi_n)]e^{i(\beta z - \omega t)} \tag{A.29}$$

outside the core, where

$$\begin{aligned} h &= (k_1{}^2 - \beta^2)^{1/2}, \qquad & p &= (\beta^2 - k_0{}^2)^{1/2} \\ k_1 &= \omega(\mu\varepsilon_1)^{1/2}, \qquad & k_0 &= \omega(\mu\varepsilon_0)^{1/2}. \end{aligned} \tag{A.30}$$

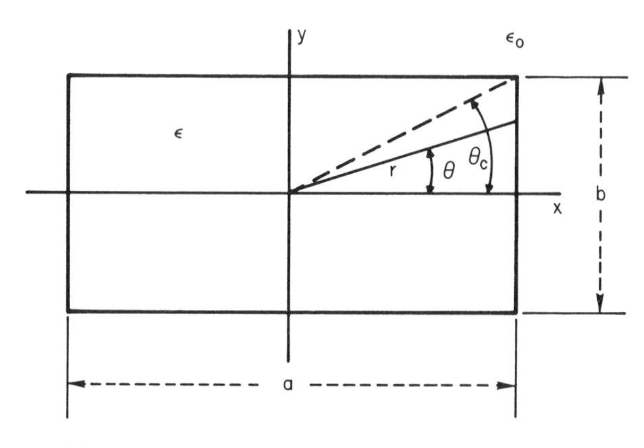

Fig. A-12. The rectangular dielectric waveguide.

a_n, b_n, c_n, and d_n are arbitrary constants. J_n and K_n are the nth-order Bessel functions and modified Bessel functions, respectively, and ψ_n and ϕ_n are arbitrary phase angles. The transverse fields can be found easily from Maxwell's equations. It is further noted that the component of the electric field tangent to the rectangular core is given by

$$E_t = \pm(E_r \sin\theta + E_\theta \cos\theta) \qquad \begin{cases} (-\theta_c < \theta < \theta_c) \\ (\pi - \theta_c < \theta < \pi + \theta_c) \end{cases} \tag{A.31}$$

or

$$E_t = \pm(-E_r \cos\theta + E_\theta \sin\theta) \qquad \begin{cases} \theta_c < \theta < \pi - \theta_c \\ \pi - \theta_c < \theta < -\theta_c \end{cases} \tag{A.32}$$

where θ_c is given in Fig. A-12. E_r and E_θ are, respectively, the radial component and the θ component of the electric field. Similar expressions exist for the tangential magnetic field.

Matching the tangential electric and magnetic fields along the boundary of the rectangular core gives, in matrix form,

$$\begin{bmatrix} G_1 & 0 & -G_2 & 0 \\ 0 & G_3 & 0 & -G_4 \\ G_5 & G_6 & -G_7 & -G_8 \\ G_9 & G_{10} & -G_{11} & -G_{12} \end{bmatrix} \begin{bmatrix} A \\ B \\ C \\ D \end{bmatrix} = 0 \tag{A.33}$$

where A, B, C, and D are N-element column matrices of the a_n, b_n, c_n, and d_n mode coefficients, respectively. The elements of the $m \times n$ matrices $G_1, G_2, \ldots, G_{12}$ are given by

$$\begin{aligned} g_{1mn} &= JS, & g_{2mn} &= KS \\ g_{3mn} &= JC, & g_{4mn} &= KC \\ g_{5mn} &= -\beta(P'SR + PCT), & g_{6mn} &= \omega\mu(PSR + P'CT) \\ g_{7mn} &= \beta(Q'SR + QCT), & g_{8mn} &= -\omega\mu(QSR + Q'CT) \\ g_{9mn} &= \omega\varepsilon_1(PCR - P'ST), & g_{10mn} &= -\beta(P'CR - PST) \\ g_{11mn} &= -\omega\varepsilon_0(QCR - Q'ST), & g_{12mn} &= \beta(Q'CR - QST) \end{aligned} \tag{A.34}$$

where

$$S = \sin(n\theta_m + \phi), \qquad C = \cos(n\theta_m + \phi), \qquad (\phi = 0 \quad \text{or} \quad \phi = \pi/2)$$

$$\begin{aligned} J &= J_n(hr_m), & K &= K_n(pr_m) \\ P &= nJ_n(hr_m)/h^2 r_m, & Q &= nK_n(pr_m)/p^2 r_m \\ P' &= J_n'(hr_m)/h, & Q' &= K_n'(pr_m)/p \end{aligned}$$

and

$$\begin{array}{llll} R = \sin\theta_m, & T = \cos\theta_m, & r_m = \tfrac{1}{2}a\cos\theta_m, & \theta < \theta_c, \\ R = -\cos\theta_m, & T = \sin\theta_m, & r_m = \tfrac{1}{2}b\sin\theta_m, & \theta > \theta_c. \end{array} \tag{A.35}$$

For $\theta = \theta_c$, $R = \cos(\theta_m + \frac{1}{4}\pi)$, $T = \cos(\theta_m - \frac{1}{4}\pi)$, and $r_m = \frac{1}{4}(a^2 + b^2)^{1/2}$. Any given mode must consist of either even harmonics or odd harmonics. For the odd harmonic cases, $\theta_m = (m - \frac{1}{2})\pi/2N$, $m = 1, \ldots, N$, where N is the number of space harmonics. For the even harmonic cases, if the aspect ratio is unity ($a/b = 1$), $\theta_m = (m - \frac{1}{2})\pi/2N$, $m = 1, \ldots, N$ for the field components with even symmetry about the x axis, and $\theta_m = (m - N - \frac{1}{2})\pi/2(N - 1)$, $m = N + 1, N + 2, \ldots, 2N - 1$ for the field components with odd symmetry about the x axis; if the aspect ratio is other than unity, the θ_m are chosen according to the first formula, except that the first and last points for the odd z component are omitted.

Modes are designated as $E^y_{s,q}$ if, in the limit of short wavelength, their electric field is parallel to the y axis, and as $E^x_{s,q}$ if, in the limit, their electric field is parallel to the x axis. The s and q subscript are used to designate the number of maxima in the x and y directions, respectively. The dominant $E^y_{1,1}$ and $E^x_{1,1}$ modes correspond, respectively, to the ${}_e\mathrm{HE}_{1,1}$ and ${}_o\mathrm{HE}_{1,1}$ modes on an elliptical dielectric guide.*

The dispersion relation of various modes is obtained by equating the determinant of (A.33) to zero. Numerical computation shows that for small aspect ratio, only a few harmonics are needed to achieve the desired three-place accuracy, but for larger aspect ratios, the convergence is not very good. According to Goell (5), the computed results are believed to be accurate to better than 2%. The normalized propagation constant Λ is plotted against the normalized waveguide height u in Figs. A-13–A-15 for various modes and various a/b ratios. The Λ and u are defined as follows:

$$\Lambda = \frac{(\lambda_0/\lambda)^2 - 1}{(\varepsilon_1/\varepsilon_0) - 1}, \qquad u = \frac{2b}{\lambda_0}\left(\frac{\varepsilon_1}{\varepsilon_0} - 1\right)^{1/2}$$

with

$$\beta = 2\pi/\lambda, \qquad k_0 = 2\pi/\lambda_0.$$

Figure A-13 shows that the $E^y_{s,q}$ and $E^x_{s,q}$ modes are degenerate as $\varepsilon_1 \to \varepsilon_0$. The degeneracy disappears when $\varepsilon_1/\varepsilon_0 = 2.25$ with $a/b = 2$ as shown in

* The subscripts (s, q) for the $E^{x,y}_{s,q}$ modes do not usually correspond to the subscripts (m, n) for the ${}_{e,o}\mathrm{HE}_{m,n}$ modes.

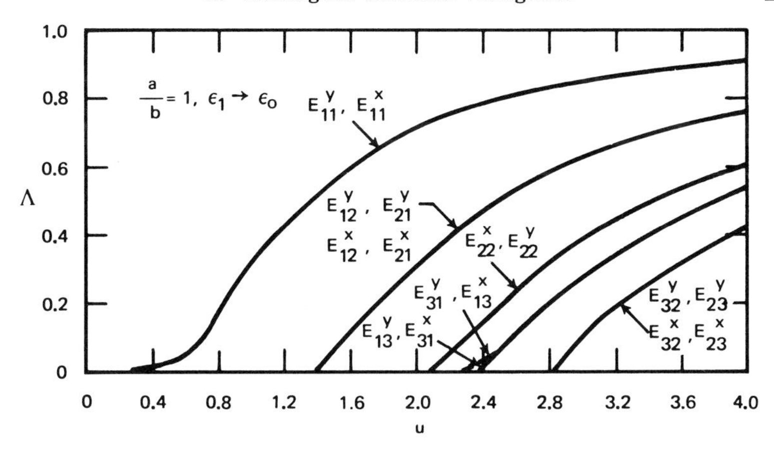

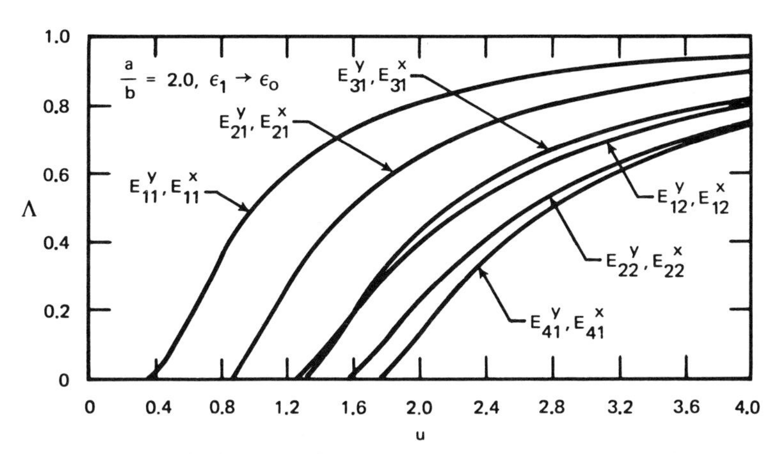

Fig. A-13. Normalized propagation constant as a function of normalized height for various modes with $\varepsilon_1 \to \varepsilon_0$. [From Goell (5).]

Fig. A-14. The effect of the change of relative dielectric constant on the propagation characteristics of the dominant modes can be seen from Fig. A-15. The curves for the dominant modes in these figures compare favorably with those presented in Section B on the elliptical dielectric waveguides. It is noted that if one plots Λ as a function of the normalized cross-sectional area (u^2a/b) for a fixed $\varepsilon_1/\varepsilon_0$, the variation between the propagation curves of a lower-order mode for various a/b ratios is quite small. This effect will be discussed in greater detail later.

A sketch of the cross sectional field configurations for various modes is given in Fig. A-16.

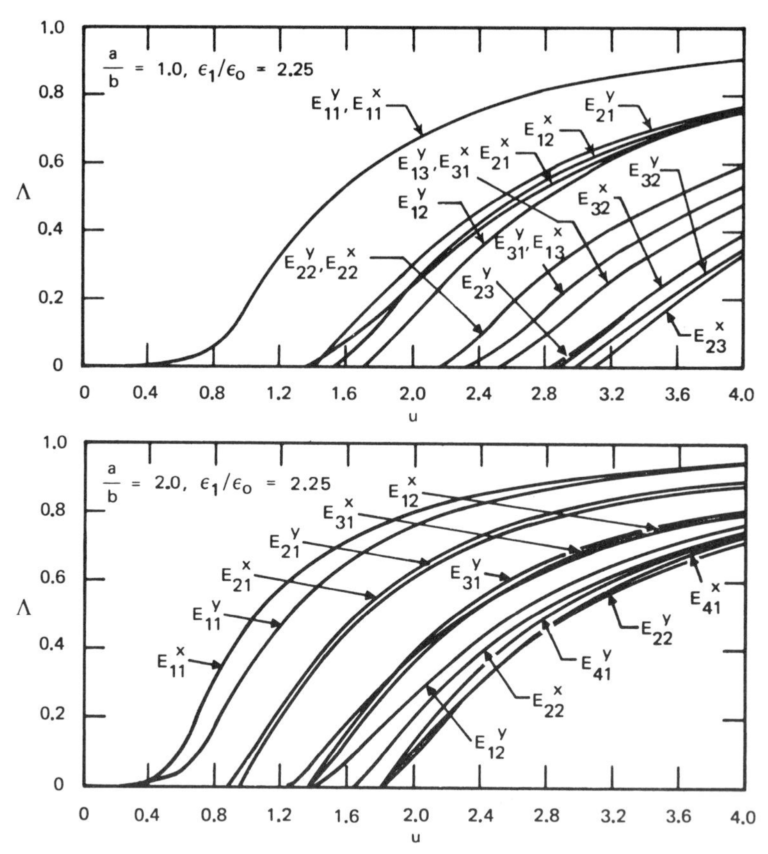

Fig. A-14. Normalized propagation constant as a function of normalized height fc various modes with $\varepsilon_1/\varepsilon_0 = 2.25$. [From Goell (5).]

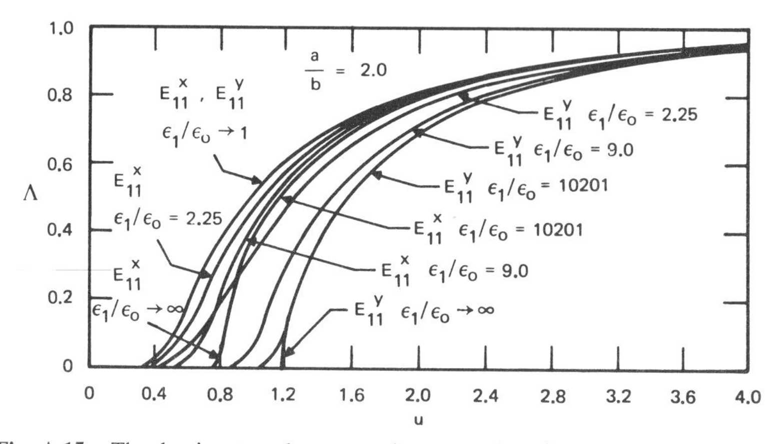

Fig. A-15. The dominant mode propagation curves for $a/b = 2.0$. [From Goell (5).]

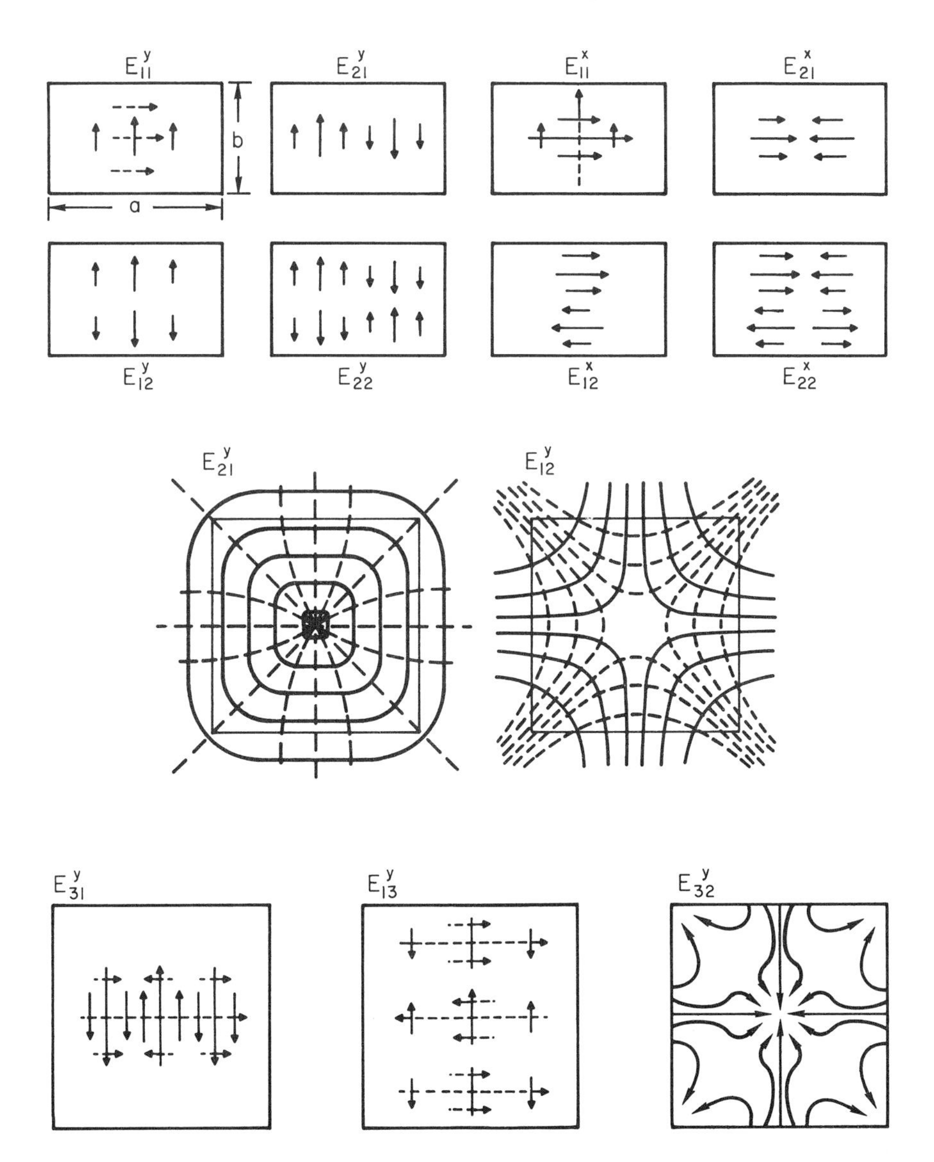

Fig. A-16. Field configuration of $E^{x,y}_{s,q}$ modes. Solid lines represent electric field lines and dashed lines represent magnetic field lines.

It is possible, with the help of a computer and a plotter, to obtain intensity pictures for various modes. These are given in Figs. A-17–A-19. The fact that most of the energy is contained within the core, even for relatively small values of Δ and $(\varepsilon_1 - \varepsilon_0)/\varepsilon_0$, can be seen from these figures.

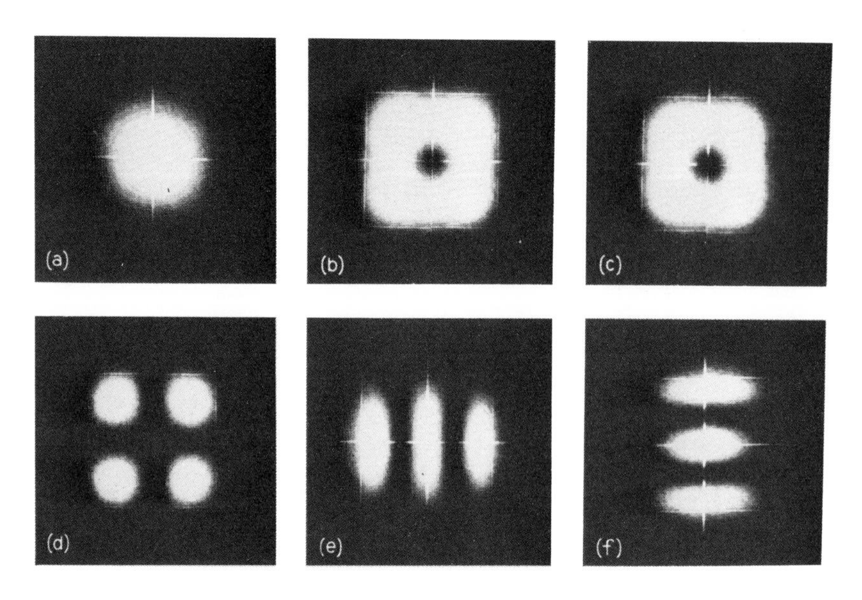

Fig. A-17. Intensity pictures for some $E^y_{s,q}$ modes with $a/b = 1.0$, $u = 3.0$, and $\varepsilon_1/\varepsilon_0 = 1.0201$: (a) $E^y_{1,1}$, (b) $E^y_{2,1}$, (c) $E^y_{1,2}$, (d) $E^y_{2,2}$, (e) $E^y_{3,1}$, and (f) $E^y_{1,3}$. [From Goell (5).]

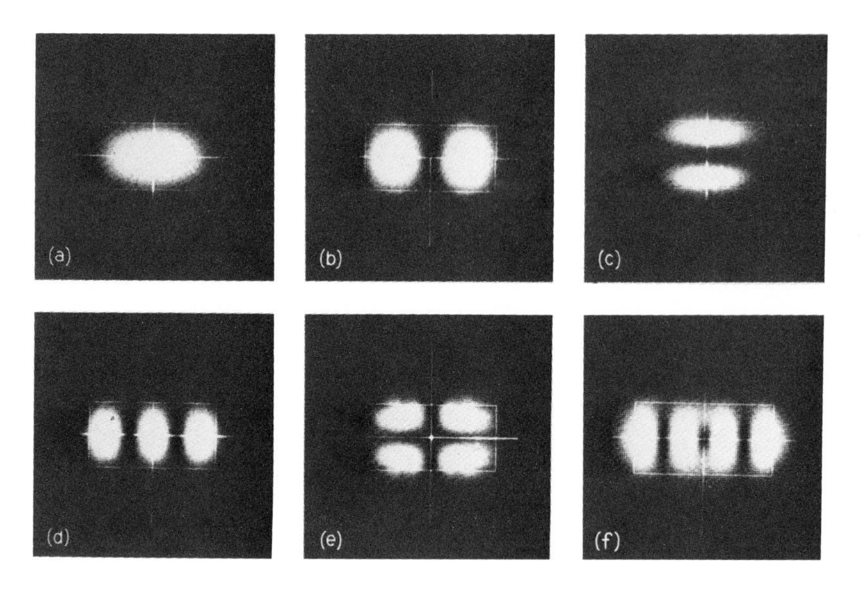

Fig. A-18. Intensity pictures for some $E^y_{s,q}$ modes with $a/b = 2.0$, $u = 2.0$, and $\varepsilon_1/\varepsilon_0 = 1.0201$: (a) $E^y_{1,1}$, (b) $E^y_{2,1}$, (c) $E^y_{1,2}$, (d) $E^y_{3,1}$, (e) $E^y_{2,2}$, and (f) $E^y_{4,1}$. [From Goell (5).]

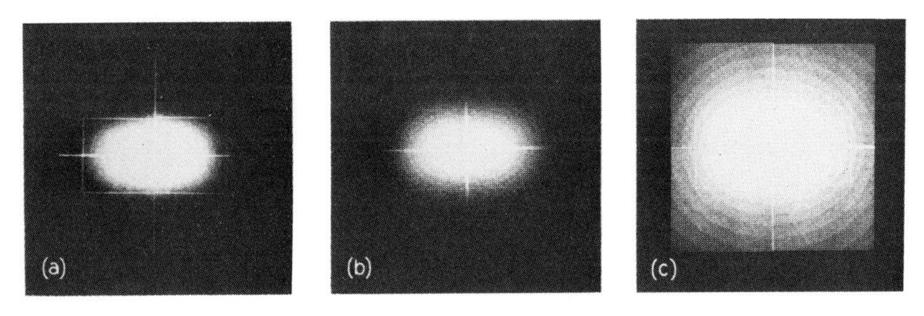

Fig. A-19. Intensity pictures for $E^y_{1,1}$ mode with $a/b = 2.0$. (a) $\Lambda = 0.81$, (b) $\Lambda = 0.50$, (c) $\Lambda = 0.02$. [From Goell (5).]

2. Approximate Analytic Solution

Since most of the energy is confined within the dielectric core for a wide range of parameters, as seen from the analysis just given, and very little energy propagates in the region of the corners of a rectangular dielectric guide, Marcatili (*6*) formulated an approximate solution to this problem by ignoring the matching of fields along the edges of the shaded area in Fig. A-20. By matching the tangential electric and magnetic fields only along the four sides of region 1 and assuming that the field components in region 1 vary sinusoidally in the x and y directions, those in 2 and 4 vary sinusoidally along x and exponentially along y, and those in regions 3 and 5 vary sinusoidally along y and exponentially along x, one obtains the following disper-

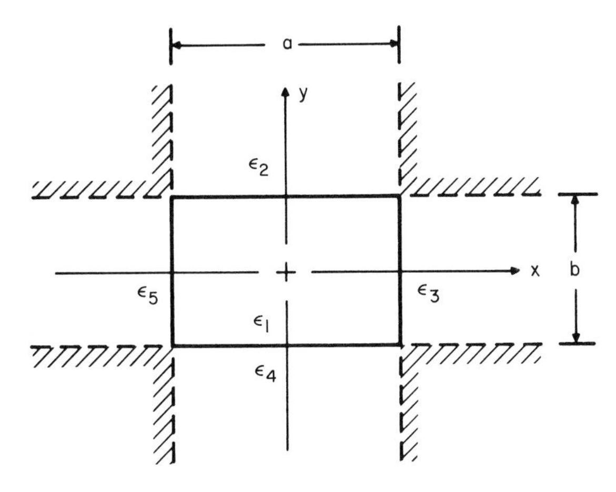

Fig. A-20. Geometry of the rectangular dielectric guide immersed in different dielectrics.

sion relations: For $E^y_{s,q}$ modes,

$$\begin{aligned}
k_x a &= s\pi - \tan^{-1}(k_x \xi_3) - \tan^{-1}(k_x \xi_5) \\
k_y b &= q\pi - \tan^{-1}[(\varepsilon_2/\varepsilon_1) k_y \eta_2] - \tan^{-1}[(\varepsilon_4/\varepsilon_1) k_y \eta_4] \\
\xi_{3,5} &= [(\pi/\chi_{3,5})^2 - k_x^2]^{-1/2} = 1/|\, k_{x3,5} \,| \\
\eta_{3,5} &= [(\pi/\chi_{2,4})^2 - k_y^2]^{-1/2} = 1/|\, k_{y2,4} \,| \\
\chi_{2,3,4,5} &= \pi(k_1^2 - k_{2,3,4,5})^{-1/2} \\
k_x &= k_{x1} = k_{x2} = k_{x4}, \qquad k_y = k_{y1} = k_{y3} = k_{y5} \\
\beta &= (k_1^2 - k_x^2 - k_y^2)^{1/2}, \quad k_{1,2,3,4,5} = \omega(\mu\varepsilon_{1,2,3,4,5})^{1/2},
\end{aligned} \tag{A.36}$$

and for $E^x_{s,q}$ modes,

$$\begin{aligned}
k_x' a &= s\pi - \tan^{-1}[(\varepsilon_3/\varepsilon_1) k_x' \xi_3'] - \tan^{-1}[(\varepsilon_5/\varepsilon_1) k_x' \xi_5'] \\
k_y' b &= q\pi - \tan^{-1}(k_y' \eta_2') - \tan^{-1}(k_y' \eta_4') \\
\xi'_{3,5} &= [(\pi/\chi_{3,5})^2 - (k_x')^2]^{1/2} = 1/|\, k'_{x3,5} \,| \\
\eta'_{2,4} &= [\pi/\chi_{2,4})^2 - (k_y')^2]^{1/2} = 1/|\, k'_{y2,4} \,| \\
k_x' &= k'_{x1} = k'_{x2} = k'_{x4}, \qquad k_y' = k'_{y1} = k'_{y3} = k'_{y5} \\
\beta' &= [k_1^2 - (k_x')^2 - (k_y')^2]^{1/2}.
\end{aligned} \tag{A.37}$$

β and β' are, respectively, the propagation constants for $E^y_{s,q}$ and $E^x_{s,q}$ modes.

It can be seen that the above approximate dispersion relations are much simpler than those obtained earlier using the point-matching technique. Comparison of the numerical results obtained according to the above approximate dispersion relations with those computed by Goell shows that for a guide and mode for which

$$\Lambda = \frac{(\lambda_0/\lambda)^2 - 1}{(\varepsilon_1/\varepsilon_0) - 1} \geq 0.5, \qquad (\varepsilon_2 = \varepsilon_3 = \varepsilon_4 = \varepsilon_5 = \varepsilon_0)$$

the approximate result is within a few per cent of the exact value. The largest discrepancy occurs for $\Lambda \simeq 0$ and for the dominant modes $E^y_{1,1}$ and $E^x_{1,1}$. The approximate theory is incapable of predicting the zero cutoff frequency of these dominant modes. Nevertheless, because of the simplicity of the approach and the fact that the dielectric media surrounding the four sides of the rectangular core may have different dielectric constants, the approximate results derived by Marcatili are very useful.

3. Experimental Results

The most thorough experimental investigation of wave propagation on a rectangular dielectric rod was carried out by Schlesinger and King (*10*). They made measurements in the microwave frequency range on 17 samples of rectangular polystyrene rod ($\varepsilon_1/\varepsilon_0 = 2.56$) using the image line configuration and the flat plate resonator. Again, only the dominant modes are summarized in Figs. A-21 and A-22. Figure A-21 shows the dependence of λ/λ_0 on the normalized cross-sectional area (NCSA $= ab/\lambda_0^2$) of the rectangular dielectric rod for various samples; a and b are shown in Fig. A-12 It can be seen that if the ratio of major axis to minor axis is not too large,

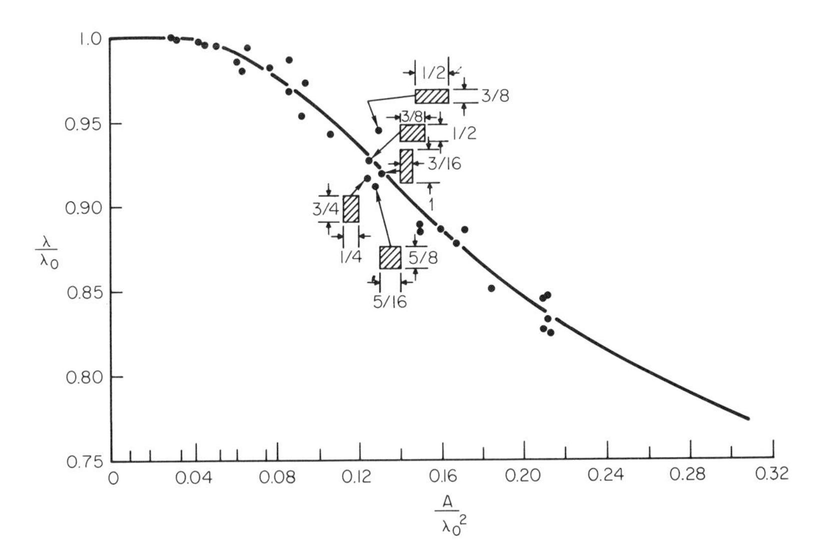

Fig. A-21. Dependence of λ/λ_0 for the dominant modes on the area of a polystyrene dielectric rod. Theoretical curve is for the round rod. Nominal diameters indicated based on 9700 mc. Experimental points for rectangular rod; $\varepsilon_1/\varepsilon_0 = 2.56$. [From Schlesinger and King (10).]

the measured points lie within 3% of a curve drawn for a circular dielectric rod with an equivalent cross-sectional area. It is further noted that given a certain area with moderate ratio of major axis to minor axis, the orientation (i.e., whether ${}_{e}HE_{1,1}$ or ${}_{0}HE_{1,1}$ mode was excited) has only a slight effect within the approximate 3%. This result is also confirmed by the analysis given in Section B on elliptical dielectric waveguide and in Section C.1 on rectangular dielectric guide. Figure A-22 shows the variation of λ/λ_0 with normalized cross-sectional area for four sets of rectangular rods. The dashed curve is the theoretical curve for a circular rod of equivalent cross-sectional area. Since the experimental points were obtained by keeping either the width or the height constant while varying the other dimension and the solid curves connecting the experimental data do not correspond to a fixed ratio of a/b, direct comparison of these curves with the results given for elliptical rods in Section B is not possible. However, the trend of these curves is as expected; i.e., since the field density for the ${}_{0}HE_{1,1}$ mode is more concentrated inside the rectangular dielectric rod than in an equivalent circular dielectric rod, the guide wavelengths are therefore correspondingly shorter.

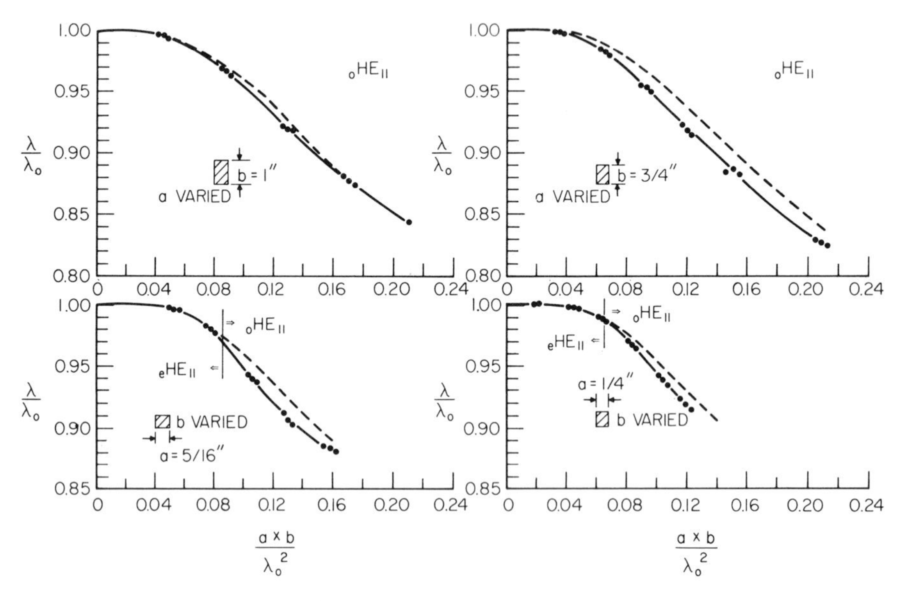

Fig. A-22. The variation of λ/λ_0 with dielectric area in number of square wavelengths for four sets of rectangular polystyrene rods ($\varepsilon_1/\varepsilon_0 = 2.56$). Dashed curves are theoretical curves for round rods of equivalent cross-sectional area. $\lambda_0 = 1.21''$.

Additional experimental data for the ${}_{e}HE_{1,1}$ and ${}_{0}HE_{1,1}$ modes were given by Schlosser and Unger (*11*) using rectangular polyethylene rods ($\varepsilon_1/\varepsilon_0=2.35$ and major axis/minor axis $= 2.0$). Results are plotted in Fig. A-23. Solid curves are calculated according to the elliptical dielectric waveguide theory given earlier and the dashed curves represent experimental results. It can be seen that very close agreement is obtained.

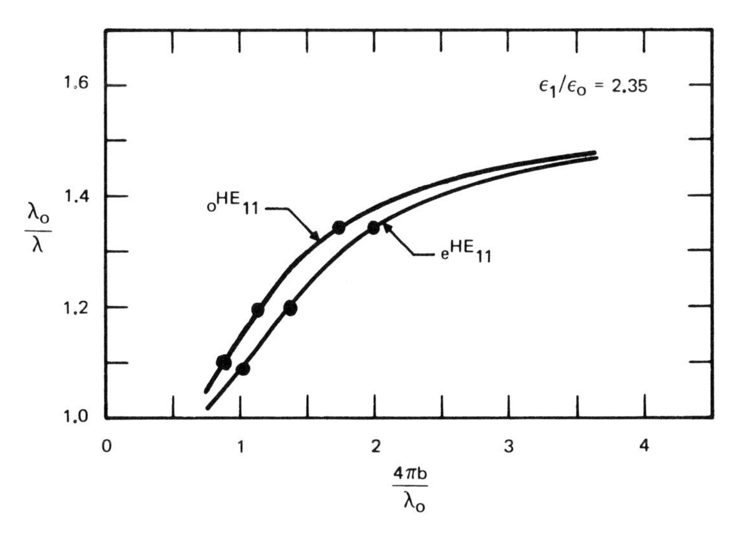

Fig. A-23. Comparison of computed, and measured λ_0/λ for the dominant modes. Ratio (major axis)/(minor axis) $= a/b = 2.0$, $_0\mathrm{HE}_{11}$ and $_e\mathrm{HE}_{11}$ measurements (rod size in mm) (○) $a = 2.5$; $b = 2.50$. (×) $a = 1.50$; $b = 3.00$. (△) $a = 2.00$, $b = 4.00$. (□) $a = 2.50$; $b = 5.00$. (♀) $a = 4.80$; $b = 9.60$. Solid curves are calculated according to the elliptical dielectric waveguide theory [Yeh (1)]. Points represent experimental results. [Schlosser and Unger (11).]

D. SUMMARY AND CONCLUSIONS

Detailed considerations have been given to the problem of wave propagation along elliptical and rectangular dielectric waveguides. It was found that the normalized guide wavelength (λ/λ_0) is strongly dependent upon the normalized cross-sectional area of the guide (A/λ_0^2). In other words, we can predict the propagation velocity of a specific mode on a dielectric rod of noncircular cross section, within a few percentage points, by the single expedient of finding the velocity of propagation for an elliptical (or rectangular) rod of equivalent ellipticity and of the same area-to-wavelength ratio. As a matter of fact, if the ellipticity of the noncircular dielectric rod is not too great, one can easily predict the propagation velocity, also within a few percentage points, based on the propagation velocity curves for a circular dielectric rod of equivalent cross-sectional area. Figure A-24 gives a plot of λ_g/λ_0 as a function of A/λ_0^2 for a circular dielectric waveguide supporting the dominant $\mathrm{HE}_{1,1}$ mode for various values of $\varepsilon_1/\varepsilon_0$.

One of the most significant differences between elliptical (or rectangular) and circular dielectric waveguides is that the twofold degeneracy of $\mathrm{HE}_{m,n}$

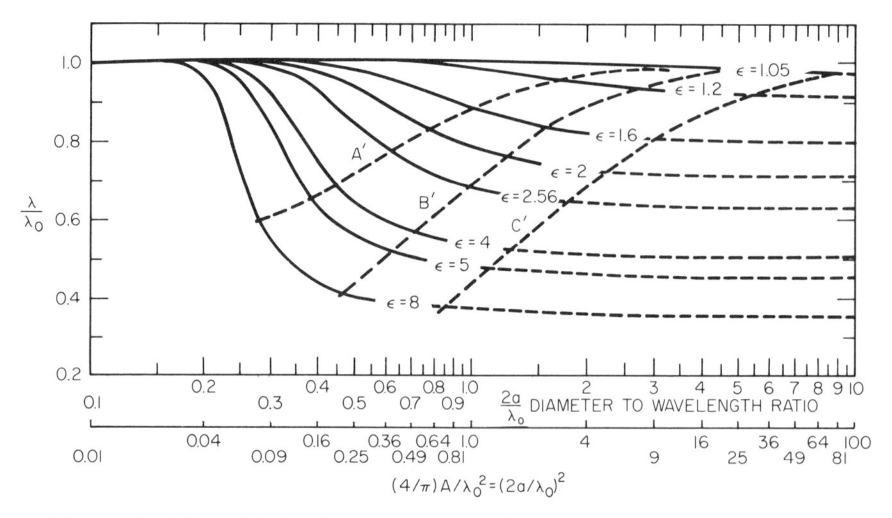

Fig. A-24. Effect of dielectric constant on the dispersion characteristics of round rod. A is the cross-sectional area of the rod, A' is the cutoff locus for $E_{0,1}$ and $H_{0,1}$ modes, B' is the cutoff locus for $HE_{1,2}$ mode, and C' is the cutoff locus for $\text{HE}_{1,3}$ mode. $\varepsilon = \varepsilon_1/\varepsilon_0$. [From Schlesinger and King (10).]

($m \geq 1$) modes on a circular dielectric waveguide no longer exists on an elliptical (or rectangular) dielectric waveguide. These nondegenerate modes on the elliptical guide have different propagation velocities. As pointed out by Snitzer and Osterberg (*12*) in their experimental investigation on dielectric waveguide modes in the visible spectrum, since an elliptical dielectric waveguide cannot support circularly symmetric modes, a slight deformation of a circular fiber carrying the circular symmetric $\text{TE}_{0,m}$ or $\text{TM}_{0,m}$ will lead to new modes which are linear combinations of the original, circular waveguide modes, i.e., the linear combination of $\text{TE}_{0,m}$ modes with $\text{HE}_{2,m}$ modes or $\text{TM}_{0,m}$ modes with $\text{HE}_{2,m}$ modes (see Fig. A-25). This observation is in agreement with the analysis given in Section B on the elliptical dielectric waveguides.

It is noted that the point-matching technique used by Goell to solve the rectangular dielectric waveguide problem can also be used to find the propagation characteristics of waves on dielectric rods of other noncircular shapes. As pointed out in Section C.1, the convergence of Goell's dispersion relation is not very good when the aspect ratio is large. This problem can be overcome by the use of elliptical harmonics as given in Section B rather than the use of circular harmonics. The fact that the approximate technique developed by Marcatili can be used to consider the problem of coupling between adjacent rectangular dielectric rods should also be noted (*6*).

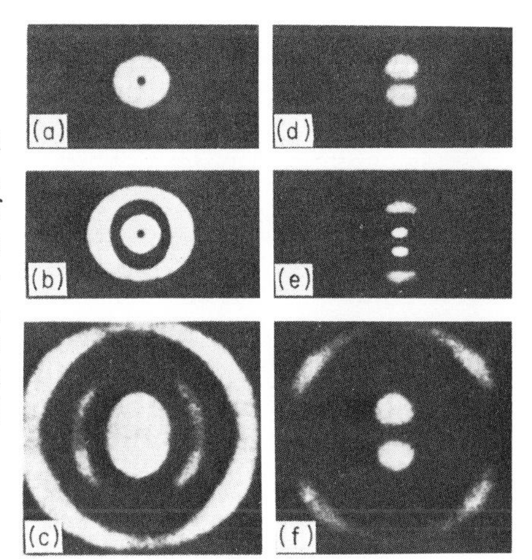

Fig. A-25. The $TE_{0,m}$, $TM_{0,m}$, and $HE_{2,m}$ modes for $m = 1, 2$, and 3; (a), (b), and (c) are the direct images of either of the sets of modes when viewed without an analyzer, and (d), (e), and (f) give the effects of admixing $n = 0$ modes with $HE_{2,m}$, or the appearance of single modes when viewed through an analyzer. [Experimental results from Snitzer and Osterberg (12).]

It is worthwhile to emphasize here that in most cases where optical fibers are used to guide light waves, the index of refraction of the core is only slightly higher than that of the cladding. In this situation, the characteristic equations for the modes on an elliptical fiber can be significantly simplified. They are given by Eqs. (A.19) and (A.20) instead of the infinite determinants. Hence the propagation constants of higher-order modes on an elliptical optical fiber can be obtained readily from these equations.

REFERENCES

1. C. Yeh, Elliptical Dielectric Waveguides, *J. Appl. Phys.* **33**, 3235 (1962).
2. C. Yeh, Attenuation in a Dielectric Elliptical Cylinder, *IEEE Trans. Ant. Prop.* **AP-11**, 177 (1963).
3. L. A. Lynbimow, G. I. Veselow, and N. A. Bei, Dielectric Waveguide with Elliptical Cross Section, *Radio Eng. Electron.* (*USSR*) (*English transl.*) **6**, 1668 (1961).
4. G. Piefke, Grundlagen zur Berechnung der Übertragungseigenschaften elliptischer Wellenleiter, *Arch. Elek. Uebertrag.* **18**, 4 (1964).
5. J. E. Goell, A Circular-Harmonic Computer Analysis of Rectangular-Dielectric Waveguides, *Bell Syst. Tech. J.* **48**, 2133 (1969).
6. E. A. J. Marcatili, Dielectric Rectangular Waveguide and Directional Coupler for Integrated Optics, *Bell. Syst. Tech. J.* **48**, 2071 (1969).
7. N. W. McLachlan, "Theory and Application of Mathieu Functions." Oxford, Univ. Press, London and New York, 1951.
8. E. Snitzer, Cylindrical Dielectric Waveguide Modes, *J. Opt. Soc. Amer.* **51**, 491 (1961).
9. J. A. Stratton, "Electromagnetic Theory." McGraw-Hill, New York, 1941.
10. S. P. Schlesinger and D. D. King, Dielectric Image Lines, *IRE Trans. Microwave Theory Tech.* **MTT-6**, 291 (1958).

11. W. Schlosser and H. G. Unger, Partially Filled Waveguides and Surface Waveguides of Rectangular Cross Section, *in* "Advances in Microwave," Academic Press, New York, 1966.
12. E. Snitzer and H. Osterberg, Observed Dielectric Waveguide Modes in the Visible Spectrum, *J. Opt. Soc. Amer.* **51**, 491 (1961).

APPENDIX B

Hollow Dielectric Waveguide *

TAKEO SAWATARI †

A. INTRODUCTION

To date, all investigations in the field of fiber optics (*1*) and basic investigations of optical waveguides (*2–8*) have made use of energy propagation along dielectric waveguides in which the refractive index of the core (N_1) is greater than that of the coating (N_2) to achieve total internal reflection. It is interesting to investigate modes excited and propagated along a hollow tubular waveguide in which the refractive index (N_0) of the inside tube is lower than that (N) of the surrounding material. The hollow waveguide can be expected to propagate the low-order mode with very small attenuation due to the grazing incidence reflection, in contrast to the total internal reflection. This phenomenon is well known to investigators in the field of microwave waveguides (*9–11*).

On the other hand, for optical waveguides, one of the central problems is the coupling of an optical source to a multimode guide in such a way as to substantially excite only one desired mode. In the case of dielectric waveguides, Snyder (*12*) has calculated mode coefficients for the $HE_{1,m}$ mode when a plane parallel beam is incident on the entrance to the guide.

* Research sponsored in part by the Air Force Office of Scientific Res., Office of Aerospace Res., U.S.Air Force, under AFOSR Contract No. AF49(638)–1626. This appendix is based in part on work reported at the 1968 Spring Meeting of the Optical Society, in *J. Opt. Soc. Amer.* **58**, 721A (1968), at the 1969 Annual Meeting of the Optical Society, in *J. Opt. Soc. Amer.* **59**, 1527A (1969), and in *J. Opt. Soc. Amer.* **60**, 132 (1970).

† This material was prepared while the author was with Optics Technology, Palo Alto, Calif. The author's present address is Bendix Research Laboratories, Southfield, Mich.

The use of a diffraction pattern of a ring aperture (*6, 7*) has been suggested to excite a desired mode into a dielectric multimode guide, and the feasibility of the technique has been verified experimentally for $HE_{1,m}$ modes, where a nominal transverse component has no azimuthal dependence.

In this appendix, we will study the characteristics of the hollow dielectric waveguide, as well as the mode excitation technique for the hollow cylindrical waveguide, from both the theoretical and experimental viewpoints.

Stratton (*11*) has determined the field components for the natural modes of the most general circularly cylindrical structure with arbitrary isotropic internal and external media. Marcatili and Schmeltzer (*13*) recently have developed the theory in the optical spectrum region. One of the simplest descriptions of the cylindrical dielectric waveguide, given by Kapany (*1*) and Kapany *et al.* (*5*), expresses transverse components of the field by right and left circularly polarized components, instead of tangential and azimuthal components. A characteristic of this expression is that each component of a mode field is proportional to a single eigenfunction, whereas in conventional expressions (*2, 11, 13*), the transverse components are proportional to linear combinations of the eigenfunctions. When this theoretical analysis is applied to the hollow, circular, dielectric waveguide, all of the modes described are in a simple symmetric form, and the various characteristic parameters of the mode can be conveniently organized to make the discussions on mode launching easier.

The synthesis technique (*6, 7*) for the circular optical waveguide mode field, which was developed to excite a desired mode in a dielectric multimode guide, has been limited to Fraunhofer diffraction patterns of ring apertures. A spatial filter (ring aperture) for generating a general field similar to the arbitrarily chosen mode field requires an amplitude and phase transmittance of $\delta(\varrho - a_0)e^{in\psi}$, where δ is a delta function, (ϱ, ψ) are polar coordinates in a pupil plane of a launching lens, a_0 is a constant, and n is an integer corresponding to the subscript of the mode to be excited. This mode excitation technique using the spatial filter can be directly applied to a hollow dielectric waveguide. However, the difficulty of this technique is the fabrication of a phase filter having the form $e^{in\psi}$. To eliminate this difficulty, we will use a Fresnel diffraction pattern of a spiral aperture and show theoretically that this pattern is almost identical to the transverse component of a hollow dielectric waveguide mode. Intensity distributions of hollow waveguides can be synthesized experimentally from the Fresnel diffraction patterns of spiral apertures. Some modes can be excited in a hollow tube (5–50 μm in diameter) using the launching technique of optical filtering and conventional mode excitation.

B. THEORY

1. Mode Field and Characteristic Parameters of a Hollow Waveguide

a. Mode Field

Consider a hollow circular waveguide, e.g., a tube of radius a and free-space dielectric constant ε_0, embedded in an external medium of dielectric constant ε. For simplicity of discussion, assumptions similar to those of Marcatili and Schmeltzer (*13*) are made for hollow tubes. The radius a is much larger than the free-space wavelength λ, and only the modes whose propagation constant h is nearly equal to that in free space k_0 and whose attenuations are consequently small are discussed. That is,

$$\lambda/a \ll 1, \qquad |\,(h/k_0) - 1\,| \ll 1. \tag{B.1}$$

Using these conditions, the approximate field components of all possible modes are represented by circularly polarized components and are derived by neglecting terms with powers of λ/a larger than one.

At all internal points, $r < a$,

$$\begin{aligned} E_{\pm} &= \pm(1 \pm A)f_{n\pm1}, \\ E_z &= -i(\beta_1/h)f_n, \\ H_{\pm} &= i(\varepsilon_0/\mu_0)^{1/2}(1 \pm A)f_{n\pm1}, \\ H_z &= i(\varepsilon_0/\mu_0)^{1/2}AE_z, \end{aligned} \tag{B.2}$$

and at external points, $r > a$,

$$\begin{aligned} E_{\pm} &= -i(1 \pm A)[u'J_n(u')/v]g_{n\pm1}, \\ E_z &= -i(\beta_2/h)[u'J_n(u')/v]g_n, \\ H_{\pm} &= \pm(\varepsilon_0/\mu_0)^{1/2}[(\varepsilon/\varepsilon_0) \pm A][u'J_n(u')/v]g_{n\pm1}, \\ H_z &= i(\varepsilon_0/\mu_0)^{1/2}AE_z, \end{aligned} \tag{B.3}$$

where the functions f_p and g_p are given by

$$\begin{aligned} f_p &= J_p(\beta_1 r)\exp(ip\phi - i\omega t + ihz) \\ g_p &= (r/a)^{-1/2}\exp[-\beta_2 i(r-a) + ip\phi - i\omega t + ihz]; \qquad p = n-1,\ n,\ n+1. \end{aligned} \tag{B.4}$$

Here, J_p is the Bessel function of the first kind of order p. The constant A (polarization parameter) determines the ratio of H_z and E_z, and has a value fixed by the boundary condition. The complex propagation constants β_1, β_2, and h satisfy the relationships

$$\begin{aligned} \beta_1^2 + h^2 &= k_0^2 = \omega^2\varepsilon_0\mu_0 = [(2\pi/\lambda)N_0]^2 \\ \beta_2^2 + h^2 &= k^2 = \omega^2\varepsilon\mu = [(2\pi/\lambda)N]^2. \end{aligned} \tag{B.5}$$

Also, it is assumed that

$$u = \beta_1 a, \qquad v = \beta_2 a. \tag{B.6}$$

Note that here the parameters h, A, β_1, β_2, u, and v are all characteristic of the (n, m) mode and should be subscripted as such. The subscripts have been omitted for convenience.

In the above expression, each field component of a mode is proportional to only one eigenfunction. It will be seen later that this expression makes the calculation of mode launching much easier than is the case in the conventional expression, where each field component is proportional to a linear combination of two eigenfunctions.

b. Characteristic Equation

To determine the propagation constant h and other characteristics of each mode, the roots of the following quadratic characteristic equation are required (*11*):

$$\begin{aligned} &\left[\frac{vH_n(v)}{H_{n-1}(v)} - \frac{\bar{\varepsilon}}{\varepsilon_0}\frac{uJ_n(u)}{J_{n-1}(u)}\right]\left[\frac{vH_n(v)}{H_{n+1}(v)} - \frac{\bar{\varepsilon}}{\varepsilon_0}\frac{uJ_n(u)}{J_{n+1}(u)}\right] \\ &= \left(\frac{\Delta\varepsilon}{2\varepsilon_0}\right)^2 \frac{u^2 J_n^2(u)}{J_{n-1}(u)J_{n+1}(u)}, \end{aligned} \tag{B.7}$$

where $\bar{\varepsilon} = (\varepsilon_0 + \varepsilon)/2$ and $\Delta\varepsilon = (\varepsilon_0 - \varepsilon)/2$. This equation is greatly simplified when the approximation used to obtain Eq. (B.3) is again applied. From Eqs. (B.5) and (B.6), $v = \beta_2 a \gg 1$. Thus, if the asymptotic expression for the Hankel function is used, the ratio becomes

$$H_n(v)/H_{n\pm1}(v) = \pm\, i. \tag{B.8}$$

Since from Eqs. (B.5) and (B.6)

$$1/v = 1/(R^2 + u^2)^{1/2} \doteqdot 1/R \ll 1, \tag{B.9}$$

where

$$R^2 = [(2\pi/\lambda)a]^2(N^2 - N_0^2), \tag{B.10}$$

powers of $1/R$ larger than $1/R^2$ can be neglected. Further, if we take

$$J_{n\pm1}(u)/uJ_n(u) = \xi_{\pm1}(u), \tag{B.11}$$

the characteristic equation then reduces to

$$\xi_{+1}\xi_{-1} - (i/R)(\bar{\varepsilon}/\varepsilon_0)(\xi_{+1} - \xi_{-1}) + (\varepsilon/\varepsilon_0)(1/R^2) = 0. \tag{B.12}$$

In the case of $n = 0$, this equation becomes

$$\xi_{n=0}(u) = -i\frac{\varepsilon}{\varepsilon_0}\frac{1}{R}, \ (\mathrm{TM}_{0,m}) \quad \text{or} \quad \xi_{n=0}(u) = -i\frac{1}{R}, \ (\mathrm{TE}_{0,m}), \tag{B.13}$$

where $\xi_{n=0}(u) = J_1(u)/uJ_0(u)$, because $\xi_{+(n=0)} = -\xi_{-(n=0)}$. In the case of $n \neq 0$,

$$\xi_{+1}(u) = -i\frac{\bar{\varepsilon}}{\varepsilon_0}\frac{1}{R}, \ (\mathrm{EH}_{n,m}^{\oplus,\ominus}) \quad \text{or} \quad \xi_{-1}(u) = i\frac{\bar{\varepsilon}}{\varepsilon_0}\frac{1}{R}, \ (\mathrm{HE}_{n,m}^{\ominus,\oplus}), \tag{B.14}$$

for $n = \pm1, \pm2, \pm3, \ldots$.

Mode designations shown in these equations will be discussed in detail in a later section.

c. Propagation Constant

In Eqs. (B.13) and (B.14), the right-hand sides are close to zero. Expanding $\xi(u)$ around zero and keeping only the first term of the expression,

$$\xi_{\pm1}(u) = \xi_{\pm1}(u') + \Delta u'\,\xi'_{\pm1}(u') = \pm\Delta u'(1/u'), \tag{B.15}$$

since

$$u = u' + \Delta u', \qquad \xi_{\pm1}(u') = 0 \quad [\text{i.e.,} \quad J_{n\pm1}(u') = 0],$$

and

$$\xi'_{\pm1}(u') = \frac{J'_{n\pm1}(u')}{u'J_n(u')} = \frac{\pm J_n(u')}{u'J_n(u')} = \pm\frac{1}{u'}.$$

Substituting Eq. (B.15) into Eqs. (B.13) and (B.14) and using Eq. (B.6) and the relation $u = u' + \Delta u'$, the transverse propagation constant β_1 is given by

$$\beta_1 = (u'/a)[1 - i(c/R)], \tag{B.16}$$

where

$$c = \begin{cases} \varepsilon/\varepsilon_0, & \text{(TM)}, \\ 1, & \text{(TE)}, \\ \bar{\varepsilon}/\varepsilon_0, & \text{(EH and HE)}. \end{cases} \tag{B.17}$$

Note here that u' is the mth root of the equation

$$J_{n\pm1}(u'_{nm}) = 0.$$

The propagation constant h can therefore be obtained from Eq. (B.5):

$$\begin{aligned} h &= k_0[1 - (\beta_1{}^2/2k_0{}^2)] \\ &= k_0[(1 - \tfrac{1}{2}(u'/ak_0)^2 + i(c/R)(u'/ak_0)^2]. \end{aligned} \tag{B.18}$$

The phase constant h_p and attenuation constant h_a of each mode are the real and imaginary parts of h, respectively; therefore,

$$h_\mathrm{p} = k_0\left[1 - \frac{1}{2}\left(\frac{u'}{ak_0}\right)^2\right] \tag{B.19}$$

and

$$h_\mathrm{a} = \frac{c}{R}\left(\frac{u'}{ak_0}\right)^2 k_0 = \frac{c}{[(\varepsilon/\varepsilon_0) - 1]^{1/2}}\left(\frac{u'}{2\pi}\right)^2 \frac{\lambda^2}{a^3}. \tag{B.20}$$

d. Polarization Parameter

Having obtained the roots of Eq. (B.7) based on the approximation, we can now investigate the permissible values of the polarization parameter A, which relates the magnitude of the z components of the magnetic and electric fields [Eqs. (B.2) and (B.3)] and is fixed by the boundary conditions. In the case of $n = 0$, modes are either transverse electric ($TE_{0,m}$) or transverse magnetic ($TM_{0,m}$), where the z component of the magnetic (or electric) field has zero value. Hence,

$$A_{0m} = 0 \quad (\mathrm{TE}_{0,m}) \quad \text{or} \quad A_{0m} \to \infty \quad (\mathrm{TM}_{0,m}). \tag{B.21}$$

When $n \neq 0$, from the boundary condition equation (B.7), $A_{n,m}$ can be found in a straightforward way (2). The results are

$$A_n = \frac{\dfrac{J_{n-1}(u)}{uJ_n(u)} + \dfrac{J_{n+1}(u)}{uJ_n(u)} - \dfrac{H_{n-1}(v)}{vH_n(v)} - \dfrac{H_{n+1}(v)}{vH_n(v)}}{\dfrac{J_{n-1}(u)}{uJ_n(u)} - \dfrac{J_{n+1}(u)}{uJ_n(u)} - \dfrac{H_{n-1}(v)}{vH_n(v)} + \dfrac{H_{n+1}(v)}{vH_n(v)}}. \tag{B.22}$$

Using the same approximations for Eqs. (B.8), (B.9), and (B.15), from Eq. (B.22), we have

$$A_{nm}^{\pm} = \pm 1 - i\frac{\Delta\varepsilon}{\varepsilon}\frac{1}{R}\frac{u_{nm}}{2n}, \quad \text{for both } n \gtrless 0 \quad (\mathrm{EH}_{n,m}^{\oplus,\ominus} \text{ and } \mathrm{HE}_{n,m}^{\ominus,\oplus}) \tag{B.23}$$

and for each eigenvalue of u_{nm} which is obtained from Eqs. (B.6) and (B.16). The superscript $\pm$ of A_{nm} shows which equation, $J_{n+1}(u') = 0$ or $J_{n-1}(u')=0$, is used to obtain the value u'_{nm} belonging to u_{nm}.

e. Mode Designation and Numerical Evaluation of Characteristic Parameters

In the preceding discussion, we have designated each mode according to its characteristics. In Table I we summarize each mode and its characteristic condition depending on the region of integer n and the value of A. It should be noted that here, except for the case of $n = 0$, two modes [$\oplus$ and $\ominus$] belonging to one eigenvalue (u_{nm}) have been defined, since the roots of the two equations $J_{n\pm1}(u') = 0$ ($n \gtrless 0$) are identical for the same absolute value of $n \pm 1$. The relationship between these two modes ($\mathrm{HE}^{\oplus}$ and $\mathrm{HE}^{\ominus}$) and the conventional mode (*2*) (HE) should be understood as $\mathrm{HE} = \mathrm{HE}^{\oplus} + \mathrm{HE}^{\ominus}$. That is, for the $\mathrm{HE}^{\oplus}$ mode, as apparent from Eq. (B.2), the right circularly polarized component E_+ is highly dominant with respect to the left circularly polarized component E_- because of $A \doteqdot 1$. On the other hand, for the $\mathrm{HE}^{\ominus}$ mode, E_- is highly dominant with respect to E_+ because of $A \doteqdot -1$. When two modes are degenerated, the conventional HE mode is generated whose transverse component is, as is well known,

Table I

Mode Designation and the Characteristic Equation

Region of n	Value of A	Mode designation	Characteristic equation
$n > 0$	$A \doteqdot 1$	$\mathrm{EH}_{n,m}^{\oplus}$	$J_{n+1}(u') = 0$
	$A \doteqdot -1$	$\mathrm{HE}_{n,m}^{\ominus}$	$J_{n-1}(u') = 0$
$n < 0$	$A \doteqdot -1$	$\mathrm{EH}_{n,m}^{\ominus}$	$J_{n+1}(u') = 0$
	$A \doteqdot 1$	$\mathrm{HE}_{n,m}^{\oplus}$	$J_{n-1}(u') = 0$
$n = 0$	$A = 0$	$\mathrm{TM}_{0,m}$	$J_1(u') = 0$
	$A \to \infty$	$\mathrm{TE}_{0,m}$	$J_1(u') = 0$

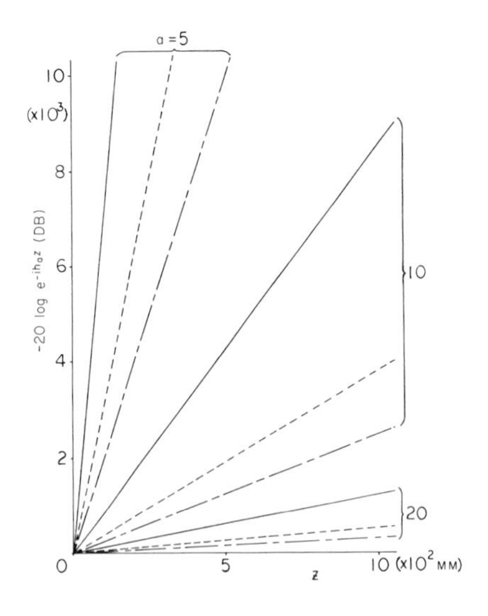

Fig. B-1. Attenuation of $TM_{0,1}$ (———), $TE_{0,1}$ (- - - - -), and $HE_{1,1}$ (— — — -) modes plotted as a function of the tube length z (mm) for different tube diameters a (μm).

almost linearly polarized as a consequence of a superposition of right and left circularly polarized light.

Figure B-1 shows the attenuation of $TM_{0,1}$, $TE_{0,1}$, and $HE_{1,1}$ modes versus length of hollow tube z, where the diemater $2a$ is used as a parameter, and Fig. B-2 shows numerically evaluated relationships connecting A, R, and u. In the calculations, the parameters are chosen as $N_0 = 1.0$, $N = 1.5$, and $\lambda = 0.62\ \mu$m. From Fig. B-1, we can see that the attenuation can be

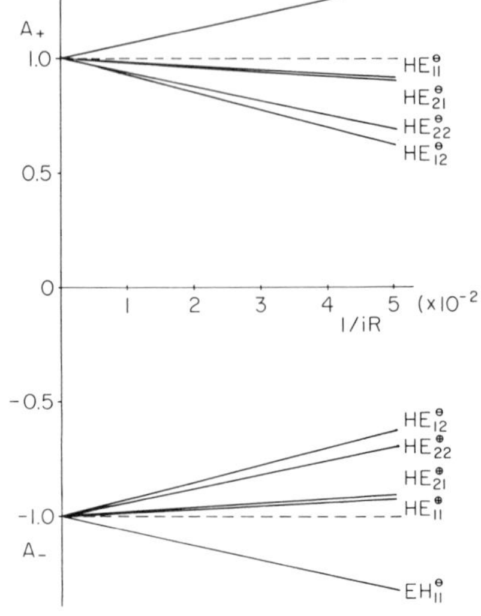

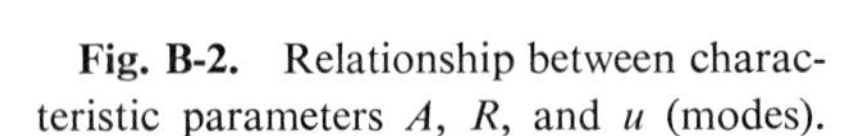
Fig. B-2. Relationship between characteristic parameters A, R, and u (modes).

arbitrarily decreased by choosing a value for a that is sufficiently large. As pointed out by Marcatili and Schmeltzer (*13*), the refractive index of tube material (N) affects the attenuation of each mode in different ways, as is apparent from Eq. (B.20). The attenuation of $TE_{0,m}$ mode increases monotonically with the increase of the refractive index. However, the attenuation of the $TM_{0,m}$ mode and the HE–EH mode is minimized when the refractive index is $\sqrt{2} = 1.414$ and $\sqrt{3} = 1.73$, respectively.

It may be interesting to enumerate some modes to compare the hollow-waveguide and dielectric-waveguide characteristics:

$$HE_{1,1},\quad TE_{0,1},\quad HE_{2,1},\quad EH_{1,1},\quad HE_{3,1},\quad HE_{1,2},\quad TE_{0,2},\quad TM_{0,1},\quad \cdots$$

for hollow waveguide;

$$HE_{1,1},\quad TE_{0,1},\quad TM_{0,1},\quad HE_{2,1},\quad EH_{1,1},\quad HE_{3,1},\quad HE_{1,2},\quad TE_{0,2},\quad \cdots$$

for dielectric waveguide.

The upper line is arranged from the low-loss mode to the high-loss mode for hollow dielectric waveguide ($N_0 = 1$, $N = 1.5$, and $a/\lambda = 10$), and the lower line is from lower-order modes to higher-order modes for normal dielectric waveguides ($N_1 = 1.9$, $N_2 = 1.47$, and $a/\lambda = 10$). In this comparison, we see that the selective positions of the $TM_{0,1}$ mode are quite different in the two guides. This is because the attenuation of a TM mode in a hollow waveguide is particularly higher than that in a conventional waveguide, as seen in Fig. B-1.

In Fig. B-2, we can see that if the diameter of the tube is greater than 10 times the wavelength ($1/R < 3.2$), the polarization parameter becomes close to plus unity or minus unity for each $\oplus$ or $\ominus$ mode. That is, either the positive or the negative transverse component of the mode field becomes dominant compared to the other ($-$ or $+$). However, if the diameter decreases or higher-order modes are launched, a lower degree of domination appears.

2. Mode Synthesis and Excitation of Hollow Waveguide

a. Mode Coefficient

Having analyzed the characteristics of hollow dielectric waveguides in the preceding section, we address here the problem of launching an arbitrary mode in a hollow dielectric tube. In order to assess the technique of mode launching described later, it is necessary to establish equations by which we can determine the relative power in those modes that are excited when a given wave is incident on the end of a hollow dielectric tube. We will adopt the assumption given in a recent paper (7) on matching incident fields and mode fields at the entrance end, $z = 0$, of the tube, namely that the tangential components of the electric (but not magnetic) fields at the end of the guide are those of the incident wave, the reaction of the guide being substantially ignored. This assumption is much more valid for the hollow tube discussed here than for conventional dielectric cylinders (fibers), because the material constant ε_0 inside the hollow tube is identical to that of free space, $z < 0$. No reflected wave exists at the plane $z = 0$ on the internal region of the tube. Furthermore, in the preceding analysis, we have limited ourselves to low-loss modes which propagate a small amount of energy along the outside of the tube. Consequently, the nonuniqueness of the excited mode (that is, the fact that the magnetic mode field resulting from equating the tangential components of the incident electric field to those of the general electric field on the guide does not match the magnetic field of the incident wave at the boundary) is minimized in the case of the hollow tube.

Based on this assumption, we have at the plane $z = 0$

$$\mathbf{E}_t' = \sum_n C_n \mathbf{E}_{nt} + \mathbf{E}_{Rt}, \tag{B.24}$$

where $\mathbf{E}_t'$, $\mathbf{E}_{nt}$, and $\mathbf{E}_{Rt}$ are the tangential vectors of the incident electric field, the nth mode field, and the radiating field, respectively. C_n is the amplitude of the nth mode (hereafter called the mode coefficient). By taking the vector cross product of each side with the tangential magnetic mode field ($\mathbf{H}_{nt}$), integrating over the plane $z = 0$, and invoking the orthogonality condition satisfied by the modes, we obtain

$$C_n = \left(\int \mathbf{E}_t' \times \mathbf{H}_{nt}^* \, d\sigma\right) \Big/ \int \mathbf{E}_{nt} \times \mathbf{H}_{nt} \, d\sigma. \tag{B.25}$$

Note that although the hollow waveguide modes are not mathematically orthogonal in the power sense, the overlap integrals between nearest modes are small enough to be ignored in the approximation used. The power P_n in the nth mode, relative to the incident power, is then given by

$$P_n = \left| \int \mathbf{E}_t' \times \mathbf{H}_{nt}^* \, d\sigma \right|^2 \Big/ \left[\left(\int \mathbf{E}_t' \times \mathbf{H}_t^{*\prime} \, d\sigma \right) \int \mathbf{E}_{nt} \times \mathbf{H}_{nt}^* \, d\sigma \right]. \tag{B.26}$$

Note that tangential components of the mode field have been given in Eqs. (B.2) and (B.3) as $E_\pm$ and $H_\pm$.

b. Synthesis of Launching Field

The purpose of this section is to find a technique to generate an appropriate incident field which can excite only a desired (arbitrarily chosen) mode in the hollow tube. A spatial filtering technique has been studied (*7*) which generates the incident field for launching an arbitrary mode on a fiber optical waveguide. The amplitude and phase filter analyzed in the literature has the following form:

$$[\delta(\varrho - a_0)/\varrho]e^{in\psi} \qquad (a_0 \text{ is a constant}). \tag{B.27}$$

This is inserted in the pupil plane (ϱ, ψ) of a launching lens which focuses incident collimated light onto the end of the fiber.

The incident field is obtained at the focal plane of the lens as a two-dimensional Fourier transform of the filter transmittance. This field is a good approximation of a transverse field component inside the core of the fiber; however, a double-path illuminating system (*6*, *7*) is generally needed to obtain the required input polarization. This technique can be applied to the hollow dielectric waveguide, and it is more feasible than for the conventional waveguide. However, from an experimental point of view, it has been quite difficult to fabricate a phase filter of $e^{in\psi}$ for $n \neq 0$, which is required when a general mode, except an $HE_{1,m}$ mode, is to be excited.

The technique described here is based on the concept of spatial filtering and eliminates the difficulty of complicated filter fabrication; e.g., no phase filter is required. Consider a spiral aperture with uniphase as a spatial filter of a launching lens for a hollow tube guide. A pupil function of the lens is given as

$$p(\varrho, \psi) = \delta(\varrho - (b_0\psi + c_0)^{1/2})/\varrho \tag{B.28}$$

where b_0 and c_0 are constants. Here, let us calculate a Fresnel diffraction pattern of the distribution instead of the Fraunhofer diffraction pattern examined in the literature.

The amplitude and phase distribution of a Fresnel diffraction pattern (*14*) formed in a plane (r, ϕ) (hereafter called as Fresnel plane) perpendicular to the optical axis and at a distance d_0 from the spatial filter of the spiral aperture is obtained as

$$F(r, \phi) = \int_0^\infty \int_0^{2\pi} \frac{\delta(\varrho - (b_0\psi + c_0)^{1/2})}{\varrho} \times \left\{\exp ik_0\left[\frac{1}{2d_0}\varrho^2 - \frac{1}{d_0} r\varrho \cos(\phi - \psi)\right]\right\}\varrho \, d\varrho \, d\psi. \tag{B.29}$$

By setting $k_0b_0/2d_0 = n'$ (n' an integer) and $k_0c_0^{1/2}/d_0 = \beta_1'$, imposing a condition of $(k_0b_0/2d_0c_0^{1/2})r \ll 1$, and performing the integral, we have

$$F(r, \phi) = c_1 \int_{\alpha}^{2\pi+\alpha} \exp[in'\psi - i\beta_1'r\cos(\psi - \phi)]\, d\psi$$
$$= c_2 J_{n'}(\beta_1'r)\exp(in'\phi), \tag{B.30}$$

where c_1 and c_2 are constants of a complex number.

Therefore, using a double-path system similar to that described in the literature (7) to obtain the required input polarization, replacing each filter of Eq. (B.27) by the spiral mask of Eq. (B.28), and shifting a focus position by a distance d_0, we can experimentally produce an incident field in the plane $z = 0$ given by

$$\begin{aligned} E_\pm' &= \pm(1 \pm A')f'_{n'\pm1}, & r < b_\pm \\ &= 0 & r > b_\pm \end{aligned} \tag{B.31a}$$

and

$$\begin{aligned} H_\pm' &= i(\varepsilon_0/\mu_0)^{1/2}(1 \pm A')f'_{n'\pm1}, & r < b_\pm \\ &= 0 & r > b_\pm, \end{aligned} \tag{B.31b}$$

where

$$f'_{p'} = J_{p'}(\beta_1'r)\exp(ip'\phi - i\omega t), \qquad p' = n' \pm 1. \tag{B.32}$$

A' is a constant and $b_\pm$ is the distance that corresponds to the zero of the Bessel function that is the first to fall outside the hollow radius a. This zero then satisfies the value of $J_{n'\pm1}(\beta_1'b_\pm) = 0$.

c. Evaluation of Launching Efficiency

Assuming the field given by Eq. (B.31) to be the incident field for a hollow tube, we can proceed to evaluate the launching efficiency theoretically.

The incident electric field is a perfect match to the transverse field on the inside of the tube. We thus expect it to be very selective in all cases of hollow tubes. We can now calculate the amplitude and power coefficients of other mismatched modes (case I and II) and of the desired mode (case III).

In the calculation, the approximations used are those given by Eq. (B.1), that is, orders higher than $O(1/R^2)$ are neglected. Furthermore, since it can be assumed in Eq. (B.31) that the distance $b_\pm$ is very close to the hollow radius a, the higher orders of $O\{[(b_\pm - a)/a]^2\}$ can also be negligible.

Substituting Eq. (B.31) into Eqs. (B.25) and (B.26) and referring to Eq. (B.1), we have the following results.

Case I. If $n \neq n'$, then

$$C_{n\pm} = 0 \quad \text{and} \quad P_{n\pm} = 0. \tag{B.33}$$

Case II. If $n = n' \neq 0$ and $\text{Re}(\beta_1) \neq \beta_1'$, then

$$\begin{aligned} C_{n\pm} &= \frac{\beta_1' J_n(\beta_1' a)}{\beta_1 J_n(\beta_1 a)} \frac{2\beta_1^2}{[(\beta_1')^2 - \beta_1^2]} \left(\frac{b_\pm - a}{a} + \frac{\bar{\varepsilon}}{\varepsilon_0} \frac{1}{R} \right) \\ P_{n\pm} &= \frac{8}{[(\beta_1')^2 - (\text{Re}(\beta_1))^2]^2} \left[\frac{(b_\pm - a)a}{b_\pm^2} \frac{\bar{\varepsilon}}{\varepsilon_0} \frac{1}{R} \right] \frac{J_n^2(\beta_1' a)}{J_n(\beta_1' b_\pm) J_{n\pm2}(\beta_1' b_\pm)}, \end{aligned} \tag{B.34}$$

and if $n = n' = 0$ and $\text{Re}(\beta_1) \neq \beta_1'$, then

$$\begin{aligned} C_{n=0} &= \frac{\beta_1' J_0(\beta_1' a)}{\beta_1 J_0(\beta_1 a)} \frac{2\beta_1^2}{[(\beta_1')^2 - \beta_1^2]} \left(\frac{b - a}{a} + \frac{\bar{\varepsilon}}{\varepsilon_0} \frac{1}{R} \right) \\ P_{n=0} &= \frac{8}{[(\beta_1')^2 - (\text{Re}(\beta_1))^2]^2} \left[\frac{(b - a)a}{b^2} \frac{\bar{\varepsilon}}{\varepsilon_0} \frac{1}{R} \right] \frac{J_0^2(\beta_1' a)}{J_0(\beta_1' b) J_2(\beta_1' b)}. \end{aligned} \tag{B.35}$$

Case III. If $n = n'$ and $\text{Re}(\beta)_1 = \beta_1'$, then

$$C_{n\pm \text{ or } n=0} = 1 \quad \text{and} \quad P_{n\pm \text{ or } n=0} = 1, \tag{B.36}$$

where the subscript $n+$ of C_n and P_n stands for coefficients of the $\text{EH}^\oplus$ mode ($n > 0$) and the $\text{HE}^\oplus$ mode ($n < 0$), $n-$ stands for those of the $\text{HE}^\ominus$ mode ($n > 0$) and the $\text{EH}^\ominus$ mode ($n < 0$), and $n = 0$ corresponds to the $\text{TE}_{0,m}$ or $\text{TM}_{0,m}$ mode. Note in the case of $n = 0$ that $b_+ = b_- = b$.

From these calculations, we can see that, by using the appropriate incident field, the power coefficient of a desired mode becomes almost unity and those of unwanted modes of two kinds (cases I and II) become zero. It is apparent that, as far as our approximations are applicable, the method should be quite effective.

C. EXPERIMENT

1. Synthesized Mode Patterns

Prior to observing actual mode patterns in a hollow dielectric waveguide, we tried to synthesize various intensity distributions similar to hollow waveguide modes using Fresnel diffraction patterns of spiral apertures. For simplicity in the experiment, we did not take into account the polarization characteristic of the field and only attempted to simulate the intensity distribution.

From Eq. (B.1) and Table I, the intensity distribution of a mode or mode combination of relatively low order at an arbitrary cross section of hollow tube is given as follows:

Single mode:

$$\begin{aligned} \mathrm{HE}_{1,m} &\propto J_0^2(\beta r) \\ \mathrm{TE}_{0,m}\ (\mathrm{TM}_{0,m}) \text{ and } \mathrm{HE}_{2,m} &\propto J_1^2(\beta r) \\ \mathrm{EH}_{1,m} \text{ and } \mathrm{HE}_{3,m} &\propto J_2^2(\beta r) \end{aligned} \tag{B.37}$$

Mode combination:

$$\begin{aligned} \mathrm{TE}_{0,m}\ (\mathrm{TM}_{0,m}) \pm \mathrm{HE}_{2,m} &\propto J_1^2(\beta r) \begin{pmatrix} \cos^2 \phi \\ \sin^2 \phi \end{pmatrix} \\ \mathrm{EH}_{1,m} \pm \mathrm{HE}_{3,m} &\propto J_2^2(\beta r) \begin{pmatrix} \cos^2 2\phi \\ \sin^2 2\phi \end{pmatrix}. \end{aligned} \tag{B.38}$$

The single-mode patterns described here can be obtained simply by generating the diffraction pattern of a ring aperture or spiral aperture as shown in Eq. (B.30). It is apparent that the absolute square of $F(r, \phi)$ given by Eq. (B.30) is proportional to the single-mode pattern.

To generate a mode combination whose intensity distribution is given in Eq. (B.38), a mask which has two spiral apertures is used. One spiral is clockwise, that is, it increases in radius from the center according to the increase of the azimuth, and the other is counterclockwise. In Eq. (B.28), the first spiral corresponds to the case where $b_0 > 0$, and the second spiral corresponds to $b_0 < 0$, with another constant c_0 in the equation assumed to be identical for both spirals.

From Eqs. (B.29) and (B.30), the intensity distribution of a diffraction pattern of the two spiral masks is therefore given as follows:

$$\begin{aligned} | F_1 + F_2 |^2 &= | c_2 J_{n'}(\beta_1' r)[\exp(in'\phi)] + c_2 J_{-n'}(\beta_1' r) \exp(-in'\phi) |^2 \\ &= | c_2 |^2 J_{n'}^2(\beta_1 r) \begin{bmatrix} \cos^2(n'\phi) \\ \sin^2(n'\phi) \end{bmatrix}, \end{aligned} \tag{B.39}$$

where F_1 represents the diffraction pattern from the first spiral and F_2 that for the second spiral.

Accordingly, by choosing a mask of appropriate size (one spiral or two spirals) and an appropriate distance from the mask to the Fresnel diffraction plane, we can synthesize intensity distributions identical to those of Eqs. (B.37) and (B.38). Examples of the spiral apertures are shown in Fig. B-3. The actual diameter of the spiral ($2\sqrt{c_0}$) is 6 mm, and the spiral ratio ($\pi|b_0|/c_0$) is 1/30. These filters have been fabricated by a normal photographic reduction technique. Diffraction patterns obtained from them are shown in Figs. B-4 and B-5.

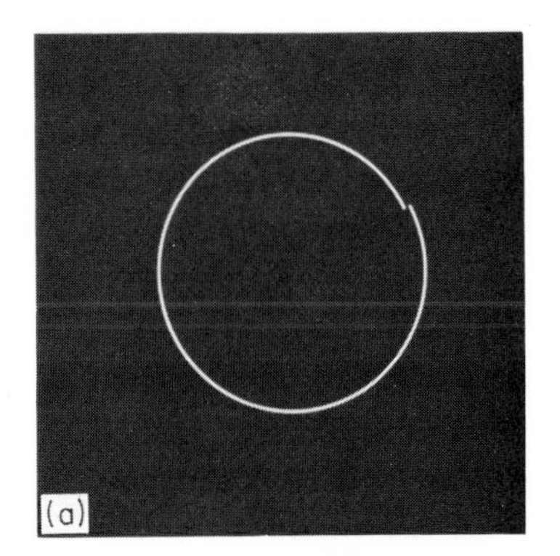

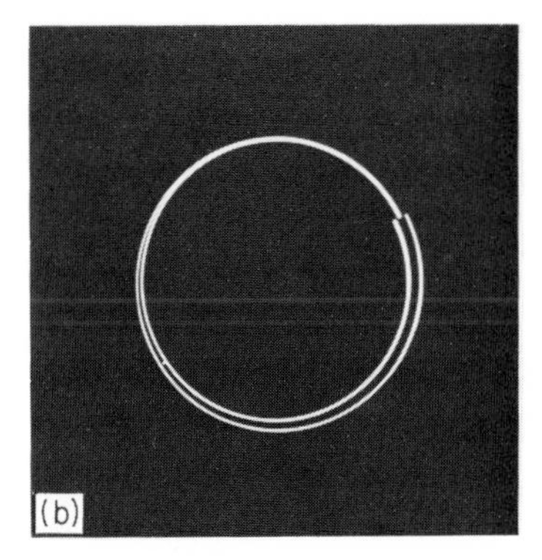

Fig. B-3. (a) Single spiral and (b) double spiral apertures used to synthesize hollow waveguide mode patterns.

Figures B-4(a–c) were obtained using the single spiral of Fig. B-3(a), corresponding to J_0^2, J_1^2, and J_2^2. To obtain these patterns for different nth orders, distances from the mask to the Fresnel plane [d_0 in Eq. (B.29)] are adjusted to satisfy the condition given in Eq. (B.30). The distances are the focal length (500 mm) of the lens for $n' = 0$, 500 mm (without lens) for $n' = 1$, and 250 mm (without lens) for $n' = 2$, respectively. The wave-number of the He–Ne laser light (k_0) is about 10^4 mm^{-1}. Figures B-5(a, b) correspond to $J_1^2 \cos^2 \phi$ and $J_2^2 \cos^2 2\phi$ in the vicinity of the center of the patterns. The distances d_0 are 50 and 25 cm for Figs. B-5(a) and B-5(b), respectively, and k_0 is the same as in Fig. B-4.

If we cut out the higher-order lobes for these diffraction patterns by placing a circular diaphragm of an appropriate diameter at the Fresnel plane and picking up only the central portion of the patterns, we obtain the synthesized mode patterns shown in Figs. B-6–B-10. An order m of the mode was determined by the size of the diaphragm, since the number of lobes through the diaphragm corresponds to the order of the mode. Figures B-6–B-8 are the single-mode patterns synthesized from the single spiral aperture diffraction patterns of Fig. B-4. The corresponding far-field patterns are obtained by focusing a camera at a plane located behind the diaphragm.

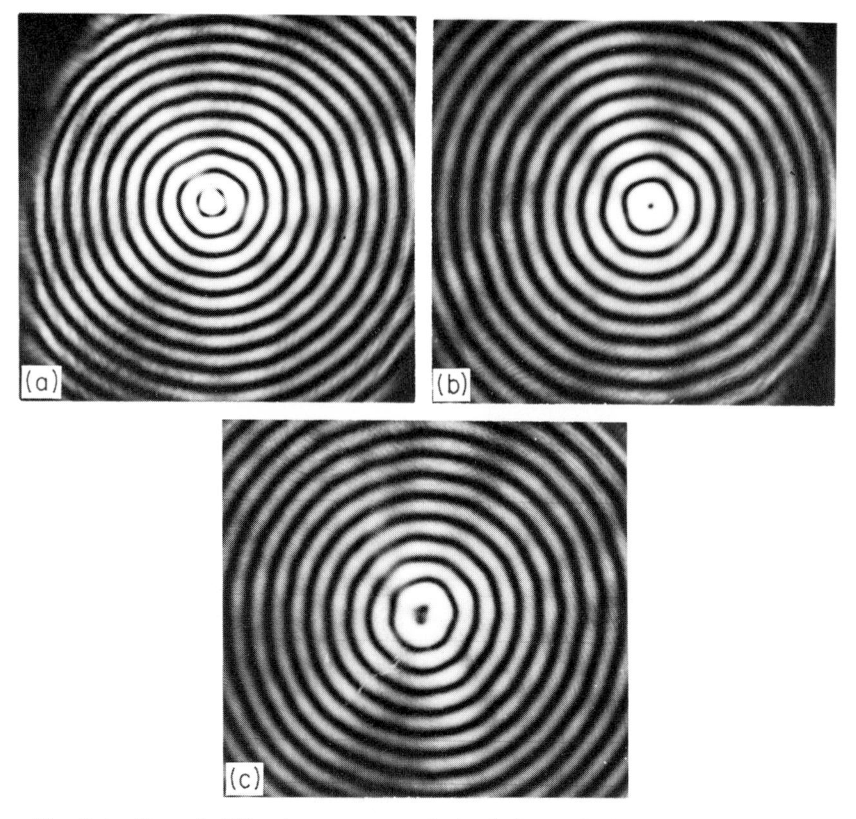

Fig. B-4. Fresnel diffraction patterns obtained from single spiral aperture: (a), (b), and (c) are proportional to J_0^2, J_1^2, and J_2^2, respectively.

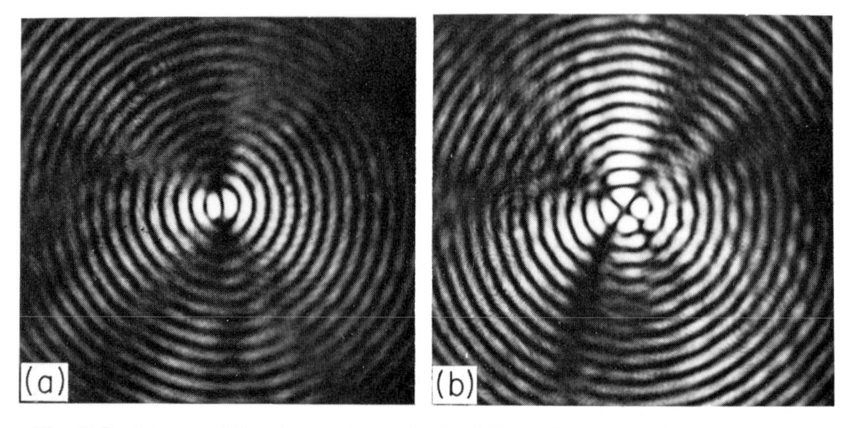

Fig. B-5. Fresnel diffraction patterns obtained from double spiral aperture; (a) and (b) are proportional to $J_1^2 \cos^2 \phi$ and $J_2^2 \cos^2 2\phi$, respectively.

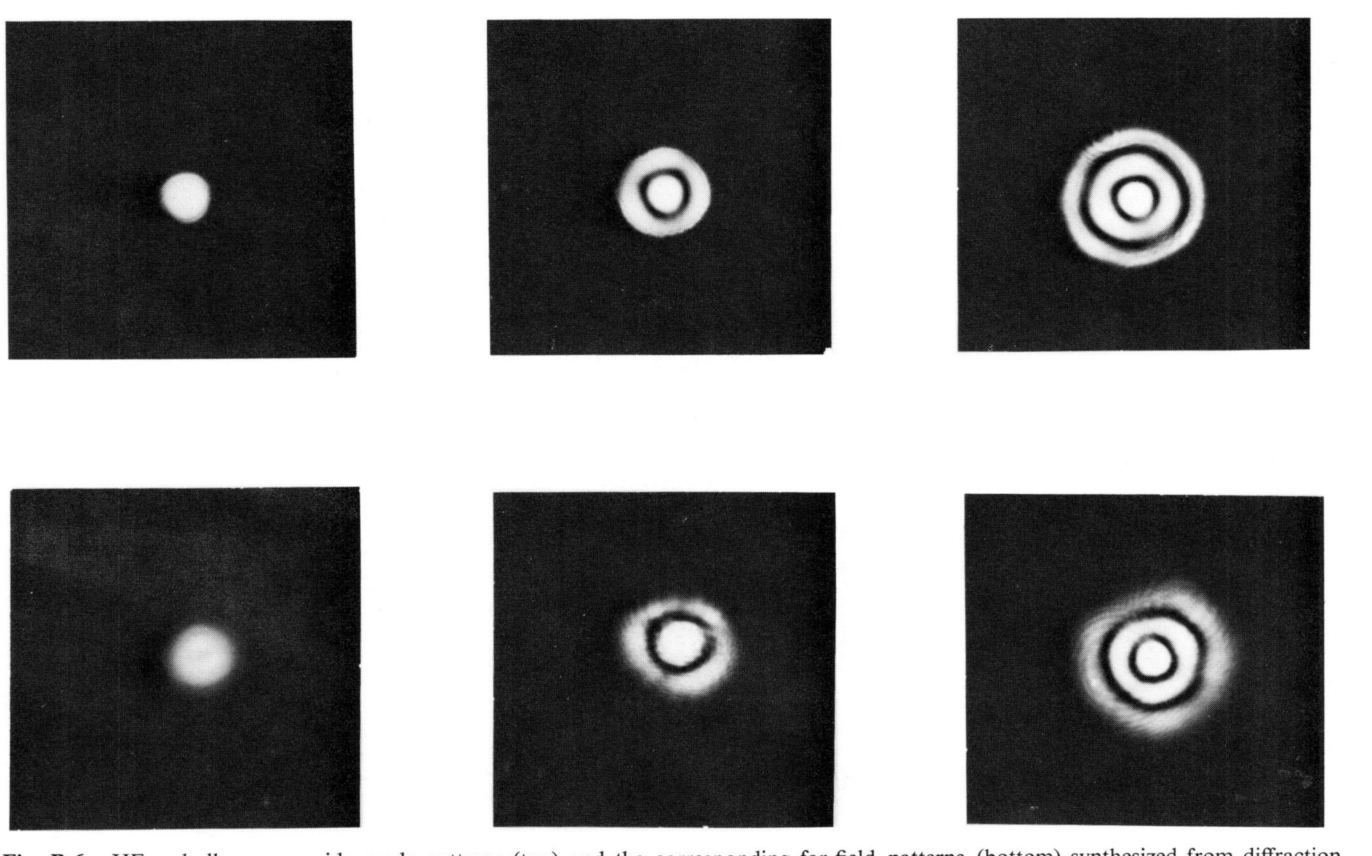

Fig. B-6. $HE_{1,m}$ hollow-waveguide mode patterns (top) and the corresponding far-field patterns (bottom) synthesized from diffraction patterns of spiral apertures. From left to right, $m = 1$, 2, and 3.

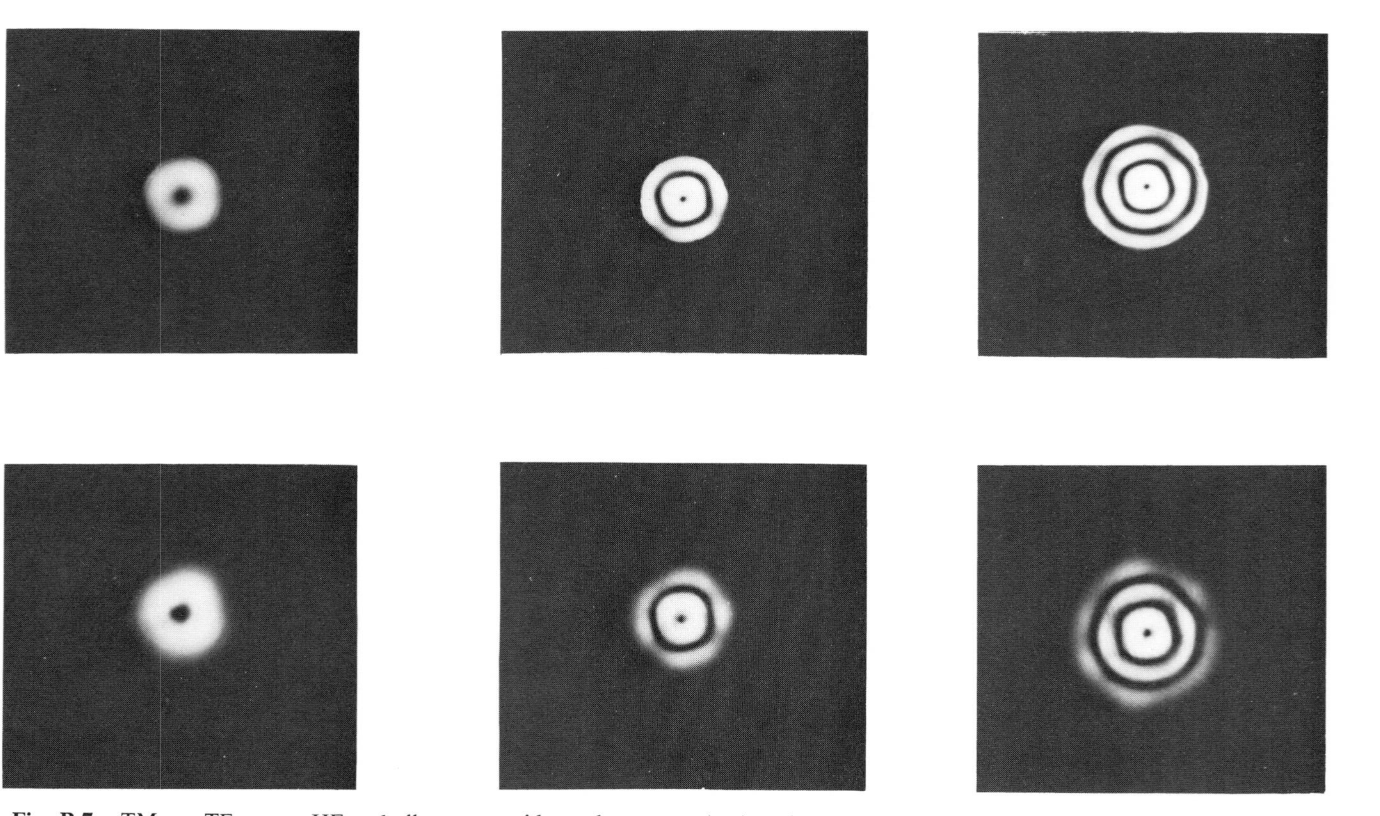

Fig. B-7. $TM_{0,m}$, $TE_{0,m}$, or $HE_{2,m}$ hollow-waveguide mode patterns (top) and the corresponding far-field patterns (bottom) synthesized from diffraction patterns of spiral apertures. From left to right, $m = 1, 2$, and 3.

The distance from the diaphragm to the focused plane is chosen to be large compared to the diameter of the diaphragm. It is obvious that these synthesized far-field patterns are identical to those of the actual modes, because it has been proven theoretically (*7*) that the far-field pattern of a mode field is approximately proportional to the Fourier transform of the mode distribution if the field outside of the waveguide is negligibly small.

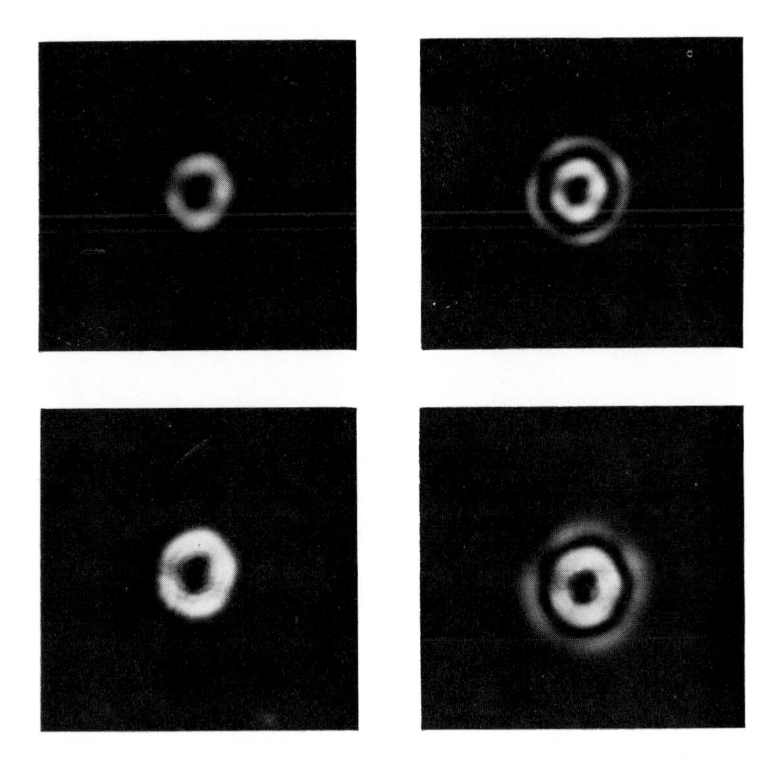

Fig. B-8. $EH_{1,m}$ or $HE_{3,m}$ hollow-waveguide mode patterns (top) and the corresponding far-field patterns (bottom) synthesized from diffraction patterns of spiral apertures. From left to right, $m = 1$ and 2.

Figures B-9 and B-10 are the synthesized mode combinations of $TE_{0,m} \pm HE_{2,m}$ and $EH_{1,m} \pm HE_{3,m}$, respectively. These are obtained from the double spiral aperture, and the far-field patterns are generated in a similar way to those of Figs. B-6–B-8. These experimentally synthesized mode patterns agree well with those expected from theory of the hollow dielectric waveguide.

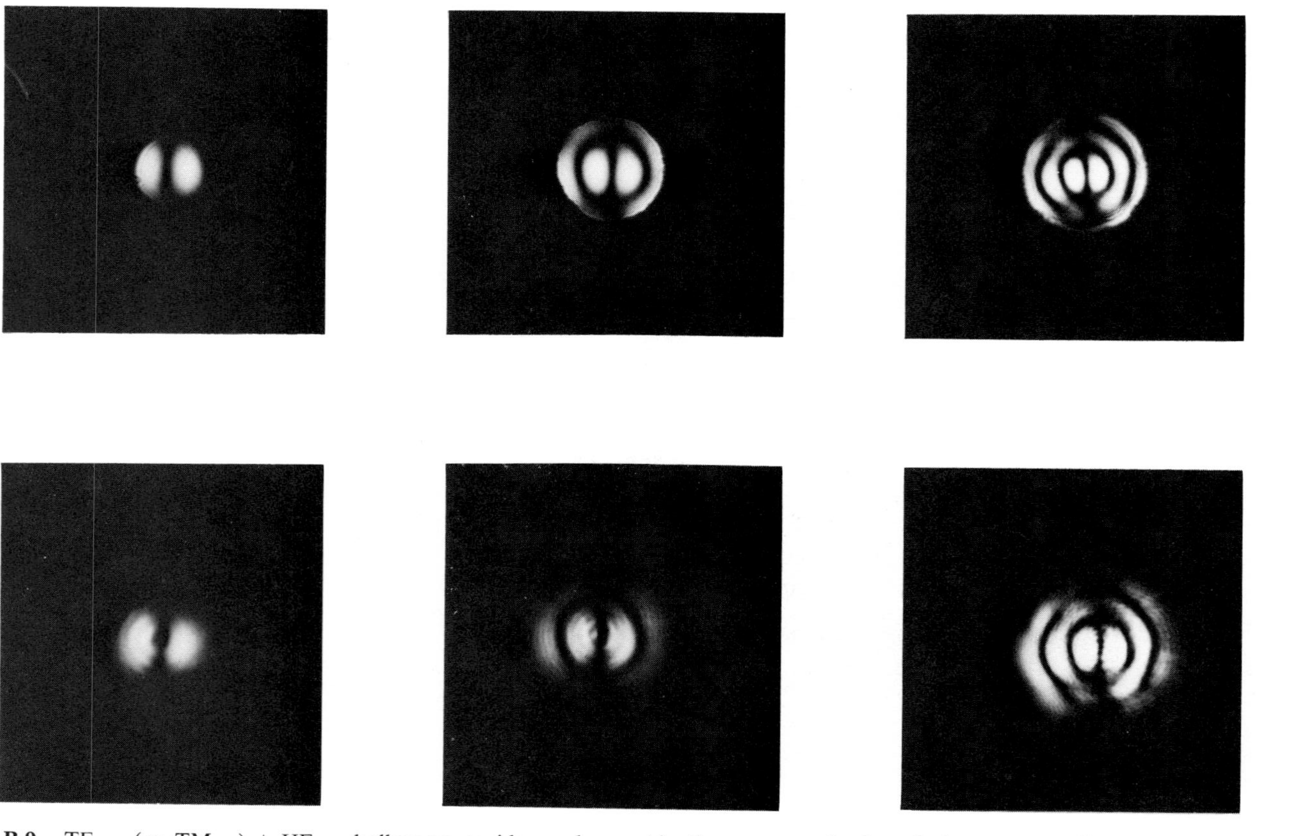

Fig. B-9. $TE_{0,m}$ (or $TM_{0,m}$) $\pm$ $HE_{2,m}$ hollow-waveguide mode combination patterns (top) and the corresponding far-field patterns (bottom) synthesized from diffraction patterns of double spiral apertures. From left to right, $m = 1$, 2, and 3.

Fig. B-10. $EH_{1,m} + HE_{3,m}$ hollow-waveguide mode combination patterns (top) and the corresponding far-field patterns (bottom) synthesized from diffraction patterns of double spiral apertures. From left to right, $m = 1$ and 2.

2. Wave Propagation along Hollow Waveguide

a. Transmission of a Hollow Tube

In Fig. B-1, the attenuation of low-order modes as a function of tube length has been calculated. In Eq. (B.20), from which we obtained the curves of Fig. B-1, we can see that the transmission for a particular wavelength depends upon the dielectric constant ε of the tube material, the diameter $2a$ of the tube, and the mode parameter u', which is related *(4)* to the characteristic angle of the mode with respect to the wall of the tube. We collected data on the angular distribution of emergent light for relatively large-diameter glass tubing, which can be described on the basis of geometric optics. The capillary tubes tested were 1 mm i.d., 3 mm o.d., and 1, 2, and 4 cm long. The refractive index of the glass was 1.5. The outer surface of each glass tube was painted black to absorb any stray light which might otherwise enter the wall of the tube. A collimated beam produced from a He–Ne gas laser was incident at an angle θ to the entrance end of the tube, and the total transmitted light was measured using a photodetector (Schottky barrier diode). The data are plotted in Fig. B-11 using the aspect ratio of the tested piece as a parameter. These curves show that when the incident angle of parallel light becomes greater than 1°, the transmitted flux is strongly dependent upon the aspect ratio. However, if the angle θ is less than 1°, the total transmitted flux does not depend quite so much on the aspect ratio, and transmittances are more than 90%.

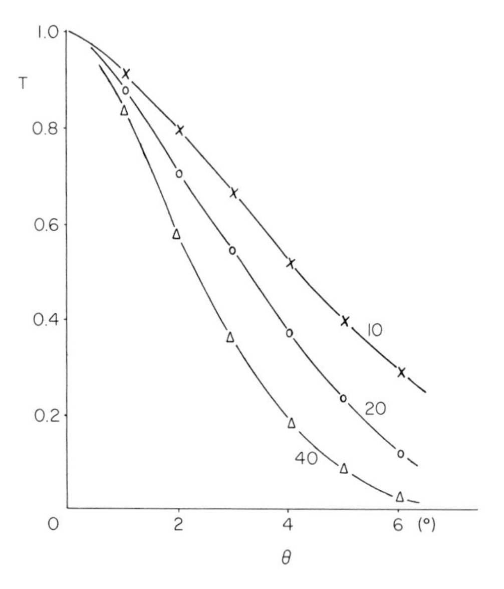

Fig. B-11. Transmittance versus angle of incidence of parallel light. The indicated numbers are aspect ratios of the tested glass tube.

Furthermore, to investigate the dependence of refractive index (dielectric constant) on the angular distribution of the transmission, we prepared three different tubes, made of quartz ($N = 1.458$), Pyrex ($N = 1.474$), and soda-lime ($N = 1.512$) glass. The experimental results for these three samples have been obtained in the same way as in the earlier measurements. We could not see any significant differences in the transmittance curves measured for the different materials. Although the reason for this is shown in Eq. (B.20) indirectly, it is obvious if we compare the reflectances of these materials for angles of incidence close to the grazing angle, for example, 85°. For the p component of the incident light, the reflectance of the high-index material is less than that of the low-index material ($R_{\parallel}{}^2 = 49.6\%$ for $N = 1.5$, and $R_{\parallel}{}^2 = 50\%$ for $N = 1.46$). On the other hand, for the s component, the results are reversed, e.g., the lower-index material has a lower reflectance ($R_{\perp}{}^2 = 73.6\%$ for $N = 1.5$, and $R_{\perp}{}^2 = 72.5\%$ for $N = 1.46$). However, the average reflectance of the s and p components for the high-index material ($N = 1.5$) is slightly higher than for the low-index material ($N = 1.46$), but this difference is only 0.3% for the same angle of incidence, 85°.

From these simple experiments, based on geometric optics, we could see that a glass capillary tube can propagate light by internal reflection with reasonably small attenuation if the light is near grazing incidence.

b. Mode Behavior in a Hollow Waveguide

In order to observe the mode behavior in hollow waveguides, we reduced the inside diameter of the tubes over the range from 5 to 50 μm. The glass material was soda lime ($N = 1.512$), and the outer surfaces of the narrow tubes were again painted black. The technique described in the previous sections is most effective for exciting a desired single mode or mode combination. However, for mode launching experiments, the technique requires quite complicated and accurate alignment of many optical components, except in the case of exciting an $HE_{1,m}$-type mode. Because of the experimental difficulty with general modes, filtering techniques were used to excite only the $HE_{1,1}$ and $HE_{1,2}$ modes. A linearly polarized $J_0{}^2(\beta_1{}'r)$ distribution (Fig. B-6) obtained as a diffraction pattern of the mask was launched at the entrance end of a hollow tube. By choosing an appropriate combination of aperture radii (1–5 mm) and lens focal length (15 cm), the size of the diffraction pattern was matched to that of the desired mode, e.g., $\beta_1 a = 2.4$ for $HE_{1,1}$ and $\beta_1 a = 5.5$ for $HE_{1,2}$.

To excite other modes, a conventional launching technique (*3*) was employed: the image of a back-lighted pinhole (500 μm in diameter) was focused onto the entrance end of tubes with 4× and 40× microscope objectives. Although, as seen in the preceding experiment, we could synthesize intensity distributions of arbitrary modes using a filtering technique, these fields are not applicable for launching purposes, because these distributions do not match the polarization and plase characteristics of the mode field.

The light source used for this experiment was a He–Ne gas laser (6328 Å). The mode patterns thus excited and propagated along the tube were viewed with a microscope (40× objective and 10× eyepiece) and photographed.

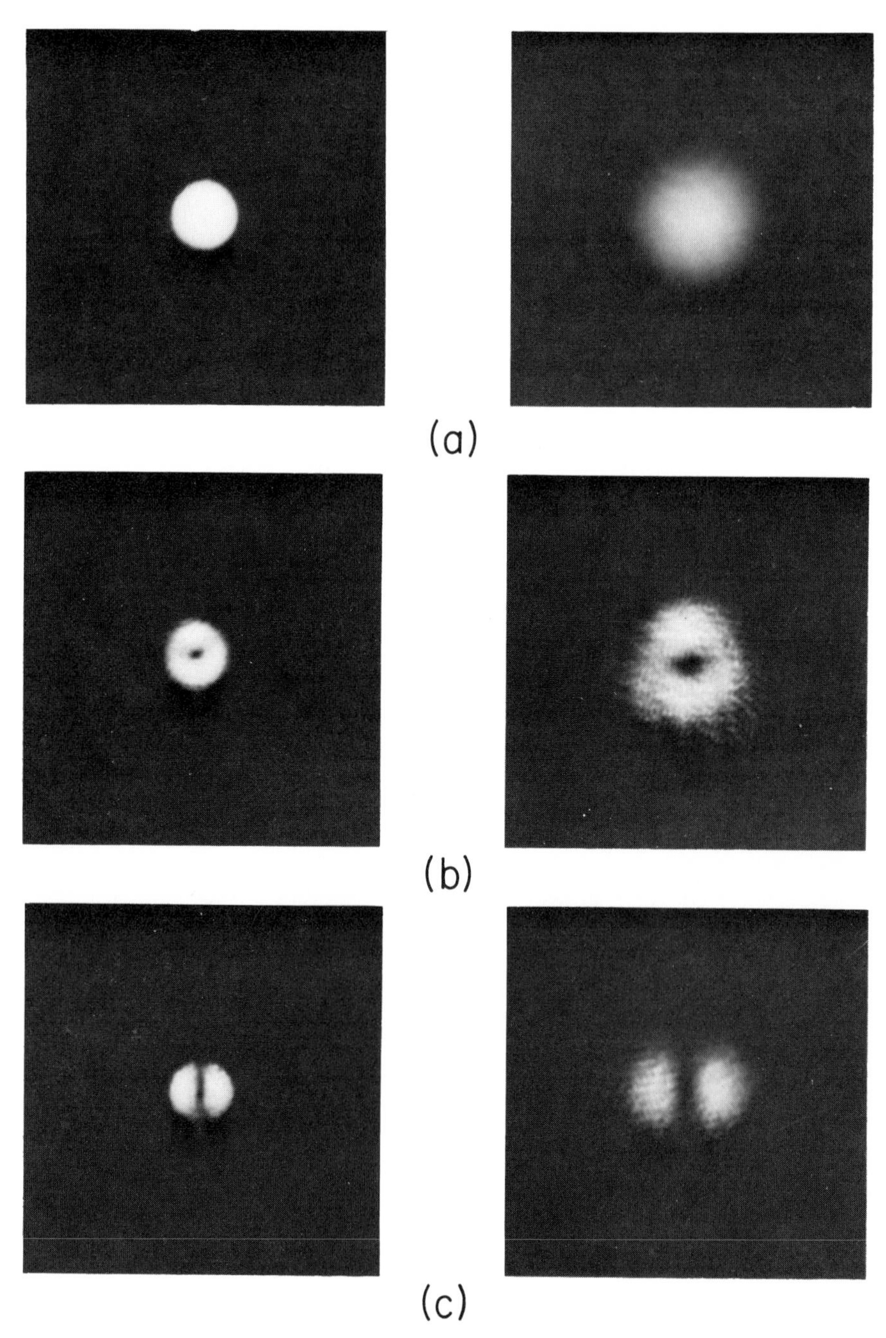

Fig. B-12. Left: hollow-dielectric waveguide modes (a) $HE_{1,1}$; (b) $TE_{0,1}$; and (c) $TE_{0,1} + HE_{2,1}$. Right: their far field patterns.

The photographs shown in Figs. B-12 and B-13 are examples of the hollow-waveguide modes observed and the corresponding far-field patterns. The mode photographs shown in Fig. B-12 were taken with a tube of approximately 30 μm i.d. and 5 cm long. The photographs in Fig. B-13 were observed in a tube of 40 μm i.d. and 5 cm long. These two modes were more difficult to excite than those of Fig. B-12, because the attenuation of these modes was greater. Furthermore, as indicated in these figures, the positive and negative modes are not distinguished, because they were considered to be excited identically. We also found that a slight bending of the tubing induced undesirable higher-order modes, and the irradiance of the output pattern decreased rapidly. This was predicted by Marcatili and Schmeltzer (*13*).

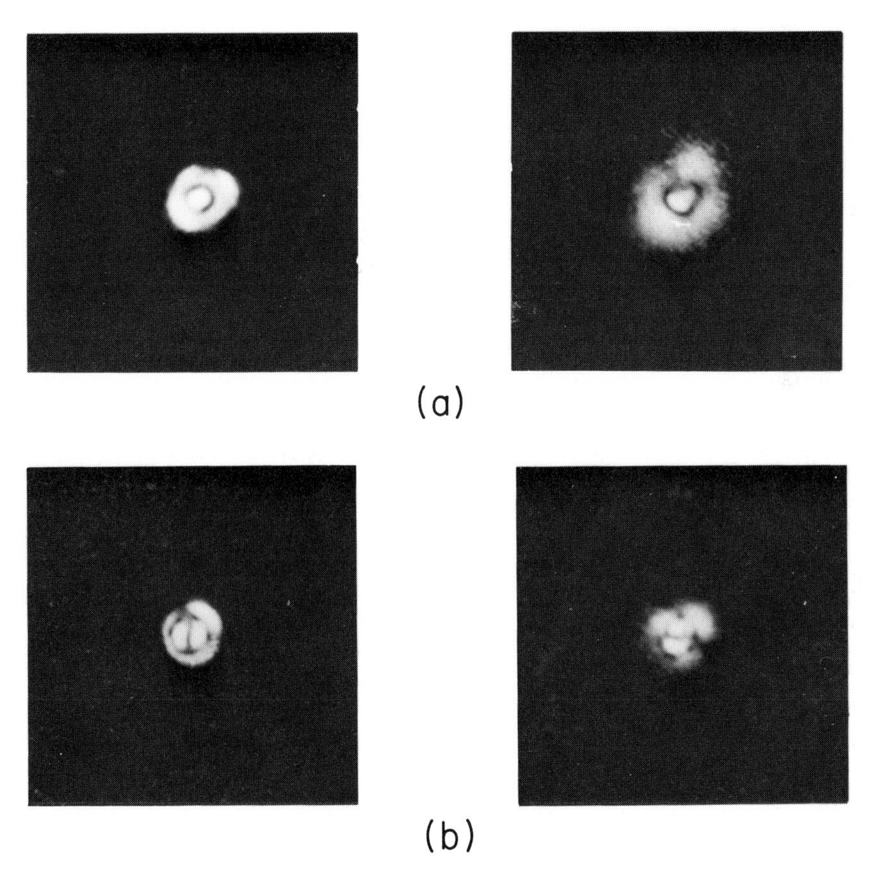

Fig. B-13. Left: hollow-dielectric waveguide modes (a) $HE_{1,2}$ and (b) $TE_{0,2} + HE_{1,2}$. Right: their far-field patterns.

D. SUMMARY

The mode field in a hollow dielectric waveguide has been analyzed and expressed in terms of right and left circular polarization components. This simple, symmetric expression is very convenient for calculating various related parameters. Important characteristic parameters of the mode field were assessed numerically using an appropriate approximation based on the geometry of the hollow tube. Numerical results show that a few low-order modes can be propagated along the hollow tube with reasonably small attenuation. Furthermore, since the polarization parameter is close to unity, each mode has one dominant transverse field. These modes were designated in a convenient manner that can be easily related to the conventional designation of a normal dielectric waveguide (fiber). By using the knowledge of the mode field of hollow waveguides, a technique for synthesizing a field equivalent to the mode field was developed using optical filtering. The launching efficiency of a desired mode for the hollow waveguide was also calculated using the synthesized field (Fresnel diffraction pattern of a spiral aperture). The results show that the technique is quite effective.

Intensity distributions of the mode patterns (single modes and mode combinations) were synthesized experimentally using optical filtering techniques. These intensity patterns show a good agreement with the theory. However, these patterns were not used for the actual launching experiment for the hollow waveguide, except in the case of the $HE_{1,m}$ mode, because of the additional experimental complexities involved.

Measurements of the transmission of a large-diameter (1 mm) hollow tube as a function of incident angle showed reasonable agreement with the theoretical predictions. Several actual modes propagated along the hollow waveguide (5–50 μm in diameter) were observed through a microscope. Some of them were excited using a technique developed here and the others were generated by a simple mode launching technique.

ACKNOWLEDGMENT

The author would like to express his sincere appreciation to Dr. N. S. Kapany and J. J. Burke, not only for their kind invitation to include this work as an appendix to their monograph, but also for their continuous encouragement and advice during the course of this work. Thanks are also extended to Professor H. Ohzu of Waseda University, Japan and Dr. R. K. Mueller of Bendix Research Laboratories for their encouragement.

REFERENCES

1. N. S. Kapany, "Fiber Optics: Principles and Applications." Academic Press, New York, 1967.
2. E. Snitzer, *J. Opt. Soc. Amer.* **51**, 491 (1961).
3. E. Snitzer and H. Osterberg, *J. Opt. Soc. Amer.* **51**, 499 (1961).
4. N. S. Kapany and J. J. Burke, *J. Opt. Soc. Amer.* **51**, 1067 (1961).
5. N. S. Kapany, J. J. Burke and K. Frame, *Appl. Opt.* **4**, 1534 (1965).
6. T. Sawatari, Doctoral Thesis, Waseda Univ., Tokyo, Japan (1969).
7. N. S. Kapany, J. J. Burke, and T. Sawatari, *J. Opt. Soc. Amer.* **60**, 1178 (1970).
8. N. S. Kapany, J. J. Burke, and T. Sawatari. *J. Opt. Soc. Amer.* **60**, 1350 (1970).
9. J. R. Carson, S. P. Mead, and S. A. Schlelkunoff, *Bell Syst. Tech. J.* **15**, 310 (1936).
10. G. Goubau and F. Schwering, *Trans. IRE* **AP-9**, 248 (1961).
11. J. A. Stratton, "Electromagnetic Theory." McGraw Hill, New York, 1941.
12. A. W. Snyder, *J. Opt. Soc. Amer.* **56**, 601 (1966).
13. E. A. J. Marcatili and R. A. Schmeltzer, *Bell Syst. Tech. J.* **43**, 1783 (1964).
14. M. Born and E. Wolf, "Principles of Optics." Pergamon Press, New York, 1964.

Author Index

Numbers in parentheses are reference numbers and indicate that an author's work is referred to, although his name is not cited in the text. Numbers in italics show the page on which the complete reference is listed.

Subject Index